NEUROTRANSMITTER RECEPTORS

New Comprehensive Biochemistry

Volume 24

General Editors

A. NEUBERGER
London

L.L.M. van DEENEN
Utrecht

ELSEVIER
Amsterdam · London · New York · Tokyo

Neurotransmitter Receptors

Editor

FERDINAND HUCHO

*Institut für Biochemie
Freie Universität Berlin,
Thielallee 63,
14195 Berlin, Germany*

1993
ELSEVIER
Amsterdam · London · New York · Tokyo

Elsevier Science Publishers B.V.
P.O. Box 211
1000 AE Amsterdam
The Netherlands

Library of Congress Cataloging-in-Publication Data
Neurotransmitter receptors / editor Ferdinand Hucho.
    p.   cm. -- (New comprehensive biochemistry ; v. 24)
   Includes bibliographical references and index.
   ISBN 0-444-89903-0 (acid-free paper)
   1. Neurotransmitter receptors.   I. Hucho, Ferdinand, 1939–
II. Series.
QD415.N48   vol. 24
[QP364.7]
574.19'2 s--dc20   93-2106
[611'.0181]   CIP

ISBN 0 444 89903 0 (Volume)
ISBN 0 444 80303 3 (Series)

© 1993 Elsevier Science Publishers B.V. All rights reserved.

No part of this publication may be reproduced, stored in a retrieval system, or transmitted in any form or by any means, electronic, mechanical, photocopying, recording or otherwise, without the prior permission of the Publisher, Elsevier Science Publishers B.V., Copyright & Permissions Department, P.O. Box 521, 1000 AM Amsterdam, The Netherlands.

No responsibility is assumed by the Publisher for any injury and/or damage to persons or property as a matter of products liability, negligence or otherwise, or from any use or operation of any methods, products, instructions or ideas contained in the material herein. Because of the rapid advances in the medical sciences, the Publisher recommends that independent verification of diagnoses and drug dosages should be made.

Special regulations for readers in the USA – This publication has been registered with the Copyright Clearance Center Inc. (CCC), Salem, Massachusetts. Information can be obtained from the CCC about conditions under which photocopies of parts of this publication may be made in the USA. All other copyright questions, including photocopying outside of the USA, should be referred to the Publisher.

This book is printed on acid-free paper.

Printed in The Netherlands

Preface

This is a good time for a book on transmitter receptors: the cloning boom is over. More than one hundred receptor sequences are known and recombinant DNA techniques, used in combination with classical biochemistry, have provided us with a wealth of information which can now be compiled in one volume of this series.

With the primary structures of most transmitter receptors at hand, it is safe to predict that the next major step will have to wait for breakthroughs in biophysical methods (spectroscopy, X-ray crystallography, ultramicroscopy). We can expect principally new insights into receptor structure–function relationships only from secondary and tertiary structures of receptor proteins. For the time being we have to live with the state reached after the major leap ahead made possible by molecular neurobiology.

It is useless to include chapters on all known receptors into a book of this kind, although the series title – New *Comprehensive* Biochemistry – may suggest this. To minimize redundancies only a few receptors (some of which are typical for a whole group of similar receptors, others which are presently of special interest) are dealt with in a full-size chapter. Others are represented in the TIPS Receptor Nomenclature Supplement which is included as a special feature in this book and which makes this volume more useful as a receptor handbook.

A major problem is the receptor nomenclature: no attempts have been made to urge the chapter authors to use a consistent system of names and terms. At this stage of 'receptorology' a logical nomenclature is not obvious. Therefore, the reader is asked to be flexible when looking in the index for 'his' receptor. As a case in point, one can take the terms *4TM* and *7TM*, which are used by some neurochemists to classify receptors into those having four and seven hydrophobic transmembrane domains, respectively. This classification is based on the assumption that receptors span the membrane with alpha helices. It may turn out that some of these transmembrane domains are actually β-strands (evidence for this is accumulating in the literature). This would mean that the twenty-two or so amino acids of the hydrophobic sequence are too long for spanning the membrane only once and the 4TM and 7TM might have to be renamed 6TM or 10TM.

It is a nice custom to use this page of a book for thanking: I have to thank the colleagues who contributed a chapter despite their many other duties, the publisher and his staff for the efficient professionality in producing this book and, last but not least, Mary Wurm, who kept a close eye on my manuscripts, making them – hopefully – readable.

Ferdinand Hucho
Berlin, Germany
June, 1993

List of contributors

Barnard, Eric A. – Molecular Neurobiology Unit, Royal Free Hospital School of Medicine, University of London, London NW3 2PF, U.K. Tel. +44-071-7940500. ext. 5445; Fax: +44-071-4311973.

Bobker, Daniel H. – Vollum Institute for Advanced Biomedical Research, Oregon Health Sciences University, L-474, 3181 S.W. Sam Jackson Park Road, Portland, Oregon 97201-3098, USA. Tel. +01-503-494-5465; Fax: +01-503-494-6972.

Böhme, Eycke – Institut für Pharmakologie, Freie Universität Berlin, Thielallee 67-73, D-14195 Berlin 33, Germany. Tel. +49-30-838 6474; Fax: +49-30-831-5954.

Bouvier, Michel – Département de Biochimie, Université de Montréal, C.P. 6128, Succ. A, Montréal, Quebec, H3C 3J7, Canada. Tel: +01-514-343-6374; Fax: +01-514-343-2210.

Darlison, Mark G. – Institut für Zellbiochemie und Klinische Neurobiologie, Universitätskrankenhaus Eppendorf, Martinistr. 52, D-20251 Hamburg, Germany. Tel: +49-40-4717-4395; Fax: +49-30-4717-4541.

Fagg, Graham E. – CIBA-Geigy Ltd., Building K-147.2.14, CH-4002 Basel, Switzerland. Tel: +41-61-6967824; Fax: +41-61-6963887.

Foster, Alan C. – Merck Sharp and Dohme Research Laboratories, Terlings Park, Harlow, Essex CM20 2QR, U.K. *Current address:* Gensia Pharmaceuticals, 11025 Roselle Street, San Diego, CA 92121, USA

Helmreich, Ernst J.M. – Physiologisch-Chemisches Institut, Universität Würzburg, Koellikerstr. 2, D-97070 Würzburg, Germany.

Hucho, Ferdinand – Institut für Biochemie der Freien Universität Berlin, Thielallee 63, D-14195 Berlin, Germany. Tel: +49-30-838 5545; Fax: +49-30-838 3753.

Järv, Jaak – Laboratory of Bioorganic Chemistry, University of Tartu, 2 Jakobi Street, 202 400 Tartu, Estonia. Tel: +07-01434-35112 or 32884; Fax: +07-01434-35440 or 33427; Telex: 173243 TAUN.

Koesling, Doris – Institut für Pharmakologie, Freie Universität Berlin, Thielallee 67-73, D-14195 Berlin, Germany. Tel: +49-30-838 6474; Fax: +49-30-831 5954.

Lohse, Martin J. – Genzentrum der Ludwig-Maximilians-Universität, Am Klopferspitz, D-82152 Martinsried, Germany. Tel: +49-89-8578-3992; Fax: +49-30-8578-3795.

Meyerhof, Wolfgang – Institut für Zellbiochemie und Klinische Neurobiologie, Universitätskrankenhaus Eppendorf, Martinistr. 52, D-20251 Hamburg, Germany, Tel: +49-40-4717-4395; Fax: +49-30-4717-4541.

Nantel, François – Département de Biochimie, Université de Montréal, C.P. 6128, Succ. A, Montréal, Quebec, H3C 3J7, Canada. Tel: +01-514-343-6374; Fax: +01-514-343-2210.

Otto, Henning – Institut für Biochemie der Freien Universität Berlin, Thielallee 63, D-14195 Berlin, Germany. Tel: +49-30-838 5545; Fax: +49-30-838 3753.

Richter, Dietmar – Institut für Zellbiochemie und Klinische Neurobiologie, Universitätskrankenhaus Eppendorf, Martinistr. 52, D-20251 Hamburg, Germany. Tel: +49-40-4717-4395; Fax: +49-30-4717-4541.

Rinken, Ago – Laboratory of Bioorganic Chemistry, University of Tartu, 2 Jakobi Street, 202 400 Tartu, Estonia. Tel: +07-01434-35112 or 32884; Fax: +07-01434-35440 or 33427; Telex: 173243 TAUN.

Schultz, Günter – Institut für Pharmakologie, Freie Universität Berlin, Thielallee 67-73, D-14195 Berlin, Germany. Tel: +49-30-838 6474; Fax: +49-30-831 5954.

Simon, Joseph – Molecular Neurobiology Unit, Royal Free Hospital School of Medicin, University of London, London NW 3 2PF, U.K. Tel. +44-071-7940500; Fax: +44-071-4311973.

Stephenson, F. Anne – The School of Pharmacy, University of London, 29/39 Brunswick Square, GD-London WC1N 1AX, U.K. Tel: +44-71-753-5877; Fax: +44-71-278-1939.

Strange, Philip G. – Biological Laboratory, The University, Canterbury, Kent, CT2 7NJ, U.K.

Strasser,, Ruth H. – Medizinische Klinik der Universität Heidelberg, D-69117 Heidelberg, Germany.

Williams, John T. – Vollum Institute for Advanced Biomedical Research, Oregon Health Sciences University, L-474, 3181 S.W. Sam Jackson Park Road, Portland, Oregon 97201-3098, USA. Tel. +01-503-494-5465; Fax: +01-503-494-6972.

Contents

Preface v

List of contributors vii

I. GENERAL TOPICS

Chapter 1. Transmitter receptors – general principles and nomenclature
Ferdinand Hucho 3

1. Historical aspects and definition 3
   - 1.1. History 3
   - 1.2. Definition 4
   - 1.3. Criteria for calling a binding site a receptor . . . 5
   - 1.4. Agonists–antagonists 5
   - 1.5. Mechanism of receptor function 6
     - 1.5.1. The triune receptor model 6
     - 1.5.2. Transmitter receptors are allosteric proteins . 7
   - 1.6. Receptor diversity is greater than transmitter diversity . 8
   - 1.7. Receptor nomenclature 9
     - 1.7.1. IUPHAR rules of receptor nomenclature . . . 9
     - 1.7.2. The TiPS receptor nomenclature supplement 1993 . 9
     - 1.7.3. Receptor classification 9
     - 1.7.4. Families and superfamilies 11
   - 1.8. Structural features 11
     - 1.8.1. Ligand-gated ion channels (type-I receptors) . 11
     - 1.8.2. G-coupled receptors (type-II receptors) . . . 12
References 13

TiPS Receptor Nomenclature Supplement 1993 15

Chapter 2. Ligand-binding studies – theory and experimental techniques
Henning Otto 61

1. Introduction 61
2. The experimental scenario 63
   - 2.1. Basic types of binding experiments 63
     - 2.1.1. Saturation experiments 63
     - 2.1.2. Kinetic experiments 64
     - 2.1.3. Thermodynamics of binding 64
     - 2.1.4. Inhibition experiments 64
   - 2.2. The receptor 65
   - 2.3. The ligand 67
     - 2.3.1. Ligand classification 67
     - 2.3.2. The radioligand's quality 69
       - 2.3.2.1. Tritiated compounds 69
       - 2.3.2.2. Iodinated compounds 69

			2.3.2.3. Homogeneity and purity of the ligand	70
			2.3.2.4. Stability of the radioligand	70
3.	Experimental strategies			72
	3.1.	Saturation experiments		73
		3.1.1.	Graphical evaluation; the Scatchard plot	74
		3.1.2.	Nonlinear Scatchard plots	79
			3.1.2.1. Nonspecific binding	79
			3.1.2.2. Cooperativity	79
			3.1.2.3. Heterogeneity of binding sites	80
			3.1.2.4. Ligand-ligand interactions	80
			3.1.2.5. Ligand heterogeneity	80
			3.1.2.6. Isomerization of the receptor-ligand complex	81
			3.1.2.7. Incomplete equilibration	81
		3.1.3.	Nonspecific binding	81
		3.1.4.	Cooperativity	83
	3.2.	The kinetic approach		85
		3.2.1.	The on-rate	86
		3.2.2.	The off-rate	88
	3.3.	Inhibition experiments		88
	3.4.	Computer-based versus graphical evaluation		91
4.	Experimental techniques			92
	4.1.	The separation of free and bound ligands		92
		4.1.1.	Filtration	92
		4.1.2.	Centrifugation	94
		4.1.3.	Gel filtration	94
		4.1.4.	Dialysis	95
		4.1.5.	Other techniques	95
	4.2.	Remarks on the assay setup		96
References				97

Chapter 3. Receptor regulation
François Nantel and Michel Bouvier 99

1.	Introduction		99
2.	Receptor regulation		100
	2.1.	G-protein-coupled receptors	100
	2.2.	Tyrosine kinase receptors	105
	2.3.	Ligand-gated ion-channel receptors	107
3.	Conclusion		108
References			108

II. PROTOTYPE RECEPTORS

Chapter 4. The nicotinic acetylcholine receptor
Ferdinand Hucho 113

1.	Introduction	113
2.	Function and occurrence	114

3.	Pharmacology and toxicology		115
4.	Biochemistry		118
	4.1.	AChR from *Torpedo* electric tissue	118
		4.1.1. The molecule, quaternary structure, electron microscopy	118
		4.1.2. Primary structure, evolutionary aspects	121
		4.1.3. Secondary structure, transmembrane folding	125
		4.1.4. Functional topography	125
	4.2.	Muscle AChR	128
		4.2.1. Biochemistry	128
		4.2.2. Development	128
	4.3.	Neuronal AChR	130
		4.3.1. Biochemistry and molecular biology	130
		4.3.2. Localization of neuronal receptor subunits in the brain	131
	4.4.	AChR from other sources	132
References			133

Chapter 5. The β-adrenoceptors
Martin Lohse, Ruth H. Strasser and Ernst J.M. Helmreich 137

1.	Introduction		137
2.	Pharmacological characterization		138
	2.1.	β_1-, β_2-, β_3-adrenoceptors	138
3.	Function and structures		140
	3.1.	Function	140
		3.1.1. Activation of G-proteins	140
		3.1.2. The catalytic function of the β-receptor	142
		3.1.3. Amplification	144
	3.2.	Structures	146
		3.2.1. Functional correlations with structural elements	148
4.	Regulation of β-adrenoceptor function		151
	4.1.	Desensitization	152
		4.1.1. Phosphorylation	153
		4.1.2. Sequestration	155
	4.2.	β-receptor expression	157
		4.2.1. Turnover	157
		4.2.2. Down-regulation	158
		4.2.3. Up-regulation	160
5.	Clinical aspects		161
	5.1.	Cardiovascular diseases	162
		5.1.1. Chronic heart failure and cardiomyopathy	162
		5.1.2. Myocardial infarction	164
	5.2.	Endocrinological diseases	166
		5.2.1. Hypo- and hyperthyroidism	166
	5.3.	Congenital diseases	167
		5.3.1. McCune Albright syndrome	168
6.	Outlook		168
Acknowledgements			169
References			169

III. RECEPTORS FOR 'CLASSIC' NEUROTRANSMITTERS

Chapter 6. GABA$_A$ and glycine receptors
F. Anne Stephenson 183

1. Introduction 183
2. GABA$_A$ and glycine receptor molecular pharmacology 183
   2.1. GABA$_A$ receptors 183
   2.2. Glycine receptors 185
3. GABA$_A$ and glycine receptor biochemistry and molecular biology 185
   3.1. GABA$_A$ receptors 185
   3.2. Glycine receptors 187
4. Oligomeric receptor structures and their distribution 189
   4.1. Glycine receptors 189
   4.2. GABA$_A$ receptors 190
5. Functional properties of cloned GABA$_A$ and glycine receptors 192
   5.1. GABA$_A$ receptors 192
   5.2. Glycine receptors 194
6. GABA$_A$ and glycine receptors in disease 195
   6.1. Glycine receptors 195
   6.2. GABA$_A$ receptors 196
7. Concluding remarks 196
References 196

Chapter 7. Muscarinic acetylcholine receptors
Jaak Järv and Ago Rinken 199

1. Introduction 199
2. Phenomena used for receptor assay 199
3. Compounds interacting with the receptor 203
   3.1. Agonists and antagonists 203
   3.2. Partial agonists 203
   3.3. Structural requirements for muscarinic drugs 204
   3.4. Snake venom toxins 205
4. Pharmacological and molecular subtypes of muscarinic receptor 205
5. Receptor molecule 206
   5.1. Purification and reconstitution 206
   5.2. Antibodies 207
   5.3. Protein structure 207
   5.4. Biological modification 208
   5.5. Functional groups 210
6. Ligand-receptor interactions 211
   6.1. Equilibrium-binding studies 211
   6.2. Kinetics of radioligand binding 212
   6.3. Kinetic studies with nonradioactive ligands 214
7. Mechanisms of signal transduction 215
8. Two-site receptor model 216

References 217

Chapter 8. Receptors for 5-hydroxytryptamine
Daniel H. Bobker and John T. Williams 221

1. Introduction 221
2. History 221
3. Nomenclature 222
4. Biochemistry of 5-HT synthesis, storage and neurotransmission 223
5. Anatomy of central 5-HT 223
6. 5-HT_{1A} receptor 225
    6.1. Pharmacology 225
    6.2. Anatomy 226
    6.3. Biochemistry 226
    6.4. Physiology 227
        6.4.1. Potassium conductance 227
        6.4.2. Inhibitory postsynaptic potential 228
        6.4.3. Calcium conductance 230
7. 5-HT_{1B} and 5-HT_{1D} receptors 230
    7.1. Pharmacology 230
    7.2. Anatomy 230
    7.3. Biochemistry 231
    7.4. Physiology 232
        7.4.1. 5-HT autoreceptor 232
        7.4.2. Terminal heteroreceptor 232
8. 5-HT_{1C} receptor 233
    8.1. Pharmacology 233
    8.2. Anatomy 233
    8.3. Biochemistry and physiology 234
9. 5-HT_2 receptors 235
    9.1. Pharmacology 235
    9.2. Anatomy 235
    9.3. Biochemistry 235
        9.3.1. PI turnover 235
        9.3.2. Desensitization 236
    9.4. Physiology 237
        9.4.1. Potassium conductance 237
        9.4.2. Excitatory postsynaptic potential 238
10. 5-HT_3 receptors 239
    10.1. Pharmacology 239
    10.2. Anatomy 239
    10.3. Physiology 240
        10.3.1. Peripheral nervous system 240
        10.3.2. Central nervous system 242
        10.3.3. Synaptic potentials 242
        10.3.4. Desensitization 242
11. 5-HT receptors that stimulate adenylate cyclase 243
    11.1. 5-HT_4 receptor 243
    11.2. Modulation of h-current (I_h) 243

12.	Molecular biology		244
	12.1. G-protein-coupled receptors		244
		12.1.1. Modulation of PI turnover: 5-HT$_{1C}$, 5-HT$_2$	244
		12.1.2. Modulation of adenylate cyclase: 5-HT$_{1A}$, 5-HT$_{1B/1D}$	245
	12.2. Ligand-gated channels: 5-HT$_3$		245
13.	Conclusion		246
References			246

Chapter 9. Dopamine receptors
Philip G. Strange 251

1. Introduction 251
2. Functions and distribution of dopamine receptors 251
3. Dopamine receptor subtypes: definitions and overall properties 252
   3.1. Subtypes defined from physiological, pharmacological, and biochemical studies . . 252
   3.2. Subtypes defined using molecular biology techniques 252
4. Biochemical characterisation of dopamine receptors 257
   4.1. Detection of dopamine receptors by ligand-binding assays 257
   4.2. Characterisation of dopamine receptor proteins 257
   4.3. Coupling of dopamine receptors to G-proteins 258
   4.4. Cellular responses linked to dopamine receptor activation 259
   4.5. Mechanism of ligand binding to dopamine receptors 260
   4.6. Regulation of dopamine receptors 262
Acknowledgement 264
References 264

Chapter 10. Glutamate receptors
Graham E. Fagg and Alan C. Foster 267

1. Introduction 267
   1.1. Glutamate receptor subtypes: evolution of the classification scheme 268
2. Function and occurrence 270
   2.1. NMDA receptor 270
   2.2. AMPA receptor 271
   2.3. Kainate receptor 272
   2.4. L-AP4 receptor 272
   2.5. Glu$_G$ receptor 273
3. Pharmacology and toxicology 274
   3.1. NMDA receptor 274
       3.1.1. Transmitter recognition site 275
       3.1.2. Glycine co-agonist site 277
       3.1.3. Open channel blockers 277
       3.1.4. Polyamine site 278
   3.2. AMPA receptor 279
   3.3. Kainate and L-AP4 receptors 279
   3.4. Glu$_G$ receptor 279
   3.5. Therapeutic utility 280
       3.5.1. Epilepsy 280
       3.5.2. Cerebral ischaemia 281

		3.5.3.	Parkinson's disease	281

```
            3.5.3.   Parkinson's disease  . . . . . . . . . . . . . . 281
            3.5.4.   Dementia  . . . . . . . . . . . . . . . . . . . 282
4.  Molecular biology . . . . . . . . . . . . . . . . . . . . . . . 282
    4.1.   NMDA receptor . . . . . . . . . . . . . . . . . . . . . 283
    4.2.   AMPA receptor . . . . . . . . . . . . . . . . . . . . . 285
    4.3.   Kainate receptor . . . . . . . . . . . . . . . . . . . . 287
    4.4.   Glu_G receptor  . . . . . . . . . . . . . . . . . . . . 287
5.  Miscellaneous effects . . . . . . . . . . . . . . . . . . . . . 288
    5.1.   Nitric oxide as a mediator  . . . . . . . . . . . . . . 288
Acknowledgements  . . . . . . . . . . . . . . . . . . . . . . . . . 290
References  . . . . . . . . . . . . . . . . . . . . . . . . . . . . 290
```

Chapter 11. Opioid receptors
Eric A. Barnard and Joseph Simon 297

```
1.  Introduction . . . . . . . . . . . . . . . . . . . . . . . . . . 297
2.  Multiple opioid receptors and their ligands . . . . . . . . . . 300
    2.1.   Opioid ligands  . . . . . . . . . . . . . . . . . . . . 300
    2.2.   Types of opioid receptors . . . . . . . . . . . . . . . 302
3.  Subtypes of the opioid receptor types . . . . . . . . . . . . . 303
4.  Cellular mechanisms of opioid actions . . . . . . . . . . . . . 306
    4.1.   Interactions with G-proteins . . . . . . . . . . . . . . 306
    4.2.   Coupling to effector systems . . . . . . . . . . . . . . 307
5.  The states of opioid receptors in the membrane . . . . . . . . 308
6.  Solubilisation and purification of opioid receptors . . . . . . 309
    6.1.   Solubilisation  . . . . . . . . . . . . . . . . . . . . 309
    6.2.   Purification  . . . . . . . . . . . . . . . . . . . . . 310
    6.3.   Affinity labelling of opioid receptors  . . . . . . . . 314
7.  Molecular biology of opioid receptors . . . . . . . . . . . . . 316
8.  Subtypes of the opioid receptor types at the molecular level . 319
References  . . . . . . . . . . . . . . . . . . . . . . . . . . . . 320
```

Chapter 12. Guanylyl cyclases as effectors of hormone and neurotransmitter receptors
Doris Koesling, Eycke Böhme and Günter Schultz 325

```
1.  Introduction . . . . . . . . . . . . . . . . . . . . . . . . . . 325
2.  Membrane-bound guanylyl cyclases . . . . . . . . . . . . . . . . 326
    2.1.   Regulation  . . . . . . . . . . . . . . . . . . . . . . 326
    2.2.   Structure . . . . . . . . . . . . . . . . . . . . . . . 329
3.  Soluble guanylyl cyclase . . . . . . . . . . . . . . . . . . . . 330
    3.1.   Regulation  . . . . . . . . . . . . . . . . . . . . . . 330
    3.2.   Structure . . . . . . . . . . . . . . . . . . . . . . . 332
4.  Related proteins . . . . . . . . . . . . . . . . . . . . . . . . 334
5.  Summary  . . . . . . . . . . . . . . . . . . . . . . . . . . . . 336
References  . . . . . . . . . . . . . . . . . . . . . . . . . . . . 336
```

Chapter 13. The elucidation of neuropeptide receptors and their subtypes through the application of molecular biology
Wolfgang Meyerhof, Mark G. Darlison and Dietmar Richter 339

1. Introduction 339
2. Receptor identification and purification 340
3. Receptor pharmacology 342
4. Receptor molecular biology 343
5. Physiological roles of neuropeptide receptors 350
6. Ontogeny of somatostatin receptors in the rat brain 352
7. Concluding remarks 353
Acknowledgements 354
References 354

Index 359

I. General Topics

CHAPTER 1

Transmitter receptors – general principles and nomenclature

FERDINAND HUCHO

Institut für Biochemie, Freie Universität Berlin, Thielallee 63, D-14195 Berlin, Germany

1. Historical aspects and definition

1.1. History

Concepts evolve, and so did the receptor concept. Concepts are based on observations, on special experimental data that are generalized when somebody overlooks enough of them and is wise enough to see the general principle in the flood of numbers, plots, and descriptions. The receptor concept emerged in the last quarter of the 19th century. The one name associated with the birth of 'receptorology' is John Newport Langley. In 1878 he published his observations on the mutually exclusive action of atropine and pilocarpin on saliva excretion by the submaxillary gland. He wrote [1]: "... we may, I think, without much rashness, assume that there is a substance or substances in the nerve endings or gland cells with which both atropine and pilocarpin are capable of forming compounds" In the same volume of the journal he wrote [2]: "On this assumption then the atropine or pilocarpin compounds are formed according to some law of which their relative mass and affinity for the substance are factors" The notion of a 'substance' forming a 'compound' was only a quarter century later turned into the more general term 'receptive substance' [3]. This concept of drug receptors evolved in parallel with Paul Ehrlich's 'Chemoreceptor theory' based on his immunological studies [4]. Actually both Langley and Ehrlich give much credit to each other and both probably would not have come to the general receptor concept without each other's thinking and special experimental lines of evidence. The third contemporary having contributed to the concept was Emil Fischer with his 'lock and key' description of proteins interacting with their substrates.

Nevertheless, 'receptor' is a technical term, just a word. The idea of substances specifically interacting with other substances dates back to Claude Bernard [5], who around the middle of the 19th century described the exact localization of the arrow poison curare interrupting (in modern terms) the nerve impulse transmission to the muscle. And to coin a word is not the end of scientific research: up to the beginning of the 1970s people were not sure whether a receptor really was more than a word. They thought it might turn out to be "...something that happens, but not a substance that can be isolated"[6]. In 1972 [7] the first receptor, the nicotinic acetylcholine receptor, *was* isolated. Nowadays, receptor research is one of the most active fields in molecular biology. A detailed history of receptorology is not, however, the aim of this Chapter. It would have to include milestones like the characterization of receptors as allosteric proteins [8,9], their first imaging by electron microscopy [lO,ll], and the introduction of recombinant DNA technology to receptor research [12,13], especially in connection with the introduction of *Xenopus* oocytes as expression systems [14]. The reader will find much of this in the chapters on special receptors. One milestone is still being awaited: crystallization and X-ray analysis of a receptor. We still have no authentic three dimensional structure of a transmitter receptor. Taking this not as a negative statement but rather as a challenge, we now turn to a summary of some principles which have emerged from almost exactly two decades of biochemical receptor research.

1.2. Definition

Receptors are proteins interacting with extracellular physiological signals and converting them into intracellular effects. Neurotransmitter receptors are integral membrane proteins; their physiological signals are neurotransmitters and neuromodulators.

Writing down this definition, problems and exceptions spring to mind, but for the time being I would like to stay with it. One should point out that it avoids discriminating against physiological signal molecules not yet defined unambiguously as transmitters. For example, many of the neuropeptides discussed in this chapter have not been classified as transmitters by the criteria summarized below. By including neuromodulators in our definition, we can react flexibly to newly published experimental data. On the other hand, it tacitly includes the so-called *orphan receptors*, receptors discovered by reverse genetics (the art of 'pulling out' clones with consensus probes homologous to sequences of well-known members of receptor families and superfamilies). Orphan receptors are proteins (actually in most cases just DNA sequences) for which the endogenous ligand, the physiological signal, has not yet been found.

The molecular definition given here is different from a more biological receptor concept that names cellular entities like the rods and cones of the retina 'photoreceptors' and the cells of the muscle spindle 'stretch receptors'. It is also different from the not-quite-past habit of some pharmacologists to call every binding site of a drug or toxin a receptor.

1.3. Criteria for calling a binding site a receptor

Most transmitter receptors are discovered by recombinant DNA and cloning techniques. Their physiological and neurochemical characterization requires expression of the protein and a detailed investigation of their biochemistry and pharmacology in the expression system as well as in the nervous system. The initial step in investigating a receptor therefore is studying its binding behavior (Chapter 2). Binding studies often are also the primary access to a receptor to be discovered in an animal. However, in this case several criteria have to be observed to call a ligand-binding site a true receptor [15]:
1. The ligand binding site has to be saturable. Half-maximal saturation should be in the same concentration range as the physiological concentration of the ligand.
2. The other essential criterium is specificity: the binding site should bind primarily the physiological ligand and structurally closely related compounds. It should definitely discriminate against the 'wrong' stereoisomer.
3. A third criterium is the location of the binding site under investigation within the organism. There are reports in the literature about high-affinity saturable insulin binding to glass beads, opiates binding to talcum (from the rubber gloves of the experimenter), and cholinergic ligands binding to macrocyclic organic compounds, interesting artifacts but obviously no receptors. Synaptic localization in the pre- or postsynaptic membrane may be a strong indication of a binding site being a true receptor, although one should be open-minded for extrasynaptic localizations as well.

1.4. Agonists - antagonists

In many cases it is feasible to investigate a receptor using not its physiological signal molecule, but rather an artificial ligand that may be more preferred because of its stability, selectivity, high affinity, or its being available in a radioactively labeled form. Historically, several receptors have been discovered by means of such ligands before the physiological ligand was even known. For example, the opiate receptors were detected before the discovery of the enkephalins, the endogenous opiates.

This pharmacological approach adds to our tools the large group of *antagonists*. All neurotransmitters discovered so far are *agonists*, molecules triggering an effect in the target cell after binding to its receptor. Antagonists, on the other hand, are compounds that bind to the same receptor, with a similar or even (in most cases) higher affinity, without a consequence other than preventing the agonist from exerting its effect. *Competitive antagonists* compete with the agonist for the same or an overlapping binding site on the receptor protein. *Noncompetitive antagonists* also block the effect of the agonist, but bind to a site distinct from the agonist-binding site. They do not inhibit and they may even promote agonist binding. A fourth class of receptor ligands is the *inverse agonists* discovered with the GABA/benzodiazepine receptors.

They exert effects opposite to those of the agonist. The transmitter GABA acts as an anxiolytic and muscle relaxant, while inverse agonists cause, depending on the receptor occupancy, arousal, anxiety, and seizures (see chapter 6). Some of these compounds are presently under investigation as nootropics. Of interest for pharmacologists are the *partial agonists* (or mixed agonists/antagonists). These are compounds that bind with high affinity but exert only weak effects. Being (weak) agonists they antagonize the effect of strong agonists.

For a detailed discussion of the criteria classifying an endogenous substance as a transmitter, the reader is referred to standard textbooks of neurochemistry [15] and neurobiology [16]. Briefly, transmitters should be synthesized, stored within, and released upon stimulation from the neuron were its action is observed. Furthermore, its subsequent removal from the synaptic cleft should be secured. Here I wish to point out again that all neurotransmitters detected so far are agonists. One could envision antagonistic neurotransmitters, a possibility indicated by cGMP, the 'transmitter' conveying the signal from rhodopsin in the photoreceptor's disk membrane to sodium channels of the rod's plasma membrane. The 'signal' (a photon) causes cleavage of cGMP. The 'message' arriving at the plasma membrane is negative, antagonistic in the exact definition of the word: its effect is to stop rather than to trigger an ion flux. (This action is not to be confused with inhibitory chemical transmission. GABA and glycine are inhibitory transmitters that act agonistically on anion instead of cation fluxes).

It remains to be seen whether or not there are antagonistic neurotransmitters. Another open question pivotal to our understanding of the mechanism of transmitter receptor functioning is: what is the structural and chemical difference between an agonist and an antagonist Perhaps the pharmacologists working with muscarinic cholinergic receptors are closest to answering this question. But, in general, we can state: by looking at the structural formula of a compound, we cannot yet predict whether it will act as an agonist or antagonist.

1.5. Mechanism of receptor function

1.5.1. The triune receptor model

Transmitter receptors are signal converters, operationally containing three functional moieties (Fig. 1): a signal-receiving (ligand-binding) part ('R'), an **effector** ('E'), and a **t**ransducer ('T'), coupling the former two and transducing the signal from the binding site to the effector.

The description of receptors as signal converters implies that the extracellular signal is different from the signal released from the membrane to the intracellular space. Transmitters are recognized and bound at the extracellular surface of the receptor protein. They do not enter the cell. Being 'first messengers' their 'message' is converted into an intracellular effect conveyed by 'second messengers', produced by the effector moiety of the receptor.

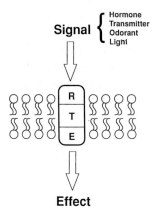

Fig. 1. Receptors as signal converters. The triune receptor concept (for R, T, E: see text).

The signal-converter concept may be considered analogous to electronic signal converters in which the input is also different from the output. An implication of this analogy is the possibility to regulate the intensity of the transmitted signal (see chapter 3). Receptors may enhance (amplify) or diminish a signal. Furthermore, they can integrate various equal or even different signals by a phenomenon called *receptor crosstalk*.

'R' is the receptor in the strict sense of the word; 'E', the effector moiety, is in many cases an ion channel, in others an enzyme. 'T', the transducer, is a protein structure coupling 'R' with 'E'. A *receptor classification* is based on the structural properties of the three functional moieties (1.7.3).

1.5.2. Transmitter receptors are allosteric proteins
One conclusion from the description of receptors given above is the allosteric receptor model [8,9]: 'R' and 'E' interact like regulatory and active sites in allosteric enzymes [17]. Both homotropic and heterotropic interactions are observed: at the neuromuscular synapse a sigmoidal dose-response curve indicates cooperative binding and channel opening by sequential binding of two acetylcholine molecules at the nicotinic acetylcholine receptor, a typical homotropic interaction. Glycine at the NMDA subtype of glutamate receptors, benzodiazepines at the $GABA_A$ receptor, and various noncompetitive antagonists at the nicotinic acetylcholine receptor act as classical heterotropic effectors, like CTP at the enzyme aspartate transcarbamylase. Attempts have been made to apply the two-state symmetry formalism to receptors [18]. According to this model the T-state would be the inactive and the R-state

$$\text{Antagonist} + T \rightleftharpoons R + \text{Agonist}$$
$$\updownarrow \qquad \updownarrow$$
$$\text{Antagonist} \quad T \quad R^*\text{Agonist}$$

the active state of the receptor, both being at equilibrium in the membrane at rest. Agonists would be compounds binding preferentially to the R-state, while competitive antagonist molecules bind to the T-state. Receptor activation in this concept is described as a reversible shift of the equilibrium between the two preformed receptor states caused by agonist binding to the R-state. Plausible as this description may be, the experimental evidence is unsatisfactory. The actual situation seems to be more in agreement with Koshland's induced fit model.

$$
\begin{array}{rl}
R + \text{Agonist} & \rightarrow R^*\text{Agonist} \xrightarrow{\uparrow \text{Agonist}} R \\
+ \text{Antagonist} & \rightarrow R\,\text{Antagonist} \xrightarrow[\downarrow \text{Antagonist}]{} R
\end{array}
$$

(R*: activated receptor)

1.6. Receptor diversity is greater than transmitter diversity

The number of neurotransmitters clearly identified and matching the accepted transmitter criteria is smaller than expected only a few years ago. The list of canonical and 'putative' neurotransmitters comprises about a dozen substances. Undoubtedly, several new neurotransmitters will be discovered, but their number and diversity will hardly explain the functional diversity of synapses in the nervous system. On the other hand, the number of newly discovered receptors and receptor genes presently seems to be in a logarithmic growth phase. (Although it seems to be approaching its end.) Most receptors for the classical transmitters have been cloned now, including their many subunits and subtypes. Only the odorant receptors may still considerably increase the number of cloned receptors. This number exceeds by far the number of identified neurotransmitters. Instead of one receptor for each transmitter, we are faced with *receptor families*, with groups of *receptor subtypes* that are based on the following principles of variablity.

The triune receptor concept introduces a parameter of variability. 'R' can be combined with different 'T' and 'E' moieties yielding completely different receptors. A dopamine receptor (D2), for example, combined with the coupling protein G_i inhibits its 'E', adenylyl cyclase. The dopamine receptor (D_1), combined with G_s, stimulates it. The receptors for adrenaline can stimulate cAMP synthesis when combined with a G_s and adenylyl cyclase (β_1, β_2); but in combination with another 'T' and 'E', they can either regulate the PI-response (α_{1A-C}) or inhibit adenylyl cyclase (α_{2A-C}). This permutation of the interacting components requires mutations in the genes coding for the respective proteins; i.e. all these molecules are iso-proteins coded by homologous genes or at least products of alternative splicing.

Another principle of variability is introduced by some receptor's quaternary structure. The GABA$_A$ receptors, for example, are composed of several homologous but different subunits (see chapter 6). Six different α, three β, three γ, and several other subunits, a total of sixteen, are presently known. Co-expression of various combinations of these in *Xenopus* oocytes results in receptors of very different properties: the effect of benzodiazepines on GABAbinding for example requires the presence of the γ_2-subunit. It is assumed that tissue-specific expression of certain genes results in different receptors and different GABA (and drug) effects in the nervous system. In the special receptor chapters of this book, more examples of receptor diversity through subunit combination can be found.

Finally, a further means of increasing receptor diversity is described in chapter 11: a glutamate receptor of the Glu R2-type has been shown to be made Ca^{2+}-impermeable by a single amino acid exchange in the second putative transmembrane helix (Glu600Arg600). Interestingly, this change does not represent an altered gene; rather, it is probably brought about by mRNA editing (21). It remains to be seen whether this mechanism of generating receptor diversity is a more general one.

1.7. Receptor nomenclature

1.7.1. IUPHAR rules of receptor nomenclature

With this diversity receptor nomenclature is a problem, especially at this early stage of research. With newly discovered receptors and subunits being published continuously, no final nomenclature is possible. The most consistent and widely accepted receptor nomenclature is compiled and updated regularly by *Trends in Pharmacological Sciences* (TIPS). This compilation has been adopted in this book and is reproduced here in its most recent version [19].

1.7.2. The TiPS receptor nomenclature supplement 1993
(see appendix to this chapter [p. 15 ff.]

1.7.3. Receptor classification

At present, neurotransmitter receptors are classified according to structural principles (Fig. 2). *Type-I receptors* are molecules with 'R', 'T', and 'E' being integral parts of one protein molecule and 'E' being an ion channel. In *type-II receptors* 'R', 'T', and 'E' are distinct and separable proteins; 'T' in these receptors is a coupling or G protein, 'E' is in most cases an enzyme, but it can also be an ion channel. In brief these two classes are called *ligand-gated ion channels* or *ionotropic receptors* (type-I) and *G-coupled* or *metabotropic receptors* (type-II), respectively. A third class, *type-III receptors*, has enzymatic activity instead of an ion channel as the integral part of its protein structure. This receptortype is better known from the field of hormone receptors like EGF, PDGF, and insulin receptors, but guanylyl cyclase as a receptor of the neurotransmitter NO and of ANF (atrial natriuretic factor) justifies its presence in our classification. We shall later see that the membranes of the three receptor types

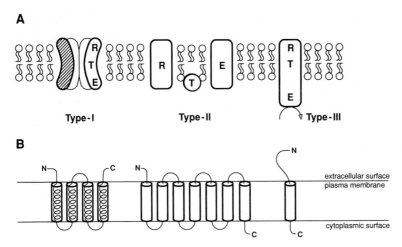

Fig. 2. Receptor classification. Structural features of receptor classes.

have significant structural features in common. Actually, they are now seen as *receptor superfamilies*, each originating from one common ancestral receptor early in the evolution.

From the distribution of hydrophobic amino acids along the primary structures, it was concluded that subunits of type-I receptors form four transmembrane helices (4TM), type-II receptors seven (7TM), and type-III receptors one (1TM). In this book the nicotinic acetylcholine receptor is described as the prototype of type-I receptors (chapter 4), the β-adrenoceptor of type-II (chapter 5), and guanylyl cyclases of type-III receptors (chapter 15). Table I summarizes this receptor classification.

Other receptor classifications, based e.g. on functional properties (excitatory vs. inhibitory) or on the type of transmitter to which it responds, are less convincing,

TABLE I

Receptor class	Type-I	Type-II	Type-III
Alternative name	Ligand-gated ion channels	G-coupled receptors	Tyrosine-kinase receptor (TKR)
Putative trans-membrane helices	4	7	1
Effector (E)	ion channel	enzyme (ion channel)	enzyme
R, T, E	integral part of receptor protein	separate protein entities	integral part of receptor protein
Examples	nAChR Glycine R GABA$_A$ R	mAChR β-adrenergic R Neuropeptide R	receptors for growth hormones (EGF, PDGF)

because receptors of one class of this definition would not have enough in common. Excitatory receptors can be G-coupled (e.g. the β-adrenoceptors) or ligand-gated ion channels (e.g. the nicotinic acetylcholine receptors). They can be 7TM or 4TM. Vice versa, the above examples show that different receptors using the same neurotransmitter can be excitatory or inhibitory, ionotropic or metabotropic. However, even within a family of metabotropic receptors, functional variability does not permit a functional classification: acetylcholine receptors of the muscarinic subtype regulate the intracellular cAMP level (M_2, M_4) or the inositol phosphate (PI) response (M_1, M_3, M_4). The main common feature of the muscarinic acetylcholine receptors seems to be their 7TM structure. Unsatisfactory as this may be, the number of putative transmembrane helices is at present the single most widely used classification criterium.

1.7.4. Families and superfamilies
The ever-growing number of receptors is grouped in *families* and *superfamilies*. Members of a family are receptors that react to the same neurotransmitter, e.g. the dopamine, serotonin, glutamate receptor families, and are pharmacologically placed in one group like the muscarinic acetylcholine receptors, the nicotinic acetylcholine receptors, and the α- and the β-adrenoceptors, etc. Structural information obtained from cDNA and gene cloning and sequencing suggests that these families may be evolutionarily related. The ligand-gated ion channels (4TM, type-I receptors) show striking sequence homologies (see below). Furthermore, a general pattern seems to be that the second transmembrane helix (M2) forms the wall of the channel structure. Based on these observations, type-I receptors are seen as a *superfamily* of homologous proteins. Similar considerations have placed the G-coupled receptors (7TM, type-II receptors) in one superfamily of homologous proteins.

1.8. Structural features

1.8.1. Ligand-gated ion channels (type-I receptors)
Primary structures of many nicotinic acetylcholine receptors (neuronal and muscle-like), several glutamate, glycine, and $GABA_A$ receptors, and one ionotropic serotonin receptor (5-HT_3) are presently known. With the exception of the glutamate receptors, these structures are significantly similar, placing them in a superfamily possibly derived by divergent evolution from a common ancestor [22–24]. The ionotropic glutamate receptors, especially their putative ion channel-forming domains, may have originated by convergent evolution from a different ancestral gene. Hydropathy plots of these sequences each disclose four hydrophobic sequences, long enough to form a membrane-spanning α-helix (Fig. 3). Therefore, they are classified as 4TM receptors. In all of them the transmembrane segment M2 has characteristics similar to the channel-forming elements found in the nicotinic acetylcholine receptor, especially several serine and threonine residues. Leaving aside the glutamate recep-

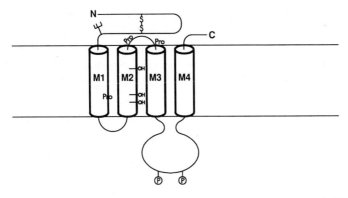

Fig. 3. Ligand-gated ion channels. General properties: -s-s-, disulfide bridge; -OH, serine and threonine residues on the putative transmembrane helix M2; Ψ, carbohydrate residue preceding M1; Pro, proline residues; and P, phosphorylation sites on the intracellular loop connecting putative transmembrane helices M3 and M4.

tors, that seem to comprise another superfamily, all ligand-gated ion channels have the following structural features in common: a disulfide bridge forming a sequence loop of thirteen amino acids; a proline residue in the middle of M1, several other proline residues, especially two in the loop connecting M2 and M3; the accumulation of serine and threonine residues in M2, already mentioned above; and potential N-glycosylation sites in the extracellular domain preceding the four transmembrane segments.

Very little can be said about the quaternary structures of the type-I receptors. Both the nicotinic acetylcholine receptors and the glycine receptors have been shown to be heteropentamers. A pentameric structure may be postulated for the other members of this class as well, but experimental proof is lacking. Primarily, it remains to be elucidated to what extent homo- or heteropolymers are expressed in vivo and what the role of subunit interactions is in receptor function. Even less can be said about their tertiary structures. Their elucidation by X-ray crystallography is still awaiting appropriate receptor crystals.

1.8.2. G-coupled receptors (type-II receptors)

At present the primary structures of more than three dozen mammalian and more than a dozen nonmammalian 7TM receptors are known (not counting the sequences of homologous receptors from different species) [25]. Including the odorant receptors, hundreds of more or less homologous sequences are to be expected. Several structural principles emerge from this wealth of information.

All these type-II receptors seem to have a common evolutionary origin. The overall sequence homology is striking, but it is strongest within the seven hydrophobic stretches presumed to form transmembrane helices. This points to a functional role of the helices beyond serving as membrane anchors (see below). The N- and C-terminal and intracellular loop sequences are the most divergent parts of the primary structures.

All these receptors consist of only one polypeptide chain on which the domains involved in ligand binding, G-protein coupling, and regulation including phosphorylation are located. They do not seem to possess a cleaved signal sequence and most of their genes do not contain introns. Although they selectively process signals ranging from light, odorants, and peptide hormones to neurotransmitters of various kinds, they all transmit, amplify, and integrate extracellular signals and activate via G-proteins effector systems releasing second messengers to the intracellular space.

The agonist-binding site is located within the bundle of transmembrane helices, buried approximately in the middle of the membrane. The antagonist-binding sites overlap, but are not identical to the agonist-binding sites.

Functional domains have been localized within the primary and the predicted secondary structures only for model receptors of this class (e.g. rhodopsin, β-receptors). Generalizations therefore must await analogous experiments with the other receptors. Typical experimental approaches are: site-directed mutagenesis; construction of chimeric receptors; photoaffinity labeling; and competition with synthetic peptides. One conclusion from this type of experiment is that there are several points of contact between receptor and G-protein. Another conclusion from chimeras produced with α_2- and β_2-receptors was that the specificity of antagonist binding changes with the transmembrane helix M7, while agonist binding changes gradually with the overall change of sequence.

References

1. Langley, J.N. (1878) J. Physiol. 1, 267–295.
2. Langley, J.N. (1878) J.Physiol. 1, 339–369.
3. Langley, J.N. (1905) J.Physiol. 33, 374–413.
4. Witkop, B.(1981) Naturwiss. Rundsch. 34, 361–379.
5. Bernard, M.C. (1857) Leçon sur les effets des substances toxiques et medicamenteuses, pp. 238–306, Baillière, Paris.
6. Silver, A. (1974) The Biology of Cholinesterases, Frontiers of Biology 36, p. 43, North-Holland Publishing Company, Amsterdam.
7. Olsen, R., Meunier, J.C. and Changeux, J.-P. (1972) FEBS Lett. 28, 96–100.
8. Karlin, A. (1967) J. Theor. Biol. 16, 306–318.
9. Changeux, J.-P. (1981) Harvey Lect. 75, 85–254.
10. Cartaud, J., Benedetti, L., Cohen, J.B., Meunier, J.C. and Changeux, J.-P. (1973) FEBS Lett. 33, 109–113.
11. Nickel, E. and Potter, L.-T. (1973) Brain Res. 57, 508–517.
12. Sumikawa, K., Houghton, M., Smith, J.C., Richards, B.M. and Barnard, E.A. (1982). Nucleic Acids Res. 10, 5809–5822.
13. Noda, M., Takahashi, H., Tanabe, T., Toyosato, M., Furutani, Y., Hirose, T., Asai, M., Inayama, S., Miyata, T. and Numa, S. (1982) Nature 299, 793–797.

14. Barnard, E.A., Miledi, R. and Sumikawa, K. (1982) Proc. R. Soc. Lond. Ser. B. 215, 241–246.
15. Hucho, F. (1986) Neurochemistry, VCH, Weinheim.
16. Hall, Z. (1992) Molecular Neurobiology, Sinauer Associates Inc. Publishers, Sunderland.
17. Changeux, J.-P. (1988/89) FIDIA Research Foundation Neuroscience Award Lecture, Vol. 4, Raven Press, New York.
18. Changeux, J.-P., Devillers-Thiéry, A. and Chemouilli, P. (1984) Science 225, 1335–1345.
19. Receptor Nomenclature Supplement, TIPS, 13 1993.
20. Green, J.-P. (1990) TIPS 11, 13–16.
21. Sommer, B., Kohler, M., Sprengel, R. and Seeburg, P.H. (1992) Cell 57, 11–19.
22. Stroud, R.M., McCarthy, M.P. and Schuster, M. (1990) Biochemistry 29, 11010–11023.
23. Betz, H. (1990) Biochemistry 29, 3597–3599.
24. Betz, H. (1990) Neuron 5, 383–392.
25. O'Dowd, B.F., Lefkowitz, R.J. and Caron, M.G. (1989) Annu. Rev. Neurosci. 12, 67–83.

TiPS Receptor Nomenclature Supplement 1993*

*The Editor of the present volume gratefully acknowledges the use of the TiPS Receptor Nomenclature Supplement, which appeared in the March 1993 issue of TiPS.

TiPS Receptor Nomenclature Supplement 1993

NO FORMAL system for the rational classification of receptors exists. This has led to inconsistencies in the use of terminology, which in turn has led to confusion.

Our decisions about nomenclature to be used in TiPS are included in this Supplement. This is not an attempt to classify receptors; we are simply reporting the most widely accepted nomenclature for receptors and their subtypes and defining them in terms of what is known of their pharmacology, biochemistry and molecular biology. Because of their particular pharmacological significance, we have also included nomenclature for subtypes of certain ion channels.

Certain ground rules have, however, been applied. A receptor is defined as a binding site with functional correlates. Cloned receptors for which endogenous expression has not been demonstrated, binding sites without an identified functional correlate and receptor subtypes proposed on the basis of preliminary functional data are included as footnotes.

Receptors are subtyped on the basis of different pharmacological profiles rather than on distinct second-messenger/signal-transduction pathways or primary structures.

Channels have been subtyped on the basis of electrophysiological properties and/or modulating ligands.

Information has been compiled with the help of several experts in each field and, where applicable, the chairperson of the relevant IUPHAR receptor nomenclature committee.

This is the fourth edition of the Supplement; the fifth edition will be published in March 1994.

Steve Watson and Debbie Girdlestone*

Department of Pharmacology
Mansfield Road
Oxford OX1 3QT

*Elsevier Trends Journals
68 Hills Road
Cambridge CB2 1LA

Contents

Key to headings	2	GABA	15
		Galanin	16
Adenosine (P$_1$ purinoceptors)	3	Glutamate (ionotropic)	17
α$_1$-Adrenoceptor	4	Glutamate (metabotropic)	18
α$_2$-Adrenoceptor	5	Glycine	19
β-Adrenoceptor	6	Histamine	20
Angiotensin	7	5-HT	21
Atrial natriuretic peptide	8	Leukotriene	23
Bombesin	9	Melatonin	24
Bradykinin	10	Muscarinic	25
Calcitonin gene-related peptide	11	Neuropeptide Y	26
Cannabinoid	11	Neurotensin	27
CCK and gastrin	12	Nicotinic	28
Dopamine	13	Opioid	29
Endothelin	14	PAF	30
		Prostanoid	31
		P$_2$ Purinoceptor	32
		Somatostatin	33
		Tachykinin	34
		Vasopressin and oxytocin	35
		VIP	36
		Ca^{2+} channels	37
		K$^+$ channels	38
		Na$^+$ channels	41
		Chemical names	42

Key to headings

Nomenclature	most commonly used nomenclature
Alternative/previous names	nomenclature that is less popular or no longer used
Potency order	approximate rank order of potency of endogenous ligands
Selective agonists	where possible, agonists have a selectivity of at least two orders of magnitude
Selective antagonists	where possible, antagonists have a selectivity of at least two orders of magnitude (values in parentheses are pA_2 estimates derived from functional studies, if available)
Channel blockers	commonly used blockers of ion channels intrinsic to the receptor
Radioligands	the most selective ligands have been included (values in parentheses are K_d estimates for antagonists derived from binding studies)
Predominant effectors	effector pathways usually associated with receptor activation
	cAMP — stimulation (↑) or inhibition (↓) of adenylyl cyclase
	cGMP — stimulation (↑) or inhibition (↓) of guanylyl cyclase
	IP_3/DG — stimulation of phosphoinositide metabolism
	int. ion channel — channel intrinsic to the receptor
	channel (G) — stimulation (↑) or inhibition (↓) of ion channel regulated by G protein
Gene	name of gene whose product displays pharmacology close to native protein
Structural information	aa — amino acids
	species — species in which the receptor has been cloned
	Pxxxxx — accession number (SwissProt) for protein sequence (accession numbers of nucleotide sequences are not shown because of their multiplicity)
	TM — predicted number of transmembrane domain(s)
Other receptors/binding sites	includes newly cloned receptor subtypes with unknown physiological significance, binding sites that have no known functional correlate and receptor subtypes proposed on the basis of preliminary functional data
Endogenous ligands	naturally occurring agonists found in mammals

Adenosine receptors*

Nomenclature	A_1	A_{2A}	A_{2B}[†]
Previous names	R_i	A_{2a}, R_a	A_{2b}, R_a
Selective agonists	N^6-cyclopentyladenosine 2-Cl-N^6-cyclopentyladenosine	CGS21680 PAPA-APEC	—
Selective antagonists	DPCPX (8.3–9.3) 8-cyclopentyltheophylline (7.4)	CP66713 (7.7)[§] KF17837 (K_d ~8 nM)[§]	—
Radioligands	[^{3}H]DPCPX (0.4 nM) [^{3}H]cyclopentyladenosine [^{125}I]APNEA	[^{3}H]CGS21680 [^{3}H]NECA[¶] [^{125}I]PAPA-APEC	[^{3}H]NECA[¶]
Predominant effectors	cAMP↓* K$^+$ channel↑ (G) Ca^{2+} channel↓ (G)	cAMP↑	cAMP↑
Gene	*a1* (*rdc7* canine)	*a2A* (*rdc8* canine; *DT35* rat)	*a2B* (*RFL9* rat)
Structural information	326 aa human 7TM 326 aa rat P25099 7TM 326 aa canine P11616 7TM 326 aa bovine P28190 7TM	409 aa human P29274 7TM 410 aa rat 7TM 412 aa canine P11617 7TM	328 aa human P29275 7TM 332 aa rat P29276 7TM

Other receptors/binding sites: An A_3 receptor with 7TM and 320 aa has been cloned in rat; the expressed receptor has been shown to inhibit adenylyl cyclase but the properties of the endogenous receptor are not known. No xanthine antagonist has been described, although CGS15943 is a nonselective antagonist; it can be radiolabelled with [^{3}H]NECA or [^{125}I]APNEA.
A novel binding site has been identified with [^{3}H]CV1808.
Adenosine binds to an intracellular site on the catalytic subunit of adenylyl cyclase (P-site) causing enzyme inhibition.

*also named P_1 purinoceptors
[†]the pharmacology of this receptor and the differences from the A_{2A} receptor are not established
[§]selectivity is no more than tenfold
[¶]NECA also binds to a ubiquitous low-affinity, high-capacity site that is similar to a heat shock protein
the A_1 receptor has also been shown to inhibit or stimulate hydrolysis of phosphoinositides depending on cell type

α_1-Adrenoceptors

Nomenclature	α_{1A}	α_{1B}	α_{1C}	α_{1D}*
Previous names	α_{1a}	α_{1b}	–	α_{1A}
Potency order	NA ≥ adrenaline	adrenaline = NA	adrenaline = NA	adrenaline = NA
Selective agonists	–	–	–	–
Selective antagonists	WB4101 (~9.2)†	CEC (irreversible)	WB4101 (~9.2)† CEC (irreversible)	WB4101 (~9.2)
Predominant effectors	IP$_3$/DG§	IP$_3$/DG	IP$_3$/DG	–
Gene	–	$\alpha 1B$-C5	$\alpha 1C$-C8	$\alpha 1A$-C20
Structural information	–	515 aa rat P15823 7TM 515 aa hamster P18841 7TM	466 aa bovine P18130 7TM	560 aa rat 7TM

Comment: Examples of agonists selective for the α_1-adrenoceptor class relative to the α_2-adrenoceptor class are phenylephrine, methoxamine and cirazoline; antagonists selective for α_1-adrenoceptors relative to α_2-adrenoceptors are prazosin (8.5–10.5) and corynanthine (6.5–7.5); radioligands are [^{3}H]prazosin (0.1 nM) and [^{125}I]HEAT (0.1 nM; also known as BE2254).

*the cloned rat α_{1D}-adrenoceptor was previously thought to correspond to the endogenously expressed α_{1A}-adrenoceptor

†other antagonists with a higher affinity for α_{1A}- and α_{1C}-adrenoceptors relative to α_{1B}- and α_{1D}-adrenoceptors include 5-methylurapidil (pA$_2$ = 9.2) and (+)-niguldipine (pA$_2$ = 10.0; also has high affinity for L-type Ca^{2+} channels)

§has also been reported to stimulate Ca^{2+} influx

α_2-Adrenoceptors

	α_{2A}	α_{2B}	α_{2C}
Nomenclature			
Potency order	adrenaline ≥ NA	adrenaline ≥ NA	adrenaline ≥ NA
Selective agonists	oxymetazoline (weak partial agonist)	–	–
Selective antagonists	–	prazosin (~7.5) ARC239 (~8.0)	prazosin (~7.5) ARC239 (~8.0)
Predominant effectors	cAMP↓ K⁺ channel↑ (G) Ca²⁺ channel↓ (G)	cAMP↓ Ca²⁺ channel↓ (G)	cAMP↓
Gene	$\alpha2A$–C10	$\alpha2B$–C2 ($rg\alpha2$ rat)	$\alpha2C$–C4 ($rg10\alpha2$ rat)
Structural information	450 aa human P08913 7TM 450 aa porcine P18871 7TM	450 aa human P18825 7TM 453 aa rat P19328 7TM	461 aa human P19328 7TM 450 aa rat P18089 7TM

Other receptors/binding sites: Evidence from binding studies in the rat submaxillary gland, bovine pineal gland and in COS cells transfected with the gene rg20 (P22909) suggests a fourth α_2-adrenoceptor, α_{2D}; rg20 may be the rat homologue of human α_{2A}. The antagonist SKF104078 has high affinity for the above receptors but also binds to a low-affinity site. Binding sites for imidazolines have been identified that have no known functional correlate; catecholamines have a low affinity for these sites.

Comment: Examples of agonists selective for the α_2-adrenoceptor class relative to the α_1-adrenoceptor class are UK14304, BHT920 and clonidine (historical); antagonists selective for α_2-adrenoceptors relative to α_1-adrenoceptors are rauwolscine (7.5–9.0) and yohimbine (7.5–9.0); radioligands are [³H]rauwolscine (1 nM), [³H]yohimbine (1 nM), [³H]UK14304 and [³H]RX821002.

β-Adrenoceptors

	β_1	β_2	β_3
Nomenclature	β_1	β_2	β_3
Previous names	–	–	atypical β
Potency order	NA ≥ adrenaline	adrenaline > NA	NA > adrenaline
Selective agonists	noradrenaline* xamoterol*	procaterol	BRL37344
Selective antagonists	CGP20712A (8.5–9.3) betaxolol (8.5) atenolol (7.0)	ICI118551 (8.3–9.2) butaxamine (6.2) α-methylpropranolol (8.5)	–†
Radioligands	[^{3}H]bisoprolol (10 nM)	[^{3}H]ICI118551 (10 nM)	[^{125}I]iodocyanopindolol (0.5 nM)
Predominant effectors	cAMP↑§	cAMP↑	cAMP↑
Gene	β1	β2	β3
Structural information	477 aa human P08588 7TM 466 aa rat P18090 7TM	413 aa human P07550 7TM 418 aa rat P10608 7TM 418 aa mouse P18762 7TM 418 aa hamster P04274 7TM	402 aa human P13945 7TM 388 aa mouse P25962 7TM

Other receptors/binding sites: There is evidence for the existence of subtypes of β$_3$-adrenoceptors, e.g. propranolol and alprenolol have weak activity at human β$_3$-adrenoceptors expressed in CHO cells in contrast to their marked antilipolytic activity in human adipocytes.

*selective relative to β$_2$-adrenoceptors
†many standard β-adrenoceptor antagonists have very low affinities (e.g. CGP20712A, ICI118551); (–)-pindolol is a partial agonist at β$_3$-adrenoceptors
§also Ca^{2+} channel↑ (G) in heart

Angiotensin receptors

Nomenclature*	AT_1	AT_2†
Previous names	AII-1, AII$_\alpha$, AII-B	AII-2, AII$_\beta$, AII-A
Potency order	AII > AIII	AII = AIII
Selective agonists	–	–
Selective antagonists	losartan (8.1) EXP31274 (10.0) SKF108566 (9.7)	PD123177 CGP42112A
Radioligands	[^{3}H]losartan (6.4 nM) [^{125}I]EXP985 (1.5 nM) [^{3}H]- or [^{125}I]AII	[^{125}I]CGP42112A [^{3}H]- or [^{125}I]AII
Predominant effectors	IP$_3$/DG cAMP↓	cGMP↓
Gene	at1	–
Structural information	359 aa human 7TM 359 aa rat P25095 7TM 359 aa mouse 7TM 359 aa bovine 25104 7TM	–

Other receptors/binding sites: Two different rat AT$_1$ receptors, AT$_{1A}$ and AT$_{1B}$ (P29089; 359 aa), with 94% homology have been sequenced; there is no evidence for functional differences, although small differences in pharmacology have been described. Receptors with similar affinity for losartan and PD123177 have been described in several species including rat, rabbit, mouse, frog and domestic fowl.

Endogenous ligands: angiotensin II (**AII**), angiotensin III (**AIII**); angiotensin I is weakly active in some systems

*nomenclature as agreed by the 'Nomenclature Committee of the Council for High Blood Pressure Research (1990)' *Hypertension*, 17, 720–721
†a functional role of the AT$_2$ receptor has only been reported in cultured neurones and in anaesthetized dogs

Atrial natriuretic peptide receptors

Nomenclature	ANP$_A$	ANP$_B$
Alternative names	GC-A	GC-B
Potency order	ANP > BNP	CNP >> ANP = BNP
Selective agonists	–	CNP
Selective antagonists	[L-α-aminosuberic acid$^{7,23'}$]-β-ANP$_{7-28}$ (7.5)	–
Radioligands	[^{125}I]ANP	[^{125}I]CNP
Predominant effectors	cGMP↑	cGMP↑
Gene	*anpA*	*anpB*
Structural information	1029 aa human P16066 1TM 1057 aa rat P18910 1TM 1057 aa mouse P18293 1TM	1025 aa human P20594 1TM

Other receptors/binding sites: Molecular cloning has defined a novel binding site (ANP$_C$; previously named ANP-R$_2$) which does not contain intrinsic guanylyl cyclase activity. This site was originally thought to be involved in the clearance of α-ANP, although recent evidence suggests that it may be linked to the cAMP or IP$_3$/DG pathways. It consists of 540 aa (human; P17342), exists as a disulfide-linked homodimer with 1TM and can be radiolabelled with iodinated α-ANP.

Endogenous ligands: α-atrial natriuretic peptide (**ANP**), brain natriuretic peptide (**BNP**), type-C natriuretic peptide (**CNP**); α-ANP may derive from γ-ANP; the physiological role and action of β-ANP is uncertain

Bombesin receptors

Nomenclature	BB$_1$	BB$_2$
Alternative names	neuromedin B-preferring	GRP-preferring
Potency order*	NMB ≥ bombesin > GRP	GRP ≥ bombesin >> NMB
Selective agonists	NMB	–
Selective antagonists	–	[DPhe^6,Cpa^{14},ψ13–14]bombesin$_{6-14}$ (10 nM)† [DPhe^6]bombesin$_{6-13}$ ethyl ester (2–10 nM)† AcGRP$_{20-26}$ ethyl ester (8.0)
Radioligands	[^{125}I]BH-NMB [^{125}I]-[Tyr^4]bombesin	[^{125}I]-[DTyr^6]bombesin$_{6-13}$ methyl ester [^{125}I]GRP [^{125}I]-[Tyr^4]bombesin
Predominant effectors	IP$_3$/DG	IP$_3$/DG
Gene	bb1	bb2
Structural information	390 aa human 7TM 390 aa rat 7TM	384 aa human 7TM 384 aa mouse P21729 7TM

Other receptors/binding sites: A BB$_3$ receptor has been cloned in guinea-pig and human; the guinea-pig receptor has 399 aa and 7TM. It can be radiolabelled with [^{125}I]GRP or [^{125}I]-[Tyr^4]bombesin. An extensive pharmacological characterization of this receptor has not been determined.

Endogenous ligands: gastrin releasing peptide (**GRP**), neuromedin B (**NMB**), GRP$_{18-27}$ (previous name - neuromedin C)§

*bombesin is not found in mammals
†K$_d$ values derived in competition studies with [^{125}I]-[Tyr^4]bombesin
§nomenclature as agreed in the symposium 'Bombesin-like Peptides in Health and Disease - Rome 1987'

Bradykinin receptors

Nomenclature	B_1	B_2
Previous names	BK_1	BK_2
Potency order	BK_{1-8} = kallidin > BK	kallidin ≥ BK >> BK_{1-8}
Selective agonists	BK_{1-8}* Sar[DPhe⁸]BK_{1-8}	[Phe⁸ψ(CH_2-NH)Arg⁹]BK [Hyp³,Tyr(Me)⁸]BK
Selective antagonists	[Leu⁸]BK_{1-8} (6.7–7.3) [des-Arg¹⁰]HOE140	HOE140 (7.5–10.5) DArg[Hyp³,Thi⁵,⁸,DPhe⁷]BK (6.0–7.5) DArg[Hyp³,Thi⁵,HypE(*trans*-propyl)⁷,Oic⁸]BK (8.5–9.6) DArg[Hyp³,DPhe⁷,Leu⁸]BK (6.8–7.9)
Radioligands	[³H]BK_{1-8}	[³H]DArg[Pro²,Hyp³,Thi⁵,DTic⁷,Tic⁸]BK [¹²⁵I]-[Tyr⁸]BK [³H]BK
Predominant effectors	–	IP_3/DG
Gene	–	*b2*
Structural information	–	364 aa human 7TM 366 aa rat (P25023) 7TM

Other receptors/binding sites: The existence of B_3 receptors in guinea-pig airways has been proposed. Certain [DPhe⁷]-substituted analogues of bradykinin are antagonists at B_2 receptors but are claimed to be inactive at B_3 receptors. The variation in pA_2 values of antagonists at B_2 receptors may reflect the existence of receptor subtypes or species homologues.

Endogenous ligands: bradykinin (**BK**), BK_{1-8}, kallidin (Lys-BK), T-kinin (Ile-Ser-BK), [Hyp³]bradykinin, [Hyp⁴]kallidin

*[des-Arg⁹]BK; the presence of B_1 receptors is most easily indicated by the activity of BK_{1-8}

Calcitonin gene-related peptide receptor

The receptor can be radiolabelled with iodinated human αCGRP and βCGRP. The predominant effector pathway is adenylyl cyclase; there are also reports of modulation of Ca^+ and K^+ channels via pertussis toxin-sensitive G proteins.

$CGRP_1$ and $CGRP_2$ receptor subtypes have been proposed, with evidence that the $CGRP_1$ receptor exists in several subtypes. The antagonist $CGRP_{8-37}$ (pA_2 = 6.0–8.0) is selective for the $CGRP_1$ receptor; [Cys(ACM)^{2,7}]CGRP is a selective agonist at the $CGRP_2$ receptor.

Other receptors/binding sites: A binding site in rat brain has similar affinities for calcitonin and CGRP; it has no known functional correlate.

Comment: Amylin (~50% homologous to CGRP) is one to two orders of magnitude less potent than CGRP at the CGRP receptor; it is unclear whether it also has its own receptor. Amylin$_{8-37}$ is claimed to be an antagonist at the amylin 'receptor'.

Endogenous ligands: human αCGRP (also named CGRP-I), human βCGRP (also named CGRP-II), amylin (see above); there is heterogeneity in the structure of CGRP between species

Cannabinoid receptor

The human (P21554) and rat (P20272) cannabinoid receptors have been cloned; they have 472 and 473 amino acids respectively, and both have seven transmembrane domains. The predominant effector pathway is inhibition of adenylyl cyclase, although inhibition of N-type Ca^{2+} channels may also be involved. Agonists include (−)-Δ⁹tetrahydrocannabinol, levonantradol, nabilone, CP55940 and WIN55212-2. Radioligands are [³H]CP55940, [³H]WIN55212-2 and [³H]HU210.

Anandamide (5,8,11,14-eicosatetraenamide (N-2-hydroxyethyl)-all-Z) has been proposed as the endogenous ligand for the cannabinoid receptor.

Cholecystokinin and gastrin receptors

Nomenclature	CCK$_A$	CCK$_B$
Previous names	CCK$_1$	CCK$_2$, gastrin
Potency order	CCK$_8$ >> gastrin = CCK$_4$	CCK$_8$ > gastrin = CCK$_4$
Selective agonists	A71623	[N-methyl-Nle^{28,31}]desulfated CCK$_8$ gastrin desulfated CCK$_8$
Selective antagonists	devazepide (9.8) lorglumide (7.2) PD140548 (7.9)	CI988 (8.1) L365260 (7.5) LY262691 (K_i vs [^{125}I]BH-CCK$_8$ = 30 nM)
Radioligands	[^{3}H]devazepide (0.2 nM) [^{3}H]- or [^{125}I]BH-CCK$_8$ (nonselective)	[^{3}H]L365260 (2.0 nM) [^{3}H]PD140376 (200 pM) [^{3}H]- or [^{125}I]gastrin
Predominant effectors	IP$_3$/DG	IP$_3$/DG
Gene	*cckA*	*cckB*
Structural information	444 aa rat 7TM	447 aa human 7TM 452 aa rat 7TM 453 aa canine 7TM

Other receptors/binding sites: Species homologues of the CCK$_B$ receptor exist; they can be distinguished by the relative affinities of L364718 and L365260.

Endogenous ligands: CCK$_4$, CCK$_8$, CCK$_{33}$, gastrin

Dopamine receptors

Nomenclature	D_1	D_2	D_4	D_5
Alternative names	D_{1A}	D_{2A}	D_{2C}	D_{1B}
Selective agonists	SKF38393 fenoldopam	N-0437 bromocriptine	–*	–†
Selective antagonists	SCH23390 SKF83566 SCH39166	(–)-sulpiride YM091512 domperidone	clozapine* (weak selectivity)	–†
Radioligands	[³H]SCH23390 (0.2 nM) [¹²⁵I]SCH23982 (0.7 nM)	[³H]YM091512 (0.1 nM) [¹²⁵I]iodosulpiride (0.5 nM)	–*	–†
Predominant effectors	cAMP↑§	cAMP↓ K⁺ channel↑ (G) Ca²⁺ channel↓ (G)	cAMP↓	cAMP↑
Gene	d1	d2¶	d4¥	d5
Structural information	446 aa human P21728 7TM 446 aa rat P18901 7TM 446 aa rhesus macaque 7TM	443 aa human P14416 7TM 444 aa rat P13953 7TM 444 aa mouse 7TM 444 aa bovine P20288 7TM	387 aa human P21917 7TM 385 aa rat 7TM	477 aa human P21918 7TM 475 aa rat P25115 7TM

Other receptors/binding sites: A D_3 receptor has been cloned; the human form has 400 aa and 7TM. Its binding profile is similar to, yet distinct from, that of the D_2 receptor; in general, antagonists at the D_2 receptor have a slightly lower affinity at the D_3 receptor while the converse is true for some agonists (notably quinpirole). No functional correlate has been described.

*the D_4 receptor generally displays similar or lower affinities for antagonists and agonists compared to the D_2 receptor
†the pharmacological profile of the D_5 receptor is similar to, yet distinct from, that of the D_1 receptor
§has also been reported to stimulate formation of IP_3/DG, possibly through a receptor subtype distinct from that coupled to cAMP
¶two protein isoforms of the D_2 receptor can be derived by alternative splicing (D_{2L} = long [number of aa shown above] and D_{2S} = short [414 aa in human]); there is preliminary evidence for very minor pharmacological and functional differences between the two forms
¥the d4 gene is polymorphic in human; genetic typing has revealed allelic variations of the protein from 387 to 483 aa

Endothelin receptors

Nomenclature*	ET$_A$	ET$_B$
Potency order	ET-1 = ET-2 > ET-3	ET-1 = ET-2 = ET-3
Selective agonists	–	[Ala^{1,3,11,15}]ET-1 sarafotoxin S6c Suc[Glu^9,Ala^{11,15}]ET-1$_{10-21}$
Selective antagonists	BE18257B (5.9) BQ123 (7.4) FR139317 (6.6) myriceron caffeoyl ester (6.7)	[Cys^{11,15}]ET-1$_{11-21}$
Radioligands	[^{125}I]ET-1 [^{125}I]ET-2	[^{125}I]ET-1, [^{125}I]ET-2 or [^{125}I]ET-3 [^{125}I]-[Ala^{11,15}]Ac-ET-1$_{6-21}$ [^{125}I]-[Ala^{1,3,11,15}]ET-1
Predominant effectors	IP$_3$/DG	IP$_3$/DG
Gene	*et$_A$*	*et$_B$*
Structural information	427 aa human P25101 7TM 426 aa rat P26684 7TM 427 aa bovine P21450 7TM	442 aa human P24530 7TM 441 aa rat P21451 7TM 441 aa bovine P28088 7TM

Other receptors/binding sites: Radioligand binding and functional studies indicate the existence of a further receptor subtype characterized by a higher potency of ET-3 relative to ET-1 and ET-2; a candidate clone for this receptor has been identified in *Xenopus*.

Endogenous ligands: endothelin 1 (**ET-1**), endothelin 2 (**ET-2**), endothelin 3 (**ET-3**); vasoactive intestinal contractor is the mouse homologue of ET-2

*nomenclature as agreed by the IUPHAR Committee on Endothelin Receptors (Tsukuba Meeting, 1990)

GABA receptors

Nomenclature	GABA$_A$		GABA$_B$
	competitive site	benzodiazepine modulatory site	
Selective agonists	isoguvacine muscimol	flunitrazepam zolpidem abecarnil	L-baclofen 3-aminopropylphosphinic acid 3-aminomethylphosphinic acid
Inverse agonists		DMCM Ro194603	
Selective antagonists	bicuculline (6.0) SR95531	flumazenil ZK93426	saclofen (5.3) CGP35348 (5.0) CGP36742 (5.1)
Radioligands	[^{3}H]muscimol [^{3}H]SR95531	[^{3}H]flunitrazepam (2 nM) [^{3}H]zolpidem (7 nM) [^{3}H]flumazenil (0.9 nM)	[^{3}H]L-baclofen [^{3}H]3-aminopropylphosphinic acid
Predominant effectors	int. Cl$^-$	modulates GABA$_A$-gated Cl$^-$ channel	cAMP↓ K$^+$ channel↑ (G) Ca^{2+} channel↓ (G)
Structural information	multisubunit*	α subunit determines recognition properties; γ subunit appears to be an absolute requirement for benzodiazepine involvement†	—

Comment: The GABA$_A$ receptor Cl$^-$ channel and the glycine receptor Cl$^-$ channel can be discriminated by strychnine. pA$_2$ values against muscimol and glycine are 5.3 and 7.0 (neonatal) or 8.0 (adult), respectively.

*sequences of six α, four β, three γ, one δ and two ρ subunits have been reported; the ρ subunits appear to be localized in the retina
†in expression systems α1β1γ2 exhibits binding characteristics of BZ$_1$ (high affinity for CL218872 and β-CCM) and α2β1γ2 and α3β1γ2 exhibit binding characteristics of BZ$_2$ subtypes; recombinant receptors containing α5 exhibit BZ$_2$ characteristics but have low affinity for zolpidem; alternative spliced versions of some subunits exist

Galanin receptor

The galanin receptor can be radiolabelled with iodinated analogues of galanin, e.g. [^{125}I]-[Tyr^3]galanin. Predominant effectors are inhibition of adenylyl cyclase, inhibition of Ca^{2+} channels and stimulation of K$^+$ channels. Galantide (a chimeric peptide made from galanin$_{1-13}$ and substance P$_{5-11}$) is claimed to be a competitive antagonist with an IC$_{50}$ of 0.1 nM. The galanin receptor has not been cloned.

Other receptors/binding sites: The antagonist, galantide, appears to distinguish between receptor subtypes in pancreas, hypothalamus and hippocampus.

Endogenous ligands: human galanin, human galanin$_{1-19}$ (there is heterogeneity in the structure of galanin between species: porcine, rat, chick and bovine forms have been described; all have 29 aa in contrast to human galanin, which has 30 aa)

Glutamate receptors (ionotropic)

Nomenclature	NMDA		AMPA	kainate
	competitive site	modulatory site		
Previous names	–	–	quisqualate	–
Selective agonists	NMDA	glycine HA966 (partial agonist)	AMPA 5-fluorowillardiine	kainate domoate
Selective antagonists	D-AP5 (5.2–5.9) CGS19755 (6.0) CGP37849	5,7-dichlorokynurenate MNQX L689560	NBQX (7.1)* GYKI52466 (noncompetitive)	–*
Channel blockers	dizocilpine (MK801) phencyclidine		–	–
Radioligands	[³H]D-AP5 [³H]CPP [³H]CGS19755 [³H]dizocilpine	–	[³H]AMPA [³H]CNQX	[³H]kainate
Predominant effectors	int. Na⁺/K⁺/Ca²⁺	occupancy required for receptor activation†	int. Na⁺/K⁺§	int. Na⁺/K⁺
Gene	nmdar1, nmdar2A–nmdar2D		glur1–glur4¶	glur5–glur7, ka1, ka2
Structural information	920–1482 aa rat 4 TM		862–889 aa rat 4 TM	884–888 aa rat 4 TM

Other receptors/binding sites: A structurally related kainate-binding protein of unknown function exists in chick (464 aa) and frog (487 aa) brain. A glutamate-binding protein (516 aa) has been cloned from rat brain that shares little homology with other cloned receptors; its function is unknown.

*other quinoxalinediones such as CNQX and DNQX show limited selectivity between AMPA and kainate receptors
†see Glycine receptors
§certain isoforms are Ca²⁺ permeable
¶also called glurA–glurD; each of these four related proteins can exist as two variants generated by alternative splicing ('flip' and 'flop') that show functional differences; several human equivalents of the glur series have been identified

Glutamate receptors (metabotropic)

Nomenclature	mGluR$_1$	mGluR$_2$	mGluR$_3$	mGluR$_4$*	mGluR$_5$
Selective agonists	–	–	–	–	–
Selective antagonists	–	–	–	–	–
Radioligands	–	–	–	–	–
Predominant effectors	IP$_3$/DG	cAMP↓	cAMP↓	cAMP↓	IP$_3$/DG
Gene	mglur1	mglur2	mglur3	mglur4	mglur5
Structural information	897–1199 aa rat 7TM†	872 aa rat 7TM	879 aa rat 7TM	912 aa rat 7TM	1171 aa rat 7TM

Other receptors/binding sites: mGluR$_6$ (with high homology to mGluR$_4$) has been cloned within the last year and their relationship to endogenously expressed receptors is unclear. The agonist 1s,3r-ACPD is selective for the metabotropic class; no antagonist or radioligand has been described.

Comment: The receptors shown above have been cloned within the last year and their relationship to endogenously expressed receptors is unclear. The agonist 1s,3r-ACPD is selective for the metabotropic class; no antagonist or radioligand has been described.

*mGluR$_4$ may be the equivalent of the 'L-AP4' receptor involved in presynaptic inhibition
†three alternatively spliced versions of mGluR$_1$ exist: α (1199 aa), β (906 aa) and c (897 aa)

Glycine receptors

The inhibitory glycine receptor is a pentamer of homologous 48 kDa and 58 kDa subunits which form an intrinsic Cl- channel. Four different isoforms of the α-subunit (α1–α4) and one variant of the β-subunit (β1) have been identified by genomic and cDNA cloning. The α-subunit bears the ligand-binding site. All subunits have four transmembrane domains. Cl- flux is stimulated by glycine and antagonized by strychnine. Adult glycine receptor has a higher affinity for strychnine ($pA_2 = 8.0$) than the neonatal glycine receptor ($pA_2 = 7.0$).

A strychnine-insensitive glycine-binding site exists on the NMDA receptor. This site needs to be occupied by glycine to enable activation of the receptor [see Glutamate receptors (ionotropic)].

Histamine receptors

Nomenclature	H_1	H_2	H_3
Previous names	–	–	–
Selective agonists	2-(*m*-fluorophenyl)histamine (partial agonist)	dimaprit impromidine (also an H_3 antagonist)	R-α-methylhistamine imetit
Selective antagonists	mepyramine (9.1) triprolidine (9.9)	ranitidine (7.2) tiotidine (7.8)	thioperamide (8.4) iodophenpropit (9.6)
Radioligands	[^{3}H]mepyramine (0.8 nM) [^{125}I]iodobolpyramine (0.1 nM)	[^{3}H]tiotidine (15 nM) [^{125}I]iodoaminopotentidine (0.3 nM)	[^{3}H]R-α-methylhistamine [^{3}H]N-α-methylhistamine [^{125}I]iodophenpropit (0.3 nM)
Predominant effectors	IP_3/DG	cAMP↑	–
Gene	h1	h2	–
Structural information	491 aa bovine 7TM	359 aa human P20521 7TM 359 aa canine P17124 7TM 358 aa rat P25102 7TM	–

5-HT receptors

Nomenclature*	5-HT$_{1A}$	5-HT$_{1B}$†	5-HT$_{1D}$†	5-HT$_{1E}$	5-HT$_{1F}$
Alternative names	–	–	–	–	5-HT$_{1E\beta}$, 5-HT$_6$
Selective agonists	8-OH-DPAT	CP93129	sumatriptan	–	–
Selective antagonists	WAY100135 (7.2–7.7)	–	–	–	–
Radioligands	[^{3}H]8-OH-DPAT	[^{125}I]GTI	[^{125}I]GTI	[^{3}H]5-HT	[^{125}I]LSD (nonselective)
Predominant effectors	cAMP↓ K$^+$ channel↑ (G)	cAMP↓	cAMP↓	cAMP↓	cAMP↓
Gene	5-ht1A	5-ht1B	5-ht1D (5-ht1Dα human) 5-ht1B (5-ht1Dβ human)	5-ht1E	5-ht1F
Structural information	421 aa human P08908 7TM 422 aa rat P19327 7TM	386 aa rat P28564 7TM 386 aa mouse P28334 7TM	377 aa human P11614 7TM (α) 390 aa human P28222 7TM (β) 374 aa rat P28565 7TM 377 aa canine P11614 7TM	365 aa human 7TM	367 aa mouse 7TM

*nomenclature as agreed by the Serotonin Club Receptor Nomenclature Committee (Houston, 1992)

†the human homologue of the rodent 5-HT$_{1B}$ receptor, 5-HT$_{1D\beta}$ (390 aa P28222 7TM) has a pharmacological profile that is similar to the human 5-HT$_{1D\alpha}$ receptor but distinct from that of the rodent 5-HT$_{1B}$ receptor; it is therefore not clear whether the 5-HT$_{1D}$ receptor identified previously in human corresponds to the 5-HT$_{1D\alpha}$ or 5-HT$_{1D\beta}$ gene product

5-HT receptors (continued)

Nomenclature*	5-HT$_{2A}$	5-HT$_{2B}$	5-HT$_{2C}$	5-HT$_3$	5-HT$_4$
Previous names	D, 5-HT$_2$	5-HT$_{2F}$	5-HT$_{1C}$	M	–
Selective agonists	α-methyl-5-HT	α-methyl-5-HT	α-methyl-5-HT	2-methyl-5-HT *m*-chlorophenyl-biguanide	5-methoxytryptamine renzapride BIMU8
Selective antagonists	ritanserin (9.5) LY53857 (8–9.5)	LY53857 (noncompetitive)	mesulergine (9.1) LY53857 (8–9.5)	tropisetron (10–11) ondansetron (8–10) granisetron (10)	GR113808 (9–9.5) RS2359190 (7.8)
Radioligands	[^{3}H]ketanserin	[^{3}H]5-HT	[^{3}H]mesulergine	[^{3}H]zacopride	[^{3}H]GR113808
Predominant effectors	IP$_3$/DG	IP$_3$/DG	IP$_3$/DG	int. cation channel	cAMP↑
Gene	5-*ht2A*	5-*ht2B*	5-*ht2C*	5-*ht3*	–
Structural information	471 aa human 7TM 471 aa rat P28223 7TM	479 aa human 7TM	458 aa human P28335 7TM 460 aa rat P08909 7TM	487 aa mouse P23979 4TM	–

Other receptors/binding sites: A further receptor with 357 aa and 7 TM has been cloned in rat (5-HT$_5$); its relationship to endogenously expressed receptors is not known.

*nomenclature as agreed by the Serotonin Club Receptor Nomenclature Committee (Houston, 1992)

Leukotriene receptors

Nomenclature*	LTB$_4$	LTC$_4$	LTD$_4$
Previous names	–	–	LTD$_4$/LTE$_4$
Potency order	LTB$_4$ > 12R-HETE	LTC$_4$ ≥ LTD$_4$†	LTD$_4$ = LTC$_4$ ≥ LTE$_4$
Selective agonists	LTB$_4$	N-methyl-LTC$_4$	–
Selective antagonists	LY255283 (8.3) SC41930 (300 nM)§ ONO-LB457 (14 nM)§ U75302 (1 μM)	–	ICI198615 (10.1) LY170680 (8.1) MK571 (9.0) SKF104353 (8.6)
Radioligands	[³H]LTB$_4$	[³H]LTC$_4$¶	[³H]LTD$_4$ [³H]ICI198615
Predominant effectors	IP$_3$/DG	–	IP$_3$/DG
Gene	–	–	–

Other receptors/binding sites: Tissue-specific subtypes of LTD$_4$ receptors have been suggested based on different affinities for antagonists. A distinct [³H]LTE$_4$ binding site has been identified as a subset of LTD$_4$ binding sites; LTE$_4$ is a partial agonist at some LTD$_4$ receptors.

Comment: Leukotrienes bind extensively to enzymes in their metabolic pathways (glutathione-S-transferase, γ-glutamyl transpeptidase and several aminopeptidases) complicating interpretation of binding and functional studies, e.g. LTC$_4$ is rapidly converted to LTD$_4$. Conversion inhibitors, e.g. serine–borate complex, reduce this problem but may also have additional nonspecific actions.

Endogenous ligands: LTB$_4$, LTC$_4$, LTD$_4$, LTE$_4$ (also 12R-HETE for LTB$_4$ receptor and lipoxin A$_4$ for LTD$_4$ receptor)

*an IUPHAR subcommittee on Receptor Nomenclature for Eicosanoids has been formed and is re-evaluating the above classification
†agonist and antagonist activity of LTE$_4$ has been reported
§IC$_{50}$ values against [³H]LTB$_4$; many LTB$_4$ antagonists have weak partial agonist activity
¶interpretation of [³H]LTC$_4$ binding is hampered by presence of a binding site for LTC$_4$ on glutathione-S-transferase (see above)

Melatonin receptor

This is a G protein-coupled receptor that has not been cloned. The effector pathway is inhibition of adenylyl cyclase. Selective agonists are melatonin and 2-iodomelatonin. Luzindole (2-benzyl-N-acetyltryptamine) is a selective antagonist (pA$_2$ = 7.7). The receptor can be radiolabelled with the agonist 2-[^{125}I]iodomelatonin.

Endogenous ligand: melatonin

Muscarinic acetylcholine receptors

Nomenclature*	M_1	M_2	M_3	M_4
Selective agonists[†]	—	—	—	—
Selective antagonists[§]	pirenzepine (8.0) telenzepine (9.1)	methoctramine (7.9) AFDX116 (7.3) gallamine (not competitive) himbacine (8.2)	hexahydrosiladifenidol (8.0) p-fluorohexahydrosiladifenidol (7.8)	tropicamide (7.8)
Radioligands[¶]	[^{3}H]pirenzepine	—	—	—
Predominant effectors	IP_3/DG	$cAMP\downarrow$ K^+ channel$\uparrow$ (G)	IP_3/DG	$cAMP\downarrow$
Gene	m1	m2	m3	m4
Structural information	460 aa human P11229 7TM 460 aa mouse P12657 7TM 460 aa rat P08482 7TM 460 aa porcine P04761 7TM	466 aa human P08172 7TM 466 aa rat P10980 7TM 466 aa porcine P06199 7TM	590 aa human P20309 7TM 589 aa rat P08483 7TM 590 aa porcine P11483 7TM	479 aa human P08173 7TM 478 aa rat P08485 7TM 490 aa chick P17200 7TM

Other receptors/binding sites: An M_5 receptor with 7TM has been cloned in human (P08912) and rat (P08911); m5 mRNA has been found in the CNS but a translation product has not been found *in vivo*. The human form has 532 aa.

*Muscarinic Receptor Nomenclature Committee (November 1992)
[†]an example of an agonist selective for muscarinic receptors relative to nicotinic receptors is oxotremorine-M
[§]no antagonist has a potency on one receptor that is more than tenfold greater than its potency on the other receptor subtypes; all receptors have K_d values for (−)-N-methylscopolamine and (−)-3-quinuclidinylbenzilate of less than 1 nM
[¶][^{3}H](−)-N-methylscopolamine and (−)-[^{3}H]3-quinuclidinylbenzilate are nonselective radioligands

Neuropeptide Y receptors

Nomenclature	Y_1	Y_2
Potency order	PYY ≥ NPY	PYY ≥ NPY
Selective agonists	[Pro^{34}]NPY [Leu^{31},Pro^{34}]NPY	NPY$_{13-36}$ NPY$_{18-36}$
Selective antagonists	–	–
Radioligands	[^{125}I]- or [^{3}H]NPY [^{125}I]PYY	[^{125}I]- or [^{3}H]NPY [^{125}I]PYY
Predominant effectors	cAMP↓*	cAMP↓ Ca^{2+} channel↓ (G)
Gene	*y1* (*fc5* rat)	–
Structural information	384 aa human P25929 7TM 382 aa rat P21555 7TM	–

Other receptors/binding sites: A third receptor, whose main characteristic is the low potency of PYY, has been proposed; a claim that the gene *lcr1* encodes for this receptor has not been substantiated.

Endogenous ligands: neuropeptide Y (**NPY**), peptide YY (**PYY**); (pancreatic polypeptide also has low potency at NPY receptors; it is unclear whether it has its own receptor)

*there is also evidence for mobilization of Ca^{2+} (the role of the phosphoinositide pathway in this response is controversial)

Neurotensin receptor

The neurotensin receptor can be radiolabelled with a number of tritiated and iodinated analogues of neurotensin. Predominant effectors are IP_3/DG and inhibition of adenylyl cyclase. SR48692 (K_i 7.5–9.0) is a nonpeptide antagonist. The rat neurotensin receptor has been cloned (P20789); it has 424 aa and 7TM.

Other receptors/binding sites: There is a second neurotensin-binding site that is selectively recognized by the antihistamine levocabastine (IC_{50} = 10 nM). Levocabastine has no agonist or antagonist properties in bioassays for neurotensin.

Endogenous ligands: neurotensin, neuromedin N

Nicotinic acetylcholine receptors

	muscle-type*	neuronal-type
Nomenclature		
Previous names	–	ganglionic
Selective agonists	–	–
Selective antagonists	α-bungarotoxin (pseudoirreversible)	κ-bungarotoxin (5 nM) methylcaconitine
Channel blockers	gallamine, decamethonium[†]	hexamethonium, mecamylamine[†]
Radioligands	[^{3}H]- or [^{125}I]α-bungarotoxin	[^{3}H]- or [^{125}I]κ-bungarotoxin [^{3}H]nicotine [^{3}H]methylcarbamylcholine
Predominant effectors	int. Na$^+$/K$^+$/Ca^{2+} (conductance 50 pS)[§]	int. Na$^+$/K$^+$/Ca^{2+} (conductance 15–55 pS)
Structural information[¶]	multisubunit pentamer: $(\alpha 1)_2 \beta 1 \epsilon \delta$ (adult muscle); $(\alpha 1)_2 \beta 1 \gamma \delta$ (electric organ/embryonic muscle) each subunit has 4TM	multisubunit probably composed of α- and β-subunits pentamer: $\alpha_2 \beta_3$ each subunit has 4TM

Other receptors/binding sites: Two α-bungarotoxin binding proteins in brain have been cloned (α7, α8). α7 is a member of the ligand-gated ion channel gene superfamily and forms a homooligomeric acetylcholine-gated ion channel in oocytes; methylcaconitine is an antagonist. It is located presynaptically. A locust α-subunit has been cloned that forms a homooligomeric nicotine-gated ion channel that is blocked by both α-bungarotoxin and κ-bungarotoxin.

*muscle-type includes data from adult and embryonic/denervated muscle, and *Torpedo* electric organ
[†]examples shown are commonly used but are not selective
[§]conductance 35 pS in embryonic or denervated muscle
[¶]α subunits are characterized by Cys192 and Cys193 (which are associated with agonist binding); in muscle, α is α1 and two distinct types (α1$_a$ and α1$_b$) have been cloned; β-subunits are designated 'non-α' in avian species; *Torpedo* and mammalian γ-subunits are not functionally identical; additional neuronal α-subunits (α2, α3, α4, α5, α6, α7, α8) and β-subunits (β2, β3, β4) have been cloned from chick and rat autonomic and central nervous systems (they are not all pharmacologically identical); there are also two species variants in rat (α4-1) and (α4-2), which are derived from the α4 gene

Opioid receptors

Nomenclature	μ	δ	κ
Potency order	β-end > dynA > met > leu	β-end = leu = met > dynA	dynA >> β-end > leu = met
Selective agonists*	DAMGO sufentanil PL017	DPDPE DSBULET [DAla²]deltorphin I or II	U69593 CI977 ICI197067
Selective antagonists	CTAP (6.4–7.9)	ICI174864 (7.5) naltrindole (9.7)	nor-binaltorphimine (10.3)
Radioligands	[³H]DAMGO	[³H]DPDPE [³H][DAla²]deltorphin II [³H]naltrindole	[³H]U69593 [³H]CI977
Predominant effectors	cAMP↓ K⁺ channel↑ (G)	cAMP↓ K⁺ channel↑ (G)	Ca²⁺ channel↓ (G)
Gene	–	δ	–
Structural information	–	372 aa mouse 7TM	–

Other receptors/binding sites: The existence of μ_1- and μ_2-opioid receptor subtypes has been suggested. μ_1-Receptors have a high affinity for many δ-receptor agonists although DPDPE is a notable exception. The pharmacological profile of μ_1-receptors may explain the suggestion that μ- and δ-receptors exist in an interacting macromolecular complex with μ-agonists allosterically inhibiting δ-agonist binding and vice versa.
The existence of δ-receptor subtypes has been suggested on the basis of *in vivo* studies using δ-selective ligands: DPDPE and the antagonist 7-benzylidine-7-dehydronaltrexone are selective for δ_1 sites; [DAla²]deltorphin II and the antagonist, naltriben, are selective for δ_2 sites.
κ-Receptor subtypes have been suggested: arylacetamides, e.g. U69593 and CI977, bind to a subset of the sites labelled by ethylketocyclazocine (κ_1); κ_2 sites have moderate affinity for ethylketocyclazocine and for other benzomorphans, but have no known functional correlate. Under κ-selective labelling conditions, naloxone benzoylhydrazone labels a further population of sites named κ_3; these sites have a similar pharmacology to μ-opioid receptors.

Endogenous ligands: [Met]enkephalin (**met**), [Leu]enkephalin (**leu**), β-endorphin (**β-end**), α-neo-dynorphin, dynorphin A (**dynA**), dynorphin B

*morphine (partial agonist), naloxone and naltrexone are weakly selective for μ-opioid receptors

PAF receptor

Agonists include PAF and its metabolically stable analogue, C-PAF. PAF receptor antagonists have been divided into four classes: i) quaternary nitrogen compounds, e.g. CV6209 (9.5); ii) sp2 nitrogen compounds, e.g. WEB2086 (7.3), SRI63072 (6.4), BB-823; iii) diaryl compounds, e.g. L659989 (8.1); iv) miscellaneous compounds, e.g. ginkgolide B (6.4). The receptor can be labelled with tritiated analogues of PAF and antagonists such as WEB2086. The predominant effector pathway involves G-protein activation of IP_3/DG. Human (P25105) and guinea-pig (P21556) PAF receptors have been cloned; both receptors have 342 aa and 7TM.

Other receptors/binding sites: Binding studies suggest that differences may exist between PAF receptors in a number of tissues. These results, however, may reflect binding to membrane transporters or intracellular sites, non-equilibrium conditions or species differences. Binding and functional studies with PAF antagonists suggest the presence of two receptors on guinea-pig eosinophils (WEB2086) and a novel receptor subtype on guinea-pig macrophages (kadsurenone, WEB2086 and CV6209).

Prostanoid receptors

Nomenclature	DP	EP$_1$, EP$_2$, EP$_3$	FP	IP	TP
Previous names	prostaglandin D	prostaglandin E	prostaglandin F	prostacyclin	thromboxane
Potency order*	D > E,F,I,T	E > F,I > D,T	F > D > E > I,T	I > D,E,F > T	T = H >> D,E,F,I
Selective agonists	BW245C ZK110841	EP$_1$: 17-phenyl-ω-trinor-PGE$_2$ EP$_2$: butaprost, AH13205 EP$_3$: enprostil, GR63799, MB28767	fluprostenol	cicaprost	U46619 STA$_2$ I-BOP
Selective antagonists	BWA868C (9.3)	EP$_1$: SC19220 (5.6) AH6809 (6.8)†	–	–	GR32191 (8.8) SQ29548 (8.7)
Radioligands	[^{3}H]PGD$_2$	[^{3}H]PGE$_2$	[^{3}H]PGF$_{2\alpha}$	[^{3}H]iloprost	[^{3}H]SQ29548 (4.5 nM) [^{125}I]SAP (0.2–1.0 nM) [^{125}I]I-BOP (0.7 nM)
Predominant effectors	cAMP↑	EP$_1$: IP$_3$/DG EP$_2$: cAMP↑ EP$_3$: IP$_3$/DG, cAMP↓	IP$_3$/DG	cAMP↑	IP$_3$/DG
Gene	–	EP$_3$: *ep3*	–	–	*tp*
Structural information	–	365 aa mouse 7TM	–	–	343 aa human P21731 7TM

Other receptors/binding sites: There is some evidence from binding and functional studies for subtypes of TP receptors on platelets and vascular tissue although it is unclear whether these results reflect species homologues or the presence of mixed receptor populations.

Comment: The absolute selectivity of most of the agonists has not been demonstrated.

Endogenous ligands: PGD$_2$ (**D**), PGE$_2$ (**E**), PGF$_{2\alpha}$ (**F**), PGH$_2$ (**H**), PGI$_2$ (**I**), TXA$_2$ (**T**)

*measurement of potency of PGH$_2$ or TXA$_2$ is hampered by rapid metabolism
†blocks DP and TP receptors at higher concentrations

P₂ Purinoceptors

Nomenclature	P_{2X}	P_{2Y}	P_{2Z}	P_{2T}	P_{2U}
Alternative names	–	–	ATP⁴⁻	ADP	pyrimidinoceptor
Potency order	ATP = ADP > AMP	ATP = ADP > AMP	ATP*	ADP†	UTP ≥ ATP > ADP > AMP
Selective agonists	α,β-methyleneATP β,γ-methylene-L-ATP	ADPβF homo-ATP 2-methylthioATP ADPβS	3′-O-(benzoyl)-benzoylATP	2-methylthioADP	UTPγS
Selective antagonists	ANAPP₃§ (photolysed; irreversible)	–¶	2-methylthio-L-ATP	2-chloro-ATP (5.2) ATP (4.6)	–
Radioligands	[³H]α,β-methyleneATP	[³⁵S]ADPβS [³⁵S]ATPαS	–	β[³²P]2-methylthioADP	[³²P]ATPγS
Predominant effectors	int. cation channel	IP₃/DG	int. cation channel	int. cation channel˅ cAMP↓	IP₃/DG
Gene	–	–	–	–	–

Comment: P_{2Z} and P_{2T} receptors have restricted distributions; the former are found on mast cells and other immune cells while the latter are found on platelets. The order of potency of the endogenous nucleotides and their analogues is complicated by their relative resistance to dephosphorylation by ectonucleotidases.

*ADP and AMP are inactive; all nucleotide agonists are more potent in the absence of divalent cations
†ATP and AMP are antagonists
§the P_{2X} receptor can also be inhibited by desensitization induced by α,β-methylene ATP or β,γ-methylene-L-ATP
¶reactive blue 2 is claimed to be selective at P_{2Y} receptors over a very narrow concentration range
˅also stimulates release of intracellular Ca²⁺, although the role of the phosphoinositide pathway in this response is uncertain

Somatostatin receptor

Nomenclature	SS_1	SS_2	SS_3	SS_4
Previous names	$SRIF_2$, SS_B	$SRIF_1$, SS_A	–	–
Potency order	$SS = SS_{28}$	$SS = SS_{28}$	$SS = SS_{28}$	$SS_{28} > SS$
Selective agonists	–	MK678	–	–
Selective antagonists	–	–	–	–
Radioligands	[^{125}I]-[Tyr^{11}]SS	[^{125}I]MK678 [^{125}I]-[Tyr^{11}]SS	[^{125}I]-[Tyr^{11}]SS	[^{125}I]-[Tyr^{11}]SS
Predominant effectors	–	K$^+$ channel↑ (G) Ca^{2+} channel↓ (G)	cAMP↓	cAMP↓
Gene	ss1	ss2	ss3	ss4
Structural information	391 aa human 7TM 391 aa mouse 7TM	369 aa human 7TM 369 aa mouse 7TM	418 aa human 7TM 428 aa mouse 7TM 428 aa rat 7TM	383 aa rat 7TM

Comment: The receptors shown above have been cloned within the last year and their relationship to endogenously expressed receptors is unclear.

Endogenous ligands: somatostatin (**SS**), somatostatin$_{1-28}$ (**SS$_{28}$**)

Tachykinin receptors

Nomenclature*	NK_1	NK_2	NK_3
Previous names	SP-P, substance P	SP-E / SP-K, substance K	SP-N / SP-E, neurokinin B
Potency order	SP > NKA > NKB	NKA > NKB >> SP	NKB > NKA > SP
Selective agonists	SP methyl ester [Sar^9,Met(O$_2$)^{11}]SP [Pro^9]SP	[β-Ala^8]NKA$_{4-10}$ GR64349 [Lys^5,MeLeu^9,Nle^{10}]NKA$_{4-10}$	senktide [MePhe^7]NKB [Pro^7]NKB
Selective antagonists	CP99994 (K_i = 0.5nM) RP67580 (7.0–9.0) GR82334 (7.2–7.6)	SR48968 (8.0–10.0) GR94800 (9.6) L659877 (6.9–7.9)	[Trp^7,β-Ala^8]NKA$_{4-10}$ (agonist at other tachykinin receptors)
Radioligands	[^{3}H]- or [^{125}I]BH-[Sar^9,Met(O$_2$)^{11}]SP [^{3}H]-[Pro^9]SP [^{3}H]- or [^{125}I]BH-SP	[^{3}H]NKA [^{125}I]iodohistidyl-NKA	[^{3}H]senktide [^{125}I]BH-eledoisin [^{125}I]-[MePhe^7]NKB
Predominant effectors	IP$_3$/DG	IP$_3$/DG	IP$_3$/DG
Gene	*spr*	*skr*	*nkr*
Structural information	407 aa human P25103 7TM 407 aa rat P14600 7TM	398 aa human P21452 7TM 390 aa rat P16610 7TM 384 aa bovine P05363 7TM	468 aa human P29371 7TM 452 aa rat P24053 7TM

Other receptors/binding sites: Species homologues of the NK$_1$ receptor exist; the antagonists CP99994 and CP96345 are selective for human and guinea-pig whereas the antagonist RP67580 is selective for rat and mouse.
Species homologues of the NK$_2$ receptor exist: the potency order of NK$_2$ antagonists in some species, e.g. rabbit (MEN10207 > L659877 > R396), differs from that in others, e.g. hamster (L659877 > R396 > MEN10207).
Functional and binding studies indicate the presence of novel receptors in bovine adrenal medulla and in the CNS that recognize substance P$_{1-7}$ and certain other N-terminal analogues. These are not tachykinin receptors since activity is dependent on the nonconserved N-terminal region of substance P.

Endogenous ligands: substance P (**SP**), neurokinin A (**NKA**; previous names substance K, neurokinin α, neuromedin L), neurokinin B (**NKB**; previous names neurokinin β, neuromedin K), neuropeptide K and neuropeptide γ (N-terminally extended forms of neurokinin A)

*nomenclature as agreed in the Symposium 'Substance P and Neurokinins - Montreal 1986'

Vasopressin and oxytocin receptors

Nomenclature	V_{1A}	V_{1B}	V_2	OT
Previous names	V_1	V_3	—	—
Potency order	VP > OT	VP > OT	VP > OT	OT ≥ VP
Selective agonists	—	—	d[DArg^8]VP d[Val^4]AVP	[Thr^4,Gly^7]OT
Selective antagonists*	OPC21268† (IC_{50} v. [^{3}H]AVP = 0.4 μM)	—	OPC31260† (K_i = 26 nM) d(CH$_2$)$_5$[DIle^2,Ile^4]AVP (8.0)	d(CH$_2$)$_5$[Tyr(Me)^2,Thr^4,Orn^8]-OT$_{1-8}$ (7.7) cyc(D1-Nal,Ile,DPip,Pip,DHis,Pro) (8.6)
Radioligands	[^{3}H]d(CH$_2$)$_5$[Tyr(Me)^2]AVP (0.1 nM) [^{125}I]Phaa,DTyr(Me),Phe,Gln,Asn,-Arg,Pro,Arg,TyrNH$_2$ (0.06 nM)	—	[^{3}H]d[Val^4]AVP [^{3}H]d[DArg^8]VP [^{3}H]d(CH$_2$)$_5$[DIle^2,Ile^4]AVP$_{1-8}$	[^{125}I]d(CH$_2$)$_5$[Tyr(Me)^2,Thr^4,Orn^8,-Tyr^9-NH$_2$]OT (0.03 nM)
Predominant effectors	IP$_3$/DG	IP$_3$/DG	cAMP↑	IP$_3$/DG
Gene	v1A	—	v2§	ot
Structural information	394 aa rat 7TM	—	371 aa human 7TM 371 aa rat 7TM	388 aa human 7TM

Comment: V_{1B} receptors can be distinguished from V_{1A} receptors by the low affinity of certain antagonists, e.g. the affinity of d(CH$_2$)$_5$[Tyr(Me)^2]AVP at the V_{1B} receptor is ~ one thousandfold lower than at the V_{1A} receptor, while [DPen^1,Tyr(Me)^2]AVP has a similar affinity at both receptors (pA$_2$ = 8.5).

Endogenous ligands: oxytocin (OT), vasopressin (AVP)

*several structurally related antagonists exist for all receptor subtypes (including V_{1B}) and examples of the most selective are shown; many antagonists have partial agonist activity
†have not been tested on oxytocin receptors
§numerous mutations in the v2 receptor gene have been identified and may underlie the disease nephrogenic diabetes insipidus

VIP and related peptide receptors

Nomenclature	VIP	GRF	PACAP	secretin
Potency order	VIP > PHI > PACAP	GRF >> VIP	PACAP > VIP	secretin >> VIP
Selective agonists	VIP	GRF	PACAP	secretin
Selective antagonists	–*	–	PACAP$_{6-27}$	–
Radioligands	[^{125}I]VIP [^{125}I]PHI	[^{125}I]GRF	[^{125}I]PACAP	[^{125}I]secretin
Predominant effectors	cAMP↑	cAMP↑	cAMP↑	cAMP↑
Gene	vip†	–	–	secretin
Structural information	359 aa rat 7TM	–	–	449 aa rat P23811 7TM

Other receptors/binding sites: The order of potency for the displacement of iodinated VIP binding to rat arteries by analogues of VIP differs from that in other rat tissues such as lung, liver and brain; e.g. [DAla⁴]VIP has low affinity in arteries.
A novel, helodermin-preferring VIP receptor has been reported in several human clonal cell lines, including SUP-T1 lymphoma cells. PACAP receptor subtypes, which recognize PACAP$_{38}$ with high or low affinity, have been proposed.

Endogenous ligands: peptide histidine isoleucineamide (**PHI**), peptide histidine methionineamide (**PHM**), vasoactive intestinal peptide (**VIP**); growth hormone releasing factor (**GRF**), pituitary adenylyl cyclase activating polypeptide (**PACAP**), secretin

*[AcTyr¹,DPhe²]GRF$_{1-29}$ and [AcTyr¹,DPhe²]GRF$_{1-29}$ are weak partial agonists
†an earlier report claiming that the protein encoded by GenBank Accession number M64749 is a human VIP receptor has not been substantiated

Ca²⁺ channels

CHANNEL TYPE*	CONDUCTANCE†	BLOCKERS	PROPERTIES	LOCATION / FUNCTION
L	~25 pS	Dihydropyridine antagonists, phenylalkylamines, benzothiazepines TOXINS: calciseptine	High-voltage activated, voltage-dependent long-lasting current responsive to dihydropyridines (agonists and antagonists) and modulated by protein kinase A-dependent phosphorylation. Slow inactivation§.	Excitation–contraction coupling; critical to cardiac muscle contraction, and involved in most forms of smooth muscle contraction. Excitation–secretion coupling in endocrine cells and some neurones.
N	~12–20 pS	TOXINS: ω-conotoxins (GIV$_A$)	High-voltage activated, voltage-dependent; moderate rate of inactivation§.	Identified only in neurones; participate in neurotransmitter release.
T	~8 pS	Octanol, flunarizine (nonselective)	Low voltage-activated, voltage-dependent transient current; channels deactivate slower than L or N channels.	Influence on SA pacemaker activity in heart, and repetitive spike activity in neurones and endocrine cells.
P	~10–12 pS	TOXINS: Funnel web spider toxin (FTX), ω-agatoxin IVA (but not sensitive to ω-conotoxin, except MVIIc dihydropyridines or ω-agatoxin IIIA)	Moderately high-voltage activated; non-inactivating.	Identified in some CNS neurones (prominent in Purkinje cells).

Structural information: The L channel from skeletal muscle is heterooligomeric and composed of α1, α2, β, γ and δ subunits. The α1 subunit provides the binding site for L channel blockers. Different types of α1 subunit, present in skeletal and cardiac muscle and brain, are encoded by distinct genes; alternative splicing may also occur. Each α1 subunit includes four homologous repeats containing six membrane-spanning regions, and shares significant homology with other voltage-gated ion channels, particularly the Na⁺ channel.

*nomenclature as agreed by the IUPHAR Committee on Calcium Channels [Spedding, M. and Paoletti, P. (1992) *J. Pharmacol. Rev.* 44, 363-376]
†with ~100 mM Ba²⁺ as charge carrier
§rate of inactivation may be greatly accelerated by [Ca²⁺]$_i$; N-type channels show slow transitions between inactivating and non-inactivating states

K+ channels

CHANNEL TYPE	CONDUCTANCE	BLOCKERS	PROPERTIES	LOCATION / FUNCTION
Voltage-dependent K+ channels				
Delayed rectifier (I_{KV}) (delayed outward K+ current, I_K; voltage-dependent K+ channel)	5–60 pS	TEA, Cs+, Ba^{2+}, Zn^{2+}, forskolin, 4AP, local anaesthetics, 9-aminoacridine, phencyclidine TOXINS: noxiustoxin, *Pandinus imperator* toxin, scorpion venom, phalloidin	Activated with some delay following membrane depolarization (threshold potential depending on tissue). Inactivation can take seconds or may occur only on repolarization.	Found in one form or another in many cells. Responsible for repolarization of AP in excitable cells (squid axon, frog node of Ranvier, skeletal muscle); slow wave activity (smooth muscle); involved in repetitive firing of spontaneously discharging cells.
A-Channel (I_A) (transient outward current)	20 pS	4AP, quinidine, tetrahydroaminoacridine, phencyclidine, TEA (weak) TOXINS: dendrotoxin, toxin I, mast cell degranulating peptide	Activated by depolarization after period of hyperpolarization. Channel activation fast, threshold being more negative than for the delayed rectifier, with rapid inactivation (<100 ms) generally complete at potentials positive to –50 mV. Structurally related to I_{KV}.	Regulation of firing frequency of spontaneously active cells by delaying return of voltage to rest from AHP, prolongs interspike interval (hippocampal and pyramidal cells, crab axons); I_A can coexist with I_{KV} in the same cell.
Slow delayed rectifier (I_{KS})	1 pS (?)	LY9241, TEA (weak)	Very slowly activating and deactivating; current increased by cAMP-dependent phosphorylation.	Heart: slow component of delayed rectifying current. Kidney epithelium: K+ homeostasis. Uterus: myometrial excitability in response to oestrogen.
Inward rectifier (I_{IR}) (anomalous rectifier, I_{AR}; I_{K1} in heart)	5–30 pS	TEA, Cs+, Ba^{2+}, Sr^{2+}, but not Zn^{2+} or 4AP TOXINS: gaboon viper venom	Channel conductance greatest at hyperpolarized potentials (e.g. open at resting potential); current reduced during depolarization probably due to [Mg^{2+}]$_i$.	Resting K+ conductance and plateau phase of cardiac AP (skeletal muscle transverse tubules, cardiac muscle).
Sarcoplasmic reticulum channel ($I_{K(SR)}$)	150 pS	Decamethonium, hexamethonium Cs+, 4AP, TEA (weak)	Strongly voltage dependent; K+/Na+ selectivity relatively low.	Skeletal and cardiac muscle SR membranes; function possibly concerned with regulation of SR membrane potential and/or allowing K+ influx to function as counter-ion during SR Ca^{2+} release.
Ca^{2+}-activated K+ channels				
High conductance Ca^{2+}-activated K+ channel ($I_{BK(Ca)}$) [maxi-K channel; BK channel; I_C (Ca^{2+}-activated)]	100–250 pS	TEA and Ba^{2+} (< mM), quinine, tubocurarine TOXINS: charybdotoxin, noxiustoxin, iberatoxin	Open probability increases with rise in [Ca^{2+}]$_i$ (0.1–10 μM) and with membrane depolarization at constant [Ca^{2+}]$_i$. Opened by K+ channel openers.	AP repolarization. May also be responsible for fast AHPs following spike activity (neurones, skeletal, smooth muscle); role in control of secretion (chromaffin cells, pancreatic β-cells, pituitary cells, salivary gland). *Contd*

K+ channels (continued)

CHANNEL TYPE	CONDUCTANCE	BLOCKERS	PROPERTIES	LOCATION / FUNCTION
Intermediate conductance Ca^{2+}-activated K+ channel ($I_{IK(Ca)}$) (IK_{Ca})	18–50 pS	TEA, quinine, Cs^+, Ba^{2+}, carbocyanine dyes (block from inside in red cells) cetiedil, nitrendipine, calmodulin antagonists (pimozide, haloperidol, trifluoperazine, calmidazolium) TOXINS: charybdotoxin	Activated by $[Ca^{2+}]_i$; may be voltage sensitive (molluscan neurone) or insensitive (red cell). Metabolic exhaustion opens red cell channel ('Gardos effect') by increasing $[Ca^{2+}]_i$ and by modifying $[Ca^{2+}]_i$ sensitivity.	Volume regulation; control of bursting activity (molluscan neurones).
Small conductance Ca^{2+}-activated K+ channel ($I_{SK(Ca)}$) (apamin-sensitive K_{Ca}, SK channel; slow AHP K_{Ca}; I_{AHP})	6–14 pS	Neuromuscular blockers (e.g. tubocurarine), quinine, mepacrine 9-aminoacridine, TEA blocks at mM concentrations (~tenfold less sensitive than $I_{BK(Ca)}$), insensitive to 4AP TOXINS: apamin, leiurotoxin I (scyllatoxin)	Sensitivity to Ca^{2+} > than $I_{BK(Ca)}$ at negative membrane potentials in myotubes. Little or no voltage dependence.	Responsible for long AHPs (neurones, non-innervated skeletal muscle); relaxation (certain intestinal smooth muscles, e.g. taenia caeci); hyperkalaemic response to α-adrenoceptor agonists (liver).
Ca^{2+}-activated nonspecific cation channel ($I_{K/NaCa}$) (CAN channels; in cardiac muscle: transient inward TTX-insensitive current, I_{ti} (or TI) in neurone somata: long-lasting inward current I_{IN}, I_N; in oocytes: slow I_{IN})	25–30 pS	Not blocked by TEA in insulinoma cells (or Schwann cells) but sensitive to 4AP and quinine	$[Ca^{2+}]_i$ increases open probability, generally voltage-insensitive with a linear current–voltage relationship and long openings (usually 100–1000 ms). Does not discriminate between Na^+ and K^+.	Cardiac pacemaker activity; pacemaker depolarizations (pituitary cells); depolarizing afterpotential and plateau potential (skeletal muscle); AP plateau (fertilization potential) in oocytes; stimulus–secretion coupling (pancreatic acinar cells, lacrimal gland, macrophages); long-lasting plateau potentials (peptide-releasing nerve terminals).

Receptor-coupled channels

CHANNEL TYPE	CONDUCTANCE	BLOCKERS	PROPERTIES	LOCATION / FUNCTION
Muscarinic-inactivated K+ channel ($I_{K(M)}$) (M current)	3–18 pS	W7, Ba^{2+}	Time- and voltage-dependent K+ current, activated at potentials more positive than −65 mV (and, in some cells, by somatostatin and β-adrenoceptors, via cAMP?), slow to activate, non-rectifying, non-inactivating. Inhibited by activators of PTX-insensitive G proteins coupled to IP_3/DG (e.g. muscarinic, bradykinin, or LHRH receptors).	Contributes steady-state component to outward K+ current that, when decreased by hormones, allows enhancement of depolarizing stimuli. Hormonal control of neuronal excitability; inhibition of $I_{K(M)}$ may underlie slow EPSP following acetylcholine release; facilitates AP discharge (sympathetic and spinal neurones, hippocampus, some smooth muscles).

Contd

K+ channels (continued)

CHANNEL TYPE	CONDUCTANCE	BLOCKERS	PROPERTIES	LOCATION / FUNCTION
Atrial muscarinic-activated K+ channel ($I_{K(ACh)}$) (atrial ACh_m channel)	25–50 pS	Cs^+, Ba^{2+}, 4AP, TEA, quinine	Muscarinic and adenosine receptors activate channel directly via α subunit of G_i-like G protein. βγ subunit may also modify channel activity via phospholipase A_2 stimulation and release of a 5-lipoxygenase metabolite. Low-threshold voltage activation; shows inward rectification.	ACh_m channel for vagal slowing of the heart (atria, AV and SA nodes). Similar channel may underlie inhibitory actions of $GABA_B$ receptors, $α_2$-adrenoceptors, opioid receptors, etc.
5-HT-inactivated channel ($I_{K(5-HT)}$) (S-K+ channel)	55 pS	Ba^{2+}, TEA (weak), 4AP (weak), Cs^+ (partial block)	Weakly voltage-dependent K+ current, open at resting potential. Insensitive to $[Ca^{2+}]_i$. 5-HT inactivates channel via cAMP-dependent phosphorylation, resulting in depolarization and neuronal excitability. FMRF amide opens the channel via a lipoxygenase metabolite (12-HPETE?).	Presynaptic facilitation by 5-HT of transmitter release in *Aplysia* sensory neurones underlies the gill withdrawal reflex.

Other K+ channels

CHANNEL TYPE	CONDUCTANCE	BLOCKERS	PROPERTIES	LOCATION / FUNCTION
ATP-sensitive K+ channel ($I_{K(ATP)}$) (ATP-dependent or ATP-regulated K+ channel)	20–200 pS	Sulfonylureas (glibenclamide at nM concentrations, tolbutamide, phentolamine, ciclazindol), AZ-DF265, lidocaine, tetracaine, 9-aminoacridine, quinine, 4AP, Ba^{2+}, TEA (weak)	$[ATP]_i$ inhibits channel opening depending on $[ATP]_i$:$[ADP]_i$ ratio. Some channels exhibit weak voltage dependence, show inward rectification, or sensitivity to intracellular/extracellular pH. Channel opened by K+ channel openers.	Present in CNS and skeletal, cardiac and smooth muscle (probably closed in physiological conditions). Control of resting potential of β-cells and glucoreceptive cells in hypothalamus and substantia nigra / regulation of GABA release (hypothalamus); possible protection against ischaemia.
Na+-activated K+ channel ($I_{K(Na)}$)	220 pS	TEA (~ 20 mM), 4AP (~ 4 mM)	Early outward K+ current activated by increase in Na+ to > 20 mM. Insensitive to voltage, $[ATP]_i$ or $[Ca^{2+}]_i$.	Opposing depolarization at high $[Na^+]_i$ (ventricular myocytes); fast repolarization during spiking activity (trigeminal ganglion neurone); reactivates currents inactivated by depolarization.
Cell-volume-sensitive K+ channel ($I_{K(vol)}$)	16–40 pS	Quinidine, lidocaine, cetiedil	Opens when cells swell.	Volume regulation (*Necturus* proximal tubule; colon epithelial cells; hepatocytes).
K+ channel opener-sensitive channel (I_{KCO})	5–20 pS	Sulfonylureas (glibenclamide at 0.1 μM concentrations)	ATP- and Ca^{2+} dependent, largely voltage independent.	Smooth muscle; function unknown.

Abbreviations: 4AP 4-aminopyridine; **AHP** afterhyperpolarization; **AP** action potential; **PTX** pertussis toxin; **TEA** tetraethylammonium; **TTX** tetrodotoxin

Na+ channels

CHANNEL TYPE*	CONDUCTANCE	BLOCKERS	PROPERTIES	LOCATION / FUNCTION
I	–	Tetrodotoxin; saxitoxin	–	Central neurones: spinal cord > brain; localized in cell bodies; also present in heart.
II	20 pS	Tetrodotoxin; saxitoxin	$V_{50} \sim -41$ mV; $V_h \sim -64$ mV	Central neurones: brain > spinal cord; localized in unmyelinated axons. Functions in axonal conduction of action potential.
III	16 pS	Tetrodotoxin; saxitoxin	$V_{50} \sim -10$ mV; $V_h \sim -40$ mV	Embryonic and neonatal central neurones.
μ1	–	Tetrodotoxin; μ-conotoxin	–	Innervated adult skeletal muscle.
h1	–	Tetrodotoxin; saxitoxin (200-fold lower affinity than other Na+ channels)	'Tetrodotoxin-resistant'; $V_h \sim -67$ mV	Heart; embryonic and denervated adult skeletal muscle.

Structural information: The Na+ channels from brain and skeletal muscle are heterooligomeric and composed of α and β subunits. The α subunit determines the major functional characteristics of Na+ channels. A sequence variant of Type II with similar or identical properties, termed IIA, results from alternative splicing. Each channel type includes four homologous repeats containing six membrane-spanning regions and shares significant homology with other voltage-gated ion channels, particularly the α1 subunit of the Ca^{2+} channel.

*all listed channel types have been identified by cDNA cloning and properties are those described in channels expressed in *Xenopus* oocytes; additional channel types exist in glia, peripheral neurones, neurosecretory cells and epithelial cells

Abbreviations: V_{50} voltage required for half-maximal activation; V_h voltage required for half-maximal inactivation.

Chemical names

Adenosine — APNEA N^6-2-(4-aminophenyl)ethyladenosine **CGS15943** 9-Cl-2-(furanyl)-5,6-dihydro-[1,2,4]-triazolo[1,5-c]quinazolin-5-iminemonomethanesulfonate **CGS21680** (2-p-carboxyethyl)phenylamino-5′-N-carboxamidoadenosine **CP66713** 4-amino-8-chloro-1-phenyl-[1,2,4]triazolo[4,3-a]-quinoxaline **CV1808** 2-phenylaminoadenosine **DPCPX** 1,3-dipropyl-8-cyclopentylxanthine **KF17837** 1,3-dipropyl-7-methyl-(3,4-dimethoxystyryl)xanthine **NECA** 5′-N-ethylcarboxamidoadenosine **PAPA-APEC** 2-(4-[2-[(4aminophenyl)methylcarbonyl]ethyl]phenyl)ethylamino-5′-N-ethylcarboxamidoadenosine

α₁-Adrenoceptors — CEC chloroethylclonidine **HEAT** 2-β-4-hydroxy-3-iodophenylethylaminomethyltetralone **WB4101** N-[2-(2,6-dimethoxyphenoxy)ethyl]-2,3-dihydro-1,4-benzodioxan-2-methanamine

α₂-Adrenoceptors — **ARC239** 2-(2,4-(O-methoxyphenyl)-piperazin)-1-yl **BHT920** 6-allyl-2-amino-5,6,7,8-tetrahydro-4H-thiazolo-[4,5-d]-azepine **RX821002** 2-(2-methoxy-1,4-benzodioxan-2-yl)-2-imidazoline **SKF104078** 6-chloro-9-(3-methyl-2-butenyl)oxyl-3-methyl-1H-2,3,4,5-tetrahydro-3-benzazepine maleate **UK14304** 5-bromo-6-[2-imidazolin-2-ylamino]quinoxaline

β-Adrenoceptors — **BRL37344** sodium-4-(2-[2-hydroxy-[3-chlorophenyl]ethylamino]propyl)phenoxyacetate **CGP20712A** 2-hydroxy-5-(2-[[2-hydroxy-3-(4-[1-methyl-4-trifluoromethyl-2-imidazolyl]phenoxy)propyl]amino]ethoxy)benzamide **ICI118551** erythro-DL-1-(7-methylindan-4-yloxy)-3-isopropyl-aminobutane-2-ol

Angiotensin — **CGP42112A** nicotinic acid-Tyr-(N-benzoylcarbonyl-Arg]Lys-His-Pro-Ile-OH **EXP31274** 2-n-butyl-4-chloro-1-[(2-[1H-tetrazol-5-yl]biphenyl-4-yl]methyl)imidazole-5-carboxylic acid **PD123177** 1-(4-amino-3-methylphenyl)methyl-3-diphenyl-acetyl-4,5,6,7-tetrahydro-1H-imidazo(4,5-O)pyridine-6-carboxylic acid **SKF108566** (E)-α-([2-butyl-1-[(4-carboxyphenyl)methyl]-1H-imidazol-5-yl]methylene)-2-thiophenepropanoic acid

Bradykinin — **HOE140** DArg[Hyp³,Thi⁵,DTic⁷,Oic⁸]BK

Cannabinoid — **CP55940** [1α,2β(R)5α]-(−)-5-(1,1-dimethylheptyl)-2-[5-hydroxy-2-(3-hydroxypropyl)-cyclohexyl]phenol **HU210** 11-OH-Δ⁸tetrahydrocannabinol, dimethyl heptyl **WIN55212-2** R-(+)-(2,3-dihydro-5-methyl-3-[(4-morpholinyl)methyl]pyrrolo[1,2,3-d,e]-1,4-benzoxazin-6-yl)(1-napthalenyl)methanone monomethanesulfonate

CCK/gastrin — **A71623** Boc-Trp-Lys(O-Me-Phe-NH)-Asp-(NMe)Phe-NH₂ **CI988** (R-[R*,R*])-4-[(3-1H-indol)-3-yl)-2-methyl-1-oxo-2-([[tricyclo(3,3,1,1.³,⁷)dec-2-yloxy]carbanoyl]amino)propylamino-1-phenylethylamino-4-oxobutanoic acid **L365260** 3R(+)-N-(2,3-dihydro-1-methyl-2-oxo-5-phenyl-1H-1,4-benzodiazepin-3-yl)-N′-3-methylphenyl urea **LY262691** 1-(4-bromophenylaminocarbonyl)-4,5,diphenyl-3-pyrazolidinone **PD140376** (L-3-[(4-aminophenyl)methyl]-N-[(tricyclo[3,3,1,1³·⁷]dec-2-yloxy)carbonyl]-D-tryptophyl-β-alanine) **PD140548** (N-[α-methyl-N-[(tricyclo[3,3,1,13,7]dec-2-yloxy)carbonyl]-L-tryptophyl-D-3-(phenylmethyl)-β-alanine)

Dopamine — **N-0437** 2(N-n-propyl-N-2-[2-thenyl]ethylamino)5-hydroxytetralin **SCH23982** 8-iodo-2,3,4,5-tetrahydro-3-methyl-5-phenyl-1H-3-benzazepin **SCH23390** 7-chloro-2,3,4,5-tetrahydro-3-methyl-5-phenyl-1H-3-benzazepine-7-ol **SCH39166** (−)-trans-6,7,7a,8,9,13b-hexahydro-3-chloro-2-hydroxy-N-ethyl-5H-benzo[d]naphto-(2,b)azepine hydrochloride **SKF38393** 2,3,4,5-tetrahydro-7,8-dihydroxy-1-phenyl-1H-3-benzazepine HCl **SKF83566** (±)7-bromo-8-hydroxy-3-methyl-1-phenyl-2,3,4,5 tetrahydro-1H-3-benzazepine hydrochloride **YM091512** cis-N-(1-benzyl-2-methylpyrrolidin-3-yl)-5-chloro-2-methoxy-4-methylamino benzamide

Endothelin — **BE18257B** cyc(DGlu-Ala-allo-DIle-Leu-DTrp) **BQ123** cyc(DTrp-DAsp-Pro-DVal-Leu) **FR139317** (R)2-{[R-2-[(S)-2-[(1-[hexahydro-1H-azepinyl]-carbonyl]amino-4-methyl-pentanoyl]amino]-3-(3-[1-methyl-1H-indodyl])propionylamino-3-(2-pyridyl)propionic acid

GABA — **β-CCM** ethyl-β-carboline-3-carboxylate methyl ester **CGP35348** (3-aminopropyl)-P-diethoxymethylphosphinic acid **CGP36742** p-(3-aminopropyl)-P-n-butyl-phosphinic acid **SR95531** 2-(3′-carboxy-2′-propyl)-3-amino-6-p-methoxyphenylpyridazinium bromide **DMCM** methyl-6,7-dimethoxy-4-ethyl-β-carboline-3-carboxylate **ZK93426** 5-isopropoxy-4-methyl-β-carboline-3-carboxylate ethyl ester

Glutamate — **1s, 3R-ACPD** 1-aminocyclopentane-1s, 3R-dicarboxylic acid **AMPA** D,L-α-amino-3-hydroxy-5-methyl-4-isoxalone propionic acid **t-AP4** L-amino-4-phosphonobutanoate **D-AP5** D-amino-5-phosphonopentanoate **CGP37755** 4-phosphonomethyl-2-piperidinecarboxylic acid **CGP37849** D,L-(E)-2-amino-4-methylphosphono-3-pentanoic acid **CNQX** 6-cyano-7-nitroquinoxaline-2,3-dione **CPP** (±)-3-(2-carboxypiperazine-4-yl)propyl-1-phosphonic acid **DNQX** 6,7-dinitroquinoxaline-2,3-dione **GYKI52466** 1-(4-aminophenyl)-4-methyl-7,8-methylenedioxy-5H-2,3-benzodiazepine HCl **HA966** 3-amino-1-hydroxypyrrolid-2-one **L689560** trans-2-carboxy-5,7-dichloro-4-phenylaminocarbonylamino-1,2,3,4-tetrahydroquinoline **MNQX** 5,7-dinitroquinoxoline-2, 3-dione **NBQX** 6-nitro-7-sulfamobenzo(f)quinoxaline-2,3-dione

Chemical names (continued)

5-HT
BIMU8 endo-N-8-methyl-8-azabicyclo[3.2.1]oct-3-yl)-2,3-dihydro-3-isopropyl-2-oxo-1H-benzimidazol-1-carboxamide hydrochloride CP93129 5-hydroxy-3(4-1,2,5,6-tetrahydropyridyl)-4-azaindole 8-CH-DPAT 8-hydroxy-2-(di-n-propylamino)tetralin GTI 5-hydroxytryptamine-5-O-carboxymethylglycyltyrosinamide GR113808 [1-[2-(methylsulphonyl)amino]ethyl]-4-piperidinyl]methyl-1-methyl-1H-indole-3-carboxylate LY53857 4-isopropyl-7-methyl-9-(2-hydroxy-1-methylpropoxycarbonyl)4,6,6A,7,8,9,10,10A-octahydroindolo(4,3FG) RS235919O 3α-tropanyl-1-yl)propyl-4-amino-5-chloro-2-methoxybenzoate hydrochloride WAY100135 N-tert-butyl 3-4-(2-methoxyphenyl)piperazin-1-yl-2-phenylpropanamide dihydrochloride

Leukotriene
ICI198615 (1-[2-methoxy-4-[((phenylsulfonylamino)carbonyl)phenyl]methyl]-1H-indazol-6-yl)carbamic acid cyclopentyl ester LY223982 (E)-5-(3-carboxybenzoyl)-2-((6-(4-methoxyphenyl)-5-hexenyl]oxy)benzenepropanoic acid LY255283 1-(5-ethyl-2-hydroxy-4-[[6-methyl-6-(1H-tetrazol-5-yl)heptyl]oxy]phenyl)ethanone LY170680 5-(3-[2(R)(carboxyethylthio)-1(S)-hydroxy-pentadeca3(E)5(Z)-dienyl]phenyl) 1H tetrazole MK571 [3-[(2-(7-chloro-2-quinolinyl)ethenyl)phenyl]][3-(dimethylamino-3-oxopropyl)thio]methyl]thio propanoic acid ONO-LB457 5-(2-[2-carboxyethyl)-3-[6-(4-methoxyphenyl)-5E-hexenyl]oxyphenoxy]valeric acid SC41930 7-[3-(acetyl-3-methoxy-2-propylphenoxy)propoxy]-3,4-dihydro-8-propyl-2H-1-benzopyran-2-carboxylic acid SKF104353 2S-hydroxy-3R-[2-carboxyethylthio]-3-[2-C8-phenyloctyl)phenyl]-propanoic acid U75302 6-[6-(3-hydroxy-1E,5Zundecadien-1-yl)-2-pyridinyl]-1,5-hexanediol

Muscarinic
AFDX116 11-([2-[(diethylamino)methyl]-1-piperidinyl]acetyl)-5,11-dihydro-6H-pyrido[2,3-b][1,4]benzodiazepine-6-one

Neurotensin
SR48692 2-([1-[7-chloro-4-quinolinyl)-5-[2,6-dimethoxyphenyl]pyrazol-3yl]carbonylamino)tricyclo(3.3.1.1.3.7) decan-2-carboxylic acid

Opioid
CI977 (5R)(5α,7α,8β)-N-methyl-N-(7-[1-pyrrolidinyl]-1-oxaspiro[4,5]dec-8-yl]-4-benzofuraneacetamide monohydrochloride CTAP D-Phe-Cys-Tyr-D-Trp-Arg-Thr-Pen-Thr-NH₂ DAMGO Tyr-DAla-Gly-NMePhe]-NH(CH₂)₂-OH DPDPE [D-Pen²,D-Pen⁵]enkephalin DSBULET Tyr-DSer(OtBu)-Gly-Phe-Leu-Thr ICI197067 (2S)-N-[2N-methyl-3,4-dichlorophenylacetamido)-3-methylbutyl]pyrrolidine hydrochloride ICI174864 N-N-diallyl-Tyr-Aib-Phe-Thr PL017 [N-MePhe³,D-Pro⁴]morphiceptin U69593 5α,7α,β-(-)-N-methyl-N-[7-(1-pyrrolidinyl)-1-oxaspiro(4,5)dec-8-yl]benzene acetamide

PAF
C-PAF 1-O-hexadecyl-2-N-methylcarbamyl-sn-glycero-3-phosphocholine U69593 5α,7α,β-(-)-N-methyl-N-[7-(1H-2-methylimidazole[4,5-c]pyridinyl]phenyl]sulphonyl-L-leucinyl ethyl ether CV6209 2-[(N-acetyl-N-(2-methoxy-3-octadecyl-carbamoyloxypropoxycarbamoyl)aminomethyl]-1-ethylpyridinium chloride L659989 trans-2-(3-methoxy-methylsulfonyl-4-propoxyphenyl)-5-(3,4,5-trimethoxyphenyl)tetrahydrofuran SRI63072 (R,S)-3-[2-[(2-octadecylaminocarbonyloxymethyl)tetrahydro-2-furanyl]methoxy]hydroxyphosphinyloxy]ethyl]thiazolium hydroxide inner salt-4-oxide WEB2086 3-(4-[2-chlorophenyl]-9-methyl-6H-thieno[3,2-f][1,2,4]-triazolo-[4,3-a][1,4]-diazepine-2-yl)1-(4-morpholinyl)-1-propanone

Prostanoid
AH13205 trans-2-(4-[1-hydroxyhexyl]phenyl)-5-oxocyclopentaneheptanoic acid AH6809 6-isopropoxy-9-oxoxanthene-2-carboxylic acid BW245C 5-(6-carboxyhexyl)-1-(3-cyclohexyl-3-hydroxypropyl)hydantoin BWA868C 3-benzyl-5-[6-carboxyhexyl]-1-[2-cyclohexyl-2-hydroxyethylamino]hydantoin GR32191 (1R-[1α(Z],2β,5α])-(+)-7-(5-[[1,1'-biphenyl]-4-ylmethoxy]-3-hydroxy-2-(1-piperidinyl)cyclopentyl)-4-heptenoic acid GR63799 [1α(Z],2β[R*],3α])-4-benzoylamino]phenyl-7-(3-hydroxy-2-[2-hydroxy-3-phenoxypropoxy]-5-oxycyclopentyl)4-heptanoate I-BOP (1S-[1α,2β(5Z],3α(1E,3s*],4α])-7-(3-[hydroxy-4-(4'-iodophenoxy]-1-butenyl]-7-oxabicyclo-[2.2.1]heptan-2-yl)-5-heptanoic acid MB28767 (±)15S-hydroxy-9-oxo-16-phenoxy-17,18,19,20-tetranor-prost-13E-enoic acid SC19220 1-acetyl-2-(8-chloro-10,11-dihydrobenz[b,f][1,4]oxazepine-10-carbonyl)hydrazine SQ29548 (1s-[1α,2β(5Z],3β,4α])-7-[3-[[2-(phenylamino)carbonyl]hydrazino]methyl]-7-oxabicyclo[2.2.1]hept-2-yl-5-heptenoic acid STA₂ 11α-carba-9α-11α-thia-TXA₂ U46619 11α,9α-epoxymethano-PGH₂ ZK110841 9β-chloro-11α,15S-dihydroxy-16,20-methano-prost-5Z,13E-dienoic acid

P₂-purinoceptor
ANAPP₃ arylazidoaminopropionylATP

Somatostatin
MK678 cyc(tr-Me-Ala-Tyr-DTrp-Lys-Val-Phe)

Tachykinin
CP96345 2S,3s-cis-3-(2-methoxybenzylamino)-2-benzhydrylquinuclidine CP99994 (+)-(2s,3s)-3-(2-methoxybenzylamino)-2-phenylpiperidine GR64349 Lys-Asp-Ser-Phe-Val-Gly-[R-γ-lactam] GR82334 [DPro⁹(spiro-γ-lactam)Leu¹⁰]physalaemin GR94800 PhCO-Ala-Ala-DTrp-Phe-DPro-Nle-NH₂ L659877 cyc(Gln-Trp-Phe-Gly-Leu-Met), MEN10207 [Tyr⁵,DTrp⁶,⁹,Arg¹⁰]NKA₃₋₁₀ R396 Ac-Leu-Asp-Gln-Trp-Phe-Gly-NH₂ RP67580 2-(1-imino-2-(2-methoxy-phenyl)-ethyl)-7,7-diphenyl-4-perhydroisoindolone (3αR, 7αR) SR48968 (s)-N-methyl-N-[4-acetylamino-4-phenylpiperidino)-2-(3,4-dichlorophenyl)butylbenzamide

Vasopressin/oxytocin
OPC21268 1-(1-[4-[3-acetylaminopropoxy]benzoyl]-4-piperidyl)-3,4-dihydro-2(1H)-quinolinone; OPC31260 5-dimethylamino-1-(4-[2-methylbenzoylamino]benzoyl)-2,3,4,5-tetrahydro-1H-benzazepine

K⁺ channels
AZ-DF265 4-((N-(α-phenyl-2-piperidino-benzyl]carbamoyl]methyl) benzoic acid; W7 N-(6-aminohexyl)-5-chloro-1-naphthalenesulfonamide hydrochloride

CHAPTER 2

Ligand-binding studies – theory and experimental techniques

HENNING OTTO

Institut für Biochemie, Freie Universität Berlin, Thielallee 63, D-14195 Berlin, Germany

1. Introduction

Cells receive information from the extracellular environment. In organisms of a higher complexity, processing this information by an individual cell is part of an informational network, that the organism uses to maintain its integrity. Each cell reacts to a subset of the organism's messages. The cell is able to react to all relevant information according to its state of differentiation.

The basis of this process is the recognition of chemically coded information by the receptor proteins that the cell expresses. The central mechanism of this recognition is the binding of an extracellular signalling molecule by a receptor protein. The highly specific binding of the natural ligand leads to a conformational change within the receptor protein. Thereby, the receptor becomes activated. This leads to the subsequent activation of a more or less complex effector system. By integrating all intracellular processes, triggered by different signalling molecules, the cell responds in a way that normally fulfils the demands of the organism.

Receptors in this sense, whether they are integral membrane proteins that couple the extracellular binding of a first messenger to an intracellular effector system (ligand-gated ion channels, G-protein-coupled receptors) or intracellular proteins that first need the uptake of a signal molecule by the cell (steroid hormone receptors), are the entry point at which (by binding the natural ligand) regulatory information normally enters the cell.

Due to this prominent task, which results in the control of cellular and physiologi-

---

Abbreviations: B, concentration of the bound ligand; B_{max}, concentration of binding sites in an assay; cDNA, complementary DNA; EDTA, ethylene diamine tetra-acetic acid; F, concentration of the free (unbound) ligand; GppNHp, guanylyl imidodiphosphate; IC_{50}, inhibitor concentration at 50% inhibition of binding; k_{+1}, rate constant of ligand binding; k_{-1}, rate constant of receptor-ligand-complex dissociation; K_a, association constant; K_d, dissociation constant; K_i, inhibition constant; n_H, Hill coefficient; P, fractional saturation of binding; PMSF, phenylmethylsulfonyl fluoride; R, receptor; SEM, standard error of mean; L, ligand; I, competitor; RL, occupied binding site in receptor-ligand complex.

cal parameters, the binding sites of receptors are major targets for many drugs. These ligands compete with the natural ligand for the same binding site. Thereby, these substances evoke, modulate, or suppress the normal cellular response. They influence the signal transduction by a cell and therefore affect the informational network by which the organism controls its subsystems.

The basic motif consists of an appropriate molecule, the ligand, and a binding site on a receptor protein. By binding this ligand to the binding site, the receptor protein is activated. The triggered biological effect can be detected in vivo using pharmacological assays.

Besides the pharmacological characterization of a ligand, which includes binding to the receptor's binding site as well as the subsequently initiated effect, the characterization of the pure binding step is a useful task. Furthermore, the examination of a receptor's binding properties in vitro is more easily achieved than its complete pharmacological characterization. Assays for the receptor's activity or the biological effect are generally much more difficult. Nevertheless, the reduction to binding studies bears the danger of misinterpretation. The differentiation between specific and nonspecific binding in some cases is impossible without testing the biological effect coupled to it (see Section 2.2.).

Binding study methods are designed to characterize the binding of a ligand to the binding site of a receptor (Table I). Binding assays provide information about the concentration of binding sites, the ligand's affinity to the binding site, or the kinetics and thermodynamics of the binding process. They provide insight into intramolecular dynamics in that cooperativity in binding is revealed. Therefore, binding assays are useful to quantitate receptors and to identify and characterize either a set of ligands, that bind a receptor (binding profile of the receptor), or, vice versa, the receptors, that are able to bind a defined ligand (binding profile of a ligand) [1–3].

This Chapter provides a short guide to the basic features of binding studies. The basic types of experiments and their limitations will be demonstrated for the most simple situations. The basic mathematical relations and their graphical presentation, the influence of nonspecific binding as well as the influence of cooperative binding will be shown. Further comments will be given on ligands, on the computerized eval-

TABLE I
Information provided by binding studies

Affinity of binding
Concentration of binding sites
Stoichiometry of binding
Concentration of the receptor
Cooperativity of binding
Binding profile of a receptor
Identification of receptor subtypes
Binding profile of a ligand

uation of the nontransformed binding data, and on aspects of the set-up of a binding assay. Nevertheless, this guide will be a short one. For a deeper understanding of this subject, the reader is referred to the references at the end of this Chapter [4–10].

2. The experimental scenario

2.1. Basic types of binding experiments

The aim of binding studies is the characterization of a ligand's binding to its accompanying receptor. For a defined ligand/receptor pair, characteristic parameters are:
a) the ligand's *affinity*, expressed as either the *dissociation constant* K_d or its reciprocal, the *association constant* K_a. In the further course of this chapter, only the dissociation constant K_d will be used;
b) the *rate constants* for the ligand's *binding* k_{+1} or *dissociation* k_1;
c) the *stoichiometry* of binding; and
d) the *interference of multiple binding sites* on a receptor molecule, that influence one another during ligand binding, i.e. the *cooperativity* of the binding. The *Hill coefficient* n_H characterizes the strength of such an intramolecular interaction.

Most receptors bind more than one ligand species. More complex intentions of binding studies are based on this feature. First, one aim is the examination of the competition of different ligands at the same binding site. Second, this leads to a characterization of a receptor, a receptor subtype, or even a receptor's functional state by the binding characteristics of a set of ligands [1,11,12]. This was a classical approach to find and distinguish either different receptors or subtypes of a receptor, that had been classified according to its natural ligand. Today, many receptor subtypes or even previously unknown receptors are first described by their cDNA sequence. This leads to the curious situation of the identification of receptors without the knowledge of their natural ligand or their function. Thus, the main task is the identification of ligands, especially of the natural ligand, and the assignment of a biological effect, which is triggered by binding an agonistic ligand.

All this information and these parameters, concerning the binding of ligands, can be obtained with only a few basic types of experiments (Table II).

2.1.1. Saturation experiments
A constant concentration of binding sites is incubated with increasing concentrations of a ligand. After equilibration the equilibrium concentrations of the free and bound ligand are determined. This analysis results in the determination of the dissociation constant (K_d) and the concentration of binding sites in the assay (B_{max}). The number of binding sites per receptor molecule, i.e. the stoichiometry of binding, is derived from B_{max}, provided that the concentration of the receptor protein is known. Nonspecific binding of the ligand, cooperative binding, and heterogeneity of binding sites in

TABLE II
Basic types of experiments and related parameters

Saturation experiments
   Dissociation constant K_d
   Association constant K_a (reciprocal of K_d)
   Concentration of binding sites B_{max}
   Stoichiometry of binding
   Concentration of the receptor molecule
Kinetic experiments
   Rate constant for binding (on-rate) k_{+1}
   Rate constant for dissociation (off-rate) k_{-1}
   Dissociation constant K_d
   Association constant K_a
Thermodynamic experiments
   Free energy, enthalpy, and entropy for the equilibrium $\Delta G°$, $\Delta H°$ and, $\Delta S°$
   Free energy, enthalpy, and entropy for the intermediate state of binding $\Delta G^\#$, $\Delta H^\#$ and, $\Delta S^\#$
Inhibition experiments
   Inhibitor concentration IC_{50} (50% inhibition)
   Inhibition constant K_i

---

the assay appear in a characteristic way. This enables us to either correct the data or characterize the more complex experimental situation.

2.1.2. Kinetic experiments

The binding of the ligand over a period of time (on-rate) or the ligand's dissociation over time (off-rate) is measured. The results are the rate constants for either the association (k_{+1}) or the dissociation (k_{-1}). The equilibrium constant K_d is derived from these constants by the division of k_{-1} by k_{+1}.

2.1.3. Thermodynamics of binding

Monitoring the temperature dependence of the dissociation constant as well as of the velocity constants allows the determination of thermodynamic parameters. These are the free energy, enthalpy, and entropy either for the equilibrium ($\Delta G°$, $\Delta H°$, $\Delta S°$) or for the formation of the intermediate state of the binding process ($\Delta G^\#$, $\Delta H^\#$, $\Delta S^\#$). The thermodynamics of ligand binding are not covered in the further course of this chapter.

2.1.4. Inhibition experiments

Inhibition experiments analyze the competition of two different ligand species for the same binding site. Constant concentrations of both the receptor and the labelled ligand (e.g. a radioligand) are incubated with increasing concentrations of the unlabelled competing ligand. The result is a curve, which demonstrates the labelled ligand's displacement from the binding site by the unlabelled ligand.

Characteristic parameters are IC_{50} and the K_i. IC_{50} is the concentration of the competing ligand, which reduces the labelled ligand's binding to 50% of the amount obtained in absence of the competing ligand. This concentration depends on the concentration of labelled ligand. The inhibition constant K_i is the dissociation constant for the binding of the competing ligand. It is related to IC_{50}; but in contrast to IC_{50}, the K_i is not dependent on the labelled ligand's concentration.

A special example of an inhibition experiment is the displacement of a labelled ligand by its unlabelled counterpart. Thereby, the reversibility of the labelled ligand's binding is demonstrated. This is one of several criteria for the definition of specific binding.

2.2. The receptor

As discussed in chapter 1, the term 'receptor' is differently and sometimes confusingly used in different contexts. In the field of signal transduction, the term 'receptor' stands for a protein structure, which activates a subsequent effector system while it is activated by binding an agonistic ligand. In the context of binding studies, this term has a much more general sense. There it is synonymously used for the term 'binding site'. It is extremely important to notice this different usage in different contexts. Very often the terminology is mixed, thus bearing the danger of losing the awareness for the definition of binding specificity.

A binding assay only gives information about a ligand's binding to binding sites. The system to be assayed consists of a concentration of binding sites and a concentration of the ligand, which is either bound to these binding sites or free in solution. The structural basis of the binding itself is neither a subject nor a result of the binding assay. Binding might be based on different receptor proteins, different states of a receptor protein, or even on nonreceptor proteins, which have a specific binding site for the ligand used, but are linked to functions different from signal transduction. The binding analysis for all these different kinds of 'specific' binding are based on the same mathematical and theoretical treatment of binding studies and will therefore appear in a similar way. Binding of a ligand might even have a very nonspecific and artificial basis. For example, the ligand might bind to fragments of receptor proteins, which contain the binding site, or it may bind because of adsorption to glass surfaces, biological membranes, or proteins, which do not have any sign of a specific binding site for the ligand.

This feature of binding assays obviously necessitates the definition of specific binding and its separation from nonspecific binding. To avoid the determination of only artificial and therefore irrelevant binding, the specific binding of a ligand has to be correlated with its pharmacological potency. As mentioned above, the biological effect of the ligand's binding unfortunately never results from a binding assay. Hence,

TABLE III
Criteria for the specific binding of a ligand to a receptor

Selectivity of binding
Saturability of the binding sites
Reversibility of binding
Correlation of binding with a biological function

---

the pharmacological effect which a ligand initiates has to be elucidated by separate functional tests, which also include the subsequent activation of the receptor-linked effector system.

Very often, the specific binding has been characterized only by the criteria of selectivity and high affinity of a ligand's binding as well as its saturability and reversibility. These criteria are not sufficient to avoid an artificial basis of binding. This was exemplified by Cuatrecasas and Hollenberg [13] with ^{125}I-insulin. They demonstrated that ^{125}I-insulin binds selectively and with high affinity to either talc, silica, protein-agarose derivatives, or glass surfaces. This binding also fulfils the criteria of saturability and reversibility, as proven by the displacement by nonradioactive insulin. This observation may also hold for other ligands, especially for peptide ligands.

Hence, the definition of more precise criteria for specific binding is absolutely inevitable. It must also include the assignment of a biological effect on the ligand binding. A receptor in the sense of a specific binding site (Table III), therefore, is characterized by the following aspects.

1. Selective binding of the ligand
The ligand should belong to a limited set of ligand species, which bind to the investigated binding site. The ligand should have a reasonably high affinity expressed by a dissociation constant in a pharmacologically relevant range.

2. Reversibility of binding
It should be possible to displace the labelled ligand from a specific binding site by a reasonable concentration of its nonlabelled counterpart. This allows the definition of a nonspecific fraction of the bound ligand's total concentration. Furthermore, it would be advantageous to check the specific binding by structurally unrelated ligands, that are also specific for this binding site. The displacement of the labelled ligand by these compounds should reveal the same fraction of nonspecific binding as the nonlabelled counterpart.

3. Saturability
In general, the predominent feature of nonspecific binding is the impossibility of saturation in a pharmacologically relevant concentration range. This is due to a virtually infinite number of nonspecific binding sites with a low affinity for the ligand. In

contrast, the finite number of the specific binding sites is readily saturable by the ligand.

4. Correlation with a biological function
The binding of the ligand should be assigned to a defined biological effect. In a functional assay this effect should appear at a range of the ligand concentration that is quantitatively correlated to the affinity of the ligand in a binding assay.

2.3. The ligand

Essential for a binding assay is the availability of a ligand that has been labelled in order to measure its distribution between the bound and the free state. Commonly, ligands, labelled by radioisotopes, are applied. These radioligands usually bear either tritium (^{3}H) or iodine (^{125}I) as a tracer. Other isotopes used as labels are ^{131}I, ^{35}S, or ^{14}C. These radioligands are labelled to a high specific activity. In addition to radioisotopes, fluorescent labels might also be used. An example of a fluorescent ligand for the ß-adrenergic receptor has been given by Levitzki [8]. Fluorescent ligands are primarily used for localizing a receptor in various tissues.

A ligand used in binding studies should fulfil the following criteria.
1. *Unchanged binding properties of the labelled ligand.* The labelled ligand has to maintain the same binding characteristics as its nonlabelled counterpart [14].
2. *Sufficient affinity for the binding site.* The off-rate limits the use of a ligand in binding assays. This parameter determines which technique is applicable for the essential separation of the bound and the free ligand. The separation is very difficult if the ligand quickly dissociates from the binding site. The separation itself needs time, and during subsequent washing steps, which are a common part of separation protocols to reduce the nonspecific binding, the receptor-ligand complex is faced with a large volume of washing solution without free ligand. A significant dissociation of the bound ligand during this procedure is obviously detrimental for the determination of the bound ligand. This also clearly affects all deduced parameters.
3. *Kinetic properties sufficient for equilibration within the assay time.* With the exception of kinetic experiments, the binding process must be rapid enough to allow equilibration within the incubation period of the assay.

Further demands covering the criteria for specific binding sites of receptors as well as technical requirements concerning the quality of a radioligand preparation will be discussed later (see also Table V).

2.3.1. Ligand classification
The ligands may be classified according to either their affinity or according to their manner of action (Table IV).

TABLE IV
Ligand classification

Affinity
  Reversibly bound ligands
  Quasi-irreversibly bound ligands
  Irreversibly bound ligands
Mode of action
  Agonists
  Antagonists
  Partial agonists

---

1. Ligands differ in their *affinity* to a receptor's binding site. The binding takes place in either a reversible, quasi-irreversible, or irreversible manner.
   a) All physiologically and most pharmacologically important ligands bind in a *reversible* manner. The dissociation constants for these ligands range down to about 10^{-9} M. Such ligands are mainly used in binding assays.
   b) Ligands which bind *quasi-irreversibly* have dissociation constants below about 10^{-10} M. Many toxic compounds, for example from snake venoms, belong to this group. They are able to block receptors virtually irreversibly. This is in accordance with their biological function to paralyze or kill a prey. Because of their extremely high affinity, the determination of their dissociation constants by saturation experiments is hardly possible. The assayed binding sites will remain in an almost fully occupied state even at the lowest ligand concentrations possible. Hence, no reliable data will be obtained in the range of 10–90% saturation of the binding sites, which the saturation data should cover to give an acceptable estimate for the dissociation constant. On the other hand, the high affinity provides a quick and safe determination of the concentration of receptor binding sites even if their concentration in tissues or crude receptor preparations is very low. The stability of the ligand-receptor-complex allows a maximal reduction of the nonspecific binding. This is achieved first by the low ligand concentrations sufficient for the saturation of the specific binding sites, since the nonspecific binding is normally proportional to the total concentration of the ligand. Secondly, this stability allows a rather extensive and stringent washing which can be performed without significant loss of the bound ligand.
   c) The third group consists of ligands that bind *irreversibly* to a receptor's binding site. These compounds cannot be used in the context of binding studies. Instead, these substances may be applied in affinity labelling of receptors, which is useful for structural investigations of these molecules.
2. A second way to classify ligands may be based on their *mode of action*. Ligands may act as agonists, partial agonists, or antagonists.

a) *Agonists* evoke the biological effect by binding to a receptor.
b) *Antagonists*, in contrast, bind solely to the receptor, but are unable to induce receptor activation. Therefore, these ligands inhibit receptor activation by an agonist by competing for the binding site.
c) *Partial agonists* stand in the middle between agonists and antagonists. They are able to induce receptor activation, but are less effective than an agonist. In competition with a true agonistic ligand, these ligands slightly inhibit the maximal receptor activation when compared to the effect of the agonist alone.

As mentioned above, the ligand's mode of action and the resulting biological effect must be known and tested in pharmacological assays, before it can be considered a ligand in binding studies. This might help prevent artificial results. Furthermore, this mode of action may give a hint to the binding affinity. In a first estimation, the dissociation constant K_d of agonists can be expected within the range of 10^{-6} to 10^{-8} M. Antagonists often show a much higher affinity with dissociation constants in the range of 10^{-9} to 10^{-11} M. Owing to the higher affinity, the use of antagonists as radioligands seems preferable. First, the higher affinity allows lowering of the radioligand's concentration. This requires a high specific radioactivity of the radioligand, which guarantees the measurement at low ligand concentrations. The benefits are 1. a mostly lower nonspecific binding and 2. a better separation of the bound and the free ligand due to a higher stability of the receptor-ligand complex. A further lowering of the nonspecific binding may be achieved for a more stable complex, because the washing procedures may be more stringent, as discussed above for quasi-irreversibly binding ligands.

2.3.2. *The radioligand's quality*

Radioligands are commonly labelled with the radioisotopes ^{3}H or ^{125}I to a high specific activity (more than 10 Ci/mmol). Because of the difficulties in handling tritiated compounds (^{3}H), they are usually produced exclusively by commercial suppliers, and are, therefore, rather expensive. Iodinated peptides or proteins (^{125}I), on the other hand, are relatively easy to produce.

2.3.2.1. Tritiated compounds. Most non-peptide ligands are tritium-labelled. Their specific activity, as supplied by commercial sources, ranges between 10 and 150 Ci/mmol. The rather low energy of the isotope's weak β-radiation (0.018 MeV) requires the detection by liquid scintillation counting. The advantages of tritiated compounds are: 1. the long half-life (12.3 years), which allows a long use of stock solutions without a loss of radioactivity by decay; and 2. a low hazard that the substitution of a hydrogen by tritium changes the ligand's function and binding properties.

2.3.2.2. Iodinated compounds. Several techniques have been developed for the iodination of peptides or proteins by ^{125}I. With [^{125}I]-sodium iodide as a source, the labelling is achieved quickly at rather low cost using either the chloramine-T method, the Bolton-Hunter reaction or the lactoperoxidase method. These methods provide high specific activities of several hundred up to about 2000 Ci/mmol. Commercial suppliers provide fresh stocks of many [^{125}I]-ligands every week with specific activities of 2200 Ci/mmol. ^{125}Iodine emits γ-rays, which can be detected without problems using a γ-counter. It has a rather short half-life (60 days). For iodinated ligands much more caution is necessary because of a possible change in the binding characteristics compared to the nonlabelled precursors.

Two last topics concerning a ligand's quality are 1. the homogeneity and purity of the ligand preparation and 2. the stability of the ligand during storage as well as during the experiment.

2.3.2.3. Homogeneity and purity of the ligand. The successful binding analysis demands a ligand preparation, which is homogenous and has a purity of more than 90–95%. Homogeneity sometimes also includes the demand for pure stereoisomers, if the investigated receptor shows stereoselectivity.

2.3.2.4. Stability of the radioligand. It is very important to regularly check the radioligand's stability. At least two main sources of instability are responsible for a decrease of the specific activity.

a) *Instability due to the radioactivity.* The radioactive decay itself causes decomposition of ligand molecules. This is due to the formation of radicals by radiation interacting either directly with ligand molecules or the solvent. Owing to its higher concentration, the formation of solvent radicals is more important in this context. In particular, the solvent H_2O is a main source for radicals produced by the radiation. The higher the radioactivity in a stock, the more important it is to prevent these events. This might be achieved by using scavengers, which absorb the destructive radicals. The scavengers commonly used are ethanol, glycerol, β-mercaptoethanol, and benzylalcohol. In aqueous solutions routinely adding ethanol (about 2%) might therefore be useful. However, one has to make sure that the use of a scavenger does not disturb the ligand's binding.

b) *Instability due to biological or chemical reasons.* Decomposition of the ligand may also take place for biological or chemical reasons. To prevent this sort of decay, the radioligand should be stored at $-20°C$, unless the biological potency is influenced by freezing. The ligand should be stored in aliquots and used soon after thawing. Deep freezing and thawing of the aliquots should proceed as quickly as possible. The addition of glycerol might be useful as a scavenger as well as a protective during freezing. One should regularly check the quality and the specific activity of a stored ligand. The ligand must maintain its binding ability as well as its pharmacological potency. A quick control of the part of the radioactivity which is still able to bind to the receptor could therefore be the titration of a low ligand concentration using increasing concentrations of the receptor.

Apart from storage, the ligand's stability must be maintained during the assay. Hence, the receptor preparation should be examined very carefully for enzymatic activities, which are able to degrade the ligand. The degradation should be prevented by the use of enzyme inhibitors. If the ligand is, for example, acetylcholine, one should pay attention to esterases. Irreversible inhibition of these enzymes could be achieved by the addition of eserine, organophosphates, or other inhibitors of serine esterases.

Very important degrading enzymes to be taken into account are peptidases. Not only peptide ligands but also the receptor proteins could be destroyed by proteolysis. Therefore, much attention should be paid to the inhibition of these activities by phenylmethylsulfonyl fluoride (PMSF), chelating agents like ethylene diamine tetraacetic acid (EDTA), or inhibiting peptides like aprotinin, leupeptin, or pepstatin. The addition of EDTA should be checked carefully. Reducing the concentration of divalent cations might influence the binding.

Often a loss, especially of peptide ligands, is observed, which is due to the adsorption of these ligands to the wall of the test tube. The result is a lower total concentration of the ligand, which will take part in the equilibrium of the specific binding. This obviously would disturb the analysis of the binding experiment. Hence, the ligand's adsorption should be prevented by siliconization of glass surfaces or by saturation of the adsorbing surfaces for example by polylysine or by 'inert' proteins like albumin.

Table V conclusively summarizes the criteria for a ligand used in binding studies, as discussed above. Some of these criteria result from the definition of specific binding to a receptor. The ligand has to be characterized by pharmacological models in vivo, which reveal its mode of action and its pharmacological potency. The ligand should bind in a reversible manner. Furthermore, the ligand should competitively bind to a receptor's binding site with regard to the natural ligand as well as to other structurally unrelated ligands specific for this site.

Other criteria come from technical reasons. The kinetic properties of the ligand have to allow equilibration during the incubation period (an exception is kinetic experiments). A safe separation of the free and the bound ligand should be supported by a sufficient affinity of the ligand.

TABLE V
Criteria for the use of a ligand in binding studies

Definition of specific binding sites
   Selectivity of binding (competition with the natural ligand, competition with structurally unrelated ligands)
   Reversibility of binding
   Correlation with a biological function (pharmacological properties)
Technical requirements
   Unchanged binding properties of the radioligand
   Unchanged pharmacological properties of the radioligand
   Sufficient affinity for the binding site
   Kinetic properties sufficient for equilibration within the incubation period of an assay
   High specific radioactivity
   Homogenous (incl. stereoselective demands)
   Pure preparation (purity more than 90–95%)
   Stability (during storage and experiment)

Labelling of the ligand should only result in a structural change, i.e. the label should alter neither the binding characteristics nor the pharmacological potency of the radioligand compared to its precursor. A highly specific radioactivity will help to reduce ligand as well as receptor concentration in the assay. This will also improve the reduction of nonspecific binding.

The ligand preparation has to be homogenous. This includes sometimes a pure preparation of a stereoisomer. A high purity of the ligand preparation (more than 90–95%) is required. The ligand has to be stable during storage as well as during the experiment. This stability includes decomposition by radioactivity, degradation by components of the receptor preparation (e.g. enzymatic degradation), or by isomerization. The quality of the ligand preparation has to be checked regularly.

3. Experimental strategies

A binding assay is performed by incubation of a tracer ligand with a concentration of binding sites. From the view of pure binding, the term 'binding sites' seems preferable, instead of 'receptor' or 'binding protein'. The result of a binding experiment is the distribution of a radioligand between the bound and the free state. As mentioned above, the basis for this binding might be partially different from specific binding sites on a receptor molecule or another binding protein. In the case of multiple binding sites on a receptor, the nonequivalence of the terms 'receptor concentration' and 'concentration of binding sites' is also obvious. In binding assays only binding sites are seen.

The incubation time of a binding assay depends on the experimental approach. Except for kinetic experiments, the incubation proceeds until equilibrium is established. For inhibition experiments (displacement experiments) a second ligand is added, which is not labelled and competes with the first, the tracer ligand, for the binding sites.

The concentration of binding sites is maintained at a constant value $[R]_T$. This concentration is determined during a saturation experiment (see below).

The ligand concentration is varied during the experiment. Hence, the *controlled parameters* are the total concentration of the tracer ligand $[L]_T$ and, in case of displacement, the total concentration of the competitor $[I]_T$. The *measured parameters* are the concentration of the bound radioligand, B, and finally the concentration of the ligand already free in solution, F (free ligand). Prerequisite for the determination of B and F is a reliable technique for the separation of the two states (see section 4.1). Some of these techniques do not allow the direct determination of F. In that case, F is obtained from the difference between the total ligand concentration $[L]_T$ and the concentration of B. Obviously, F includes then the error of $[L]_T$ as well as the error of the determination of B. Because the determination of B by these techniques also might be erroneous, for a number of samples F should be checked independently by using other separation techniques like centrifugation or dialysis.

3.1. Saturation experiments

This is the most widely used type of binding experiment: a constant concentration of binding sites is incubated with increasing concentrations of the radioligand. The incubation is continued until the binding equilibrium has been established. At this point, the ligand's association to the binding sites and its dissociation proceed with equal velocity and no more alterations in the concentrations of the bound and the free ligand occur. Now, both states of the ligand are quickly separated and determined.

The following demonstrates for the simplest case the mathematical relations between measured and controlled concentrations as well as their visualization. Assume a receptor preparation which contains a homogenous population of receptor protein. The receptor complexes bear one or more binding sites, which bind the ligand independently from each other, i.e. no cooperativity occurs. Hence, the assay contains a concentration of equal binding sites $[R]_T$, which do not influence one another. Furthermore, the nonspecific binding may be negligible. Finally, neither receptor or ligand degradation nor isomerization of the receptor-ligand complex occurs during the incubation period.

The equilibrium may then be expressed in the direction of the dissociation process as:

$$(RL) \rightleftharpoons R + L$$

where L stands for the free ligand, R for an unoccupied binding site, and (RL) for an occupied binding site in a receptor-ligand complex. The concentration ratio of all participating states in the equilibrium then is given by the equilibrium constant according to the mass action law. This is expressed in eq. 1 as the dissociation constant K_d:

$$K_d = \frac{[L]\cdot[R]}{[(RL)]} \qquad \text{(eq. 1)}$$

Now the concentrations in eq.1 are substituted by measured parameters: F equals [L] and B is substituted for the complex between the ligand and the binding sites [(RL)]. At each concentration $[L]_T$ a pair of B and F is obtained. For each $[L]_T$ tested, the ratio of the concentrations of the receptor-ligand complex [(RL)] and the unoccupied binding sites [R] is changed. The sum of both concentrations equals the total concentration of binding sites $[R]_T$, which is maintained at a constant value. This is described by eq.2:

$$[R]_T = [R] + [(RL)] \qquad \text{(eq. 2)}$$

Isolating [R] and substituting [(RL)] by B the equation changes to eq. 3:

$$[R] = [R]_T - B \qquad \text{(eq. 3)}$$

Finally, the total concentration of binding sites $[R]_T$ has to be substituted by measured concentrations. This is achieved by determining the maximal concentration of ligand, which is able to bind the finite number of binding sites in the assay volume. If the total ligand increases, the bound ligand's concentration reaches a maximal concentration B_{max}, which equals the total concentration of binding sites $[R]_T$. At higher ligand concentrations the bound ligand's concentration will not change, because all binding sites are now in the occupied state (RL). The concentration of free binding sites equals zero. This is expressed in eq. 4:

$$B_{max} = [R]_T = [(RL)]_{([R]=0)} \qquad \text{(eq. 4)}$$

Therefore, $[R]_T$ in eq. 3 is substituted by B_{max}:

$$[R] = B_{max} - B \qquad \text{(eq. 5)}$$

These substitutions lead to eq. 6, which describes the relationship between the dissociation constant Kd and the measured concentrations of the bound and the free ligand:

$$K_d = \frac{F \cdot (B_{max} - B)}{B} \qquad \text{(eq. 6)}$$

The saturation experiment provides the determination of K_d and B_{max}. Historically, this was done by the linearization of the data according to Scatchard, which allows linear regression analysis. Nowadays, the direct analysis of the binding data, which includes weighting the error of measurement, is usually achieved by a computerized nonlinear least-squares fit. Therefore, the next section should be seen more as a demonstration of the customary visualization of the data as well as a method to obtain initial estimates of the binding model and of the binding parameters for a subsequent numerical refinement.

3.1.1. Graphical evaluation; the Scatchard plot
The first way to visualize the result of a saturation experiment is to plot B against F. This direct binding curve is shown in Fig. 1a. The graph is based on eq. 7, which is a transformation of eq. 6:

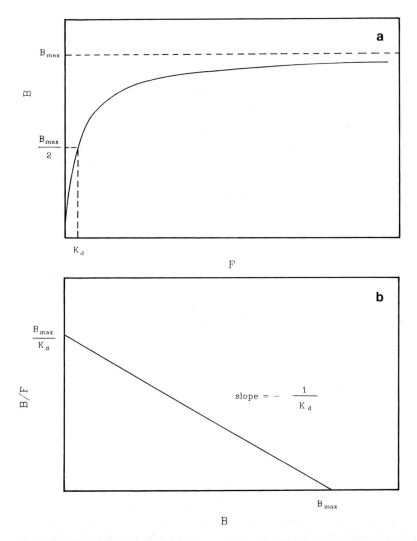

Fig. 1. Visualization of a binding experiment. a) Direct binding plot; b) Scatchard plot.

$$B = \frac{B_{max} \cdot F}{K_d + F} \qquad \text{(eq. 7)}$$

As Fig. 1a shows, the result is a typical saturation curve. Increasing concentrations of the free ligand's concentration F first leads to a rapid increase of the bound ligand's concentration B. Then, the slope of the curve decreases and the curve approaches asymptotically the total concentration of binding sites B_{max}, which indicates the complete occupation of the binding sites in the assay. A first estimate of the dissociation constant K_d might be derived from this plot. According to eq. 7, the value of Kd equals the concentration of F at a point where B equals half the value of B_{max}.

$$K_d = F_{\left(B = \frac{B_{max}}{2}\right)} \qquad \text{(eq. 8)}$$

Graphical analysis of the binding data on the basis of direct plotting is not recommended. Fig. 2a shows the experimental error for a fictive experiment (B_{max}: $1 \cdot 10^{-7}$ M; K_d: $2 \cdot 10^{-7}$ M). The experimental error results from an assumed constant proportional error of 10% for the determination of B and 2% for pipetting of the ligand ($[L]_T$). A nonspecific binding component of 2% of F was assumed and corrected, as discussed in section 3.1.3. The influence of these errors increases with increasing concentrations of F and the total ligands concentration $[L]_T$, respectively. At least, a small constant error in the determination of B (1% of B_{max}) was supposed. This error has no influence in the direct plot but does in the Scatchard plot. It is obvious that a good estimation of B_{max} from the direct plot is hardly possible. The K_d is also not very reliable due to the difficulty of correctly fitting the saturation curve visually.

For these reasons, the linearization according to Scatchard was mainly used for the determination of K_d as well as B_{max} [15,16]. This transformation made linear regression analysis available for the analysis of binding data. The Scatchard plot is shown in Fig. 1b. This graph is based on eq. 9, which is another transformation of eq. 6:

$$\frac{B}{F} = -\frac{1}{K_d} \cdot B + \frac{B_{max}}{K_d} \qquad \text{(eq. 9)}$$

$$y = m \cdot x + c \qquad \text{(eq. 10)}$$

The quotient of B divided by F is plotted against B. The resulting plot shows a linear correlation, as is indicated by the normal form of a straight line represented by eq. 10. The value for K_d can be derived from the slope of the straight line, whereas the intercept on the x-axis directly shows the concentration of binding sites B_{max} in the assay.

slope of eq. 9: $\qquad m = -\dfrac{1}{K_d}$

X-axis intercept: $\qquad B_{\left(\frac{B}{F} = 0\right)} = B_{max}$

Y-axis intercept: $\qquad \left(\dfrac{B}{F}\right)_{(B=0)} = \dfrac{B_{max}}{K_d}$

Despite the easy use of this linearization, the numerical analysis of the nontransformed data is preferable. The Scatchard plot then should be used only as a visualization or as a source of initial estimates of K_d and B_{max} for the numerical approach [17,18].

If Scatchard analysis is used for the determination of K_d and B_{max}, some pitfalls, which come from features of this transformation, have to be avoided very carefully [7,19–24]. The uncritical use of this linearization has led in many cases to wrong results due to a unrestrained extrapolation. The data never cover the whole range of receptor saturation. Hence, K_d and B_{max} are obtained by a more or less extensive extrapolation of the straight line. The main source for misleading results in this context is the use of a too narrow segment of this line for extrapolation. Even for the most simple situation of binding, a minimum of 7–10 ligand concentrations should be measured in replicate. These concentrations should ideally cover a range of B, which is between 10 and 90% of B_{max}. The occupation of binding sites should reach at least 80%. The more complex the binding situation is, the better the data must be for a reliable result. In these situations an even higher number of ligand concentrations has to be tested with ever-improved precision.

The influence of the experimental error on the Scatchard transformation is shown in Fig. 2b. The error there corresponds to the assumptions made for the graph in Fig. 2a. The error of the quotient B/F dramatically increases at low values of B, whereas the error of B increases with increasing B. Therefore, the extreme segments of this plot should not be used to judge the linearity of the plot. It is recommended to fit the line by using the data points between about 20 and 80% saturation of the binding sites. Furthermore, the determination of B_{max} by using this extrapolation is based on the assumption that the simple model of equal and noninteracting binding sites is valid over the whole range of the ligand's concentration.

For a reliable separation of the bound and the free ligand, a moderate disparity between B and F is required. If both concentrations, B and F, are in the same range, an incomplete separation will not automatically have a catastrophic influence on the result of the assay. Therefore, the concentration of binding sites B_{max} in the assay should be within the range of the dissociation constant. This is demonstrated by eq. 11. It shows the relation between the free ligand's concentration F and the fractional saturation of the binding sites P, which equals B divided by B_{max} ($P = B/B_{max}$).

$$F = \left(\frac{P}{1-P}\right) \cdot K_d \qquad \text{(eq. 11)}$$

This expression clearly shows the independence of F from B_{max}. At half-saturation of B_{max}, $F_{(P=0.5)}$ equals K_d. Furthermore, the values of F, which cover 10–90% saturation of the binding sites, then range from $F_{(P=0.1)} = K_d/9$ to $F_{(P=0.9)} = 9 \cdot K_d$.

As recommended by Hulme [25], the optimal concentration of binding sites for a saturation experiment is in the range of 0.1 to 0.5 times the K_d. In this case, for an

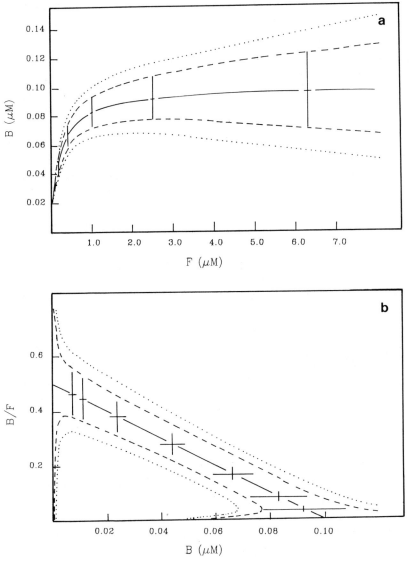

Fig. 2. Result of a fictive binding experiment. a) Direct binding plot; b) Scatchard plot. A concentration of binding sites B_{max} of $1 \cdot 10^{-7}$ M and a K_d of $2 \cdot 10^{-7}$ M was assumed. Both plots show representative points which range from 7–97% occupation of the binding sites. The error bars show the SEM and were calculated based on the following assumptions: a nonspecific binding of 2% of the free radioligand concentration; a pipetting error of $[L]_T$ of 2%; an error for the determination of total bound ligand B_{s+ns}; as well as non-specific bound ligand B_{ns} of 10% and a small constant overestimation of the bound ligand concentrations of $1 \cdot 10^{-9}$ M (1% of B_{max}). The dashed curves mark the boundary given by the SEM, the dotted curves limit the area of 90% confidence. While the proportional errors in the determination of the bound ligand predominantly influence the right end of both plots, the little constant error leads to a marked uncertainty at the left end of the Scatchard plot.

occupation of 10–90% of B_{max} the bound ligand's concentration B will range between one hundredth and one half the concentration of the free ligand F. If only the determination of B_{max} is intended, a higher concentration of binding sites in the assay might be acceptable. In this special case one has to insure an almost complete saturation of the binding sites by the application of the ligand at a total concentration exceeding at least 10 times the value of the dissociation constant K_d.

For ligands which bind with very high affinity, a limiting factor might be the maximal specific radioactivity. At 10% saturation the bound ligand's concentration should at least be represented by a radioactivity of 1000 cpm. According to Hulme [25], in a saturation experiment covering 10–90% saturation in an assay volume of 1 ml, the minimal concentration of binding sites, which will just give a reliable signal (1000 cpm) at 10% saturation, is limited by the specific radioactivity of the ligand. For a ^{3}H-labelled ligand with a specific activity of about 80 Ci/mmol, this minimal value of B_{max} is about $5 \cdot 10^{-11}$ M and for a ^{125}I-labelled ligand with approximately 2000 Ci/mmol $2 \cdot 10^{-12}$ M.

It might be impossible to fulfil the requirements for a saturation experiment due to a very high affinity of the ligand. This might be caused by an insufficient specific activity of the radioligand or by the inability to handle highly diluted solutions of either the receptor or the ligand. In this case, the dissociation constant K_d may alternatively be determined by a kinetic approach (see section 3.2.).

3.1.2. Nonlinear Scatchard plots

So far, we have just considered the simplest case of equal and noninteracting binding sites. The resulting Scatchard plot gave a marvelous straight line. On the other hand, a more complex type of binding results in a nonlinear Scatchard plot. For this nonlinearity some causes must be considered.

3.1.2.1. Nonspecific binding. Since the nonspecific binding is normally proportional to the free ligand's concentration F, it appears in the Scatchard plot as a straight line, which parallels the x-axis. The curve of total binding reaches this line asymptotically. This characteristic appearance of nonspecific binding may be used to correct the raw data. If the nonspecific binding is underestimated, nonlinearity remains. This might be incorrectly interpreted as negative cooperativity (see section 3.1.3.).

3.1.2.2. Cooperativity. This is a feature depending on the protein structure of the receptor under investigation. A receptor, which shows cooperativity, possesses two or more interacting ligand-binding sites. Binding of a ligand may cause a conformational change within the receptor protein, which either facilitates (positive cooperativity) or weakens (negative cooperativity) binding of the next ligand. The characteristics of nonlinearity allow discrimination between positive and negative cooperativity. The binding of acetylcholine to the nicotinic actylcholine receptor, for example, shows positive cooperativity [26].

Owing to the variety of reasons for nonlinearity, one should carefully exclude some more artificial causes. Very useful in this context might be to consider the stoichiometry of binding: 1 mol receptor has to bind at least 2 mol of the ligand to make it a candidate for cooperative binding (see section 3.1.4.).

3.1.2.3. Heterogeneity of binding sites. If a receptor preparation contains more than one type of binding sites with different binding characteristics for the tracer ligand, the Scatchard plot will be nonlinear. The heterogeneity can have several causes. It might be due, for example, to different protein species that possess a binding site for the ligand, to subtypes of a receptor class, or even to different states of one receptor species, which may result from the interaction with other proteins (e.g. G-coupled receptors). In some cases, the nonspecific binding to materials, which are used in the binding assay, may falsely suggest specific binding by a receptor, as discussed above for ^{125}I-insulin (see section 2.2.). Furthermore, heterogeneity of a receptor-preparation may result from an insufficient selectivity of the ligand.

To tackle the problem of receptor heterogeneity, one has several options. Up to three different binding sites might be separated by a numerical analysis, providing excellent quality binding data. Otherwise, either the receptor preparation or the selectivity of the ligand has to be improved. A better selectivity might be achieved by either another more selective tracer ligand or the addition of unlabelled ligands, which specifically suppress the binding of the tracer ligand to a subset of the different types of binding sites in the receptor preparation.

An example for the latter strategy is the [^{3}H]-spiperone binding assay for the solubilized D_2-dopamine receptor [27]. In addition to the D_2-receptor, spiperone also binds to the α_1-adrenergic receptor, the $5HT_2$-serotonin receptor, and an acceptor site, which is called spirodecanone site. To prevent binding to the serotonin receptor, specifically binding ligands like mianserin or ketanserin are used. The same principle applies for the inhibition of the adrenergic receptor. This receptor is blocked by prazosin. Most difficult is the elimination of the spirodecanone site. For this purpose, stereospecific binding by the dopamine receptor and the lack of stereospecificity of the spirodecanone site are used. Specific binding is measured as the difference of spiperone binding in the presence of either (+)-butaclamol or (−)-butaclamol.

3.1.2.4. Ligand-ligand interactions. One example is given by Cuatrecasas and Hollenberg [13]. These authors describe an apparent 'negative cooperativity' of insulin binding to placenta membranes. Negative cooperativity of insulin binding was deduced from a concave Scatchard plot. The same result was seen in absence of the receptor for nonspecific binding sites which show some of the criteria of specific binding (selective and reversible binding with high affinity, saturability, see section 2.2.). Since these binding sites are not able to interact, the reason for this behavior might be dimerization of insulin. This dimerization probably leads to a less affine binding. The same feature of insulin might also be responsible for an enhanced dissociation of ^{125}I-insulin by nonlabelled insulin, an effect also counted for 'negative cooperativity'. A similar behavior shows ^{125}I-NGF, which is also known to aggregate at neutral pH.

3.1.2.5. Ligand heterogeneity. Deviation of the Scatchard plot from linearity due to a change in the binding characteristics of the radioligand compared to its precursor are theoretically discussed by Hollemans and Bertina [14]. In particular, the iodination of peptide hormones might be problematical. However, Hulme [25] comments that ligand heterogeneity as well as the isomerization of ligands or ligand-ligand interactions seem to be relatively infrequent phenomena.

3.1.2.6. Isomerization of the receptor-ligand complex. If the receptor-ligand complex slowly isomerizes into a form which nearly irreversibly binds the ligand, curves are obtained in the Scatchard plot. The detection of these curves might be difficult. Within the experimental error a linear relationship of data might erroneously be taken into account. The absolute value of the slope in the apparent linear Scatchard plot depends on the velocity constant of isomerization as well as on incubation time. Hence, an apparent dissociation constant is determined which might be very different from the dissociation constant of the pure binding. This phenomenon is theoretically discussed by Ketelslegers et al. [22]. An example of this phenomenon given by these authors is the binding of 125I-human chorionic gonadotropin (125I-hCG) to adult rat testis homogenate. The data of the binding assay seemed to be linearly related, but measurement of the time course of dissociation revealed the existence of a receptor-ligand complex which quasi-irreversibly bound the hormone.

3.1.2.7. Incomplete equilibration. Incomplete equilibration at low but not at high ligand concentrations will appear as apparent 'positive cooperativity', whereas apparent 'negative cooperativity' might be detected in case of an incomplete equilibration of a receptor fraction which is not freely accessible for the ligand. In the latter case, the bound fraction is underestimated due to an apparently reduced number of binding sites at low but not at high ligand concentrations.

At the end of this short overview, the requirement of high-quality data shall again be stressed. Nonlinearity of the Scatchard plot should only be considered proven, if high quality data show nonlinearity in the range of 20–80% saturation of the binding sites. In the marginal segments of the curve, the probability increases that a deviation from the straight line is due merely to experimental errors.

3.1.3. Nonspecific binding

Often the total concentration of the bound ligand consists of a fraction due to specific binding and a more or less distinct fraction due to nonspecific binding. As mentioned above, nonspecific binding may be caused, for example, by ligand adsorption to components of the receptor preparation or to materials that are used in the assay. Lipophilic ligands may be integrated into biological membranes or ligands might be trapped by membrane vesicles. Investigating binding sites on intact cells, the nonspecific component of binding might be partially due to an uptake of ligand molecules by endocytosis.

Optimizing a binding assay includes minimizing nonspecific binding. If nonspecific binding exceeds about 10% of the specific component, the concentration dependence of this fraction must be determined independently to correct the raw binding data.

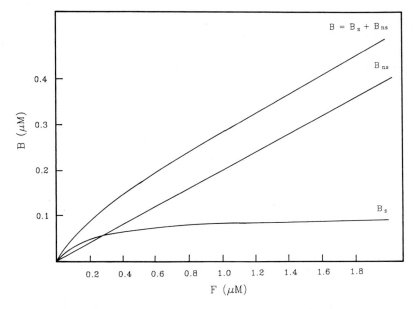

Fig. 3. The effect of nonspecific binding in a fictive binding experiment. The curve indicated as B shows the superimposition of specific binding B_s (saturation curve) and nonspecific binding B_{max} (linearly dependent on the free ligand concentration F). Assumed parameters: B_{max} $1 \cdot 10^{-7}$ M; K_d $2 \cdot 10^{-7}$ M; B_{ns} 20% of F.

The model for nonspecific binding is based on the assumption of a very large concentration of binding sites, which bind the ligand with low affinity. This has two consequences. First, these binding sites are never saturated by the ligand. Second, the nonspecific binding is linearly dependent on the concentration of the free ligand. Thus, total binding is the sum of a specific component which is saturable and a nonspecific component which increases linearly. This relation is shown in Fig. 3 in the direct binding curve. The resulting total binding is represented by superimposing the straight line of nonspecific binding onto the saturation curve of specific binding.

Nonspecific binding is evaluated by inhibiting the specific binding of the tracer ligand without significantly affecting the nonspecific component of binding. This is achieved either by the addition of an excess of the radioligand's unlabelled counterpart (i.e. the reduction of the specific radioactivity) or by the use of a second, structurally unrelated ligand which competes for the specific binding sites. The nonlabelled ligand should have a concentration of about 100–1000 times the value of K_d. This will ensure a total blockade of the specific sites without a significant reduction of non-specific binding. If specific binding occurs with very high affinity, the nonlabelled ligand should be added a short time before the addition of the tracer ligand. Without this preincubation, it could be difficult to displace the ligand in a reasonable time.

The fraction of nonspecific binding might also be determined from the Scatchard

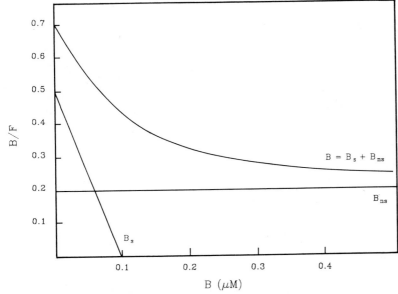

Fig. 4. Scatchard plot corresponding to Fig. 3. B represents the sum of specific binding B_s and nonspecific binding B_{ns}. The value at which the straight line B_{ns} intersects the y-axis equals the constant C which correlates the concentration of the nonspecific bound ligand B_{ns} to the free ligand concentration F.

plot due to its characteristic appearance [23]. This is shown in Fig. 4. The Scatchard plot corresponds to the direct binding curve in Fig. 3. Due to the amount of nonspecific binding, the curve resulting from the total binding asymptotically approaches a parallel to the x-axis. The quotient of B/F, which characterizes this parallel, can be taken as a constant C, which equals the slope of the straight line of nonspecific binding in the direct binding plot (Fig. 3).

Finally, it should be mentioned again that an underestimation of nonspecific binding may give rise to misinterpretation. The difference between the total binding and the underestimated nonspecific binding results again in a nonlinear Scatchard plot. This curve might then be misinterpreted as negative cooperativity or heterogeneity of binding sites. On the other hand, the overestimation of nonspecific binding is less critical. In this case, the difference of total and nonspecific binding would result in negative values for specific binding.

3.1.4. Cooperativity

Cooperativity, as mentioned above, is based on at least two interacting binding sites in a complex receptor. Binding of a ligand molecule to a first binding site leads, via conformational changes within the protein, to a different apparent dissociation constant for the binding of a second ligand molecule.

Positive cooperativity is characterized by the facilitation of a next molecule's bind-

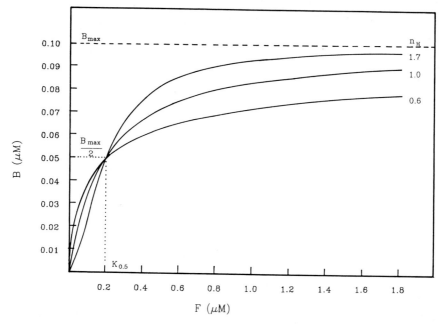

Fig. 5. Positive and negative cooperativity in comparison to binding without cooperativity in the direct binding plot. The Hill coefficients n_H indicate the type of cooperativity: $n_H > 1$ positive cooperativity, $K_{0.5} \neq K_d$; $n_H = 1$ no cooperativity, $K_{0.5} = K_d$; $n_H < 1$ negative cooperativity, $K_{0.5} \neq K_d$. The curves are calculated for B_{max} $1 \cdot 10^{-7}$ M, K_d $2 \cdot 10^{-7}$ M, and n_H as indicated, no nonspecific binding.

ing by the previously bound ligand molecule. For negative cooperativity, the binding of a next ligand molecule is weakened. The consequence of these intramolecular events for the direct binding plot and the corresponding Scatchard plot is shown in Figures 5 and 6: positive cooperativity results in a convex, negative cooperativity in a concave shape in the Scatchard plot. In both cases, the curve approaches B_{max} on the x-axis.

The strength of cooperativity is described by the Hill coefficient n_H. This constant is obtained as the slope of the linear segment of the curve resulting from the Hill transformation (eq. 12).

$$\log \frac{B}{(B_{max} - B)} = n_H \cdot \log F - n_H \cdot \log K_{0.5} \quad \text{(eq. 12)}$$

The corresponding Hill plot, which is obtained by plotting $\log (B/(B_{max} - B))$ against $\log F$, is shown in Fig. 7. Prerequisite for this plot is the independent determination of B_{max} by a saturation experiment.

Positive cooperativity is characterized by a Hill coefficient higher than 1, whereas a value lower than 1 indicates negative cooperativity. The more extreme this value is,

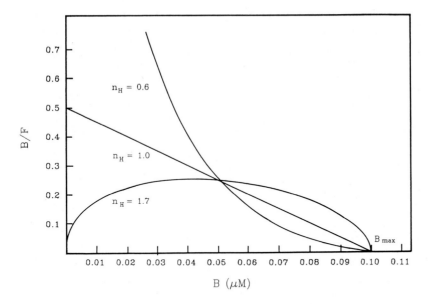

Fig. 6. Scatchard plot corresponding to Fig. 5. Hill coefficients indicate cooperativity: positive cooperativity ($n_H > 1$); no cooperativity ($n_H = 1$); and negative cooperativity ($n_H < 1$).

the stronger the cooperative interference of the binding sites. A coefficient n_H equalling 1 indicates independent binding sites, i.e. no cooperativity is detected (linear Scatchard plot).

The intercept on the x-axis of the Hill plot equals log $K_{0.5}$. $K_{0.5}$ is an integrative constant composed of the apparent dissociation constants of the distinct binding steps. It equals the free ligand's concentration at 50% saturation. If no cooperativity is seen ($n_H = 1$), this constant is identical to K_d.

3.2. The kinetic approach

Alternatively, the dissociation constant can be determined by measuring the rate constants k_{+1} for the ligand's association (on-rate) and k_{-1} for the ligand's dissociation from the receptor-ligand complex (off-rate). The change in the concentration of this complex over the time ($d[(RL)]/dt$) might be expressed as:

$$\frac{d[(RL)]}{dt} = k_{+1} \cdot [R] \cdot [L] - k_{-1} \cdot [(RL)] \qquad (eq.\ 13)$$

In this equation the term $k_{+1} \cdot [R] \cdot [L]$ describes the rate of the association process, whereas the term $k_{-1} \cdot [(RL)]$ stands for the rate of the dissociation process.

In the dynamic equilibrium, both reactions proceed with the same rate. Under this

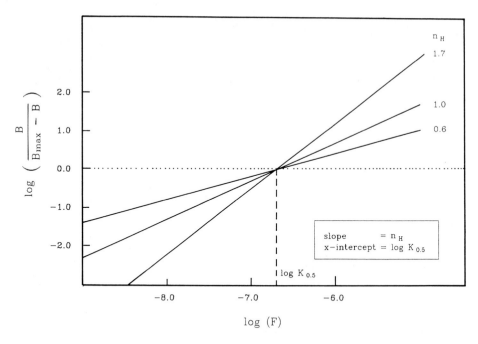

Fig. 7. Hill plot corresponding to Fig. 5. Hill coefficients indicate cooperativity: positive cooperativity ($n_H > 1$); no cooperativity ($n_H = 1$); and negative cooperativity ($n_H < 1$). The Hill coefficients equal the slope of the different straight lines.

condition, no change in the concentration of the receptor-ligand complex occurs over time. Incubation in a saturation experiment has to proceed longer than the time that is minimally needed for equilibration. Therefore, the time interval, until equilibration is finished, has to be determined before performing a saturation experiment.

From both rate constants the dissociation constant K_d can be deduced. On the assumption that the equilibrium has been established ($d[(RL)]/dt = 0$), eq. 13 can be transformed into a form that describes the ratio of the concentrations at the equilibrium. Eq. 14 expresses this ratio in a way that the dissociation constant K_d results:

$$\frac{k_{-1}}{k_{+1}} = \frac{[R] \cdot [L]}{[(RL)]} = K_d \qquad \text{(eq. 14)}$$

By the kinetic approach, only the equilibrium constant K_d is determined. B_{max} has to be obtained by an independent saturation experiment.

3.2.1. The on-rate

For the determination of k_{+1}, the ligand and the receptor are mixed at time $t = 0$. The concentration of bound ligand $B_{(t)}$ is then monitored over the time t until no further change occurs. The concentration of $B_{(t)}$ at this point equals B, the bound ligand's concentration in the equilibrium.

One approach to determine k_{+1} is based on assuming a pseudo-first order reaction. It is supposed that the concentration of the free ligand, F, is nearly unchanged over the whole period of measurement and equals approximately $[L]_T$. From this follows that F clearly exceeds the bound ligand's concentration. B should always remain below 10% of the ligand's total concentration. k_{+1} is then obtained by using the bound ligand's concentration at time t ($B_{(t)}$) and at equilibrium (B), which serves as reference point. Due to these assumptions, eq. 13 is changed to eq. 15:

$$\left(\frac{dB}{dt}\right)_{(t)} = k_{+1} \cdot (B_{max} - B_{(t)}) \cdot [L]_T - k_{-1} \cdot B_{(t)} \qquad \text{(eq. 15)}$$

The same equation describing the equilibrium is substracted from eq. 15 (at equilibrium the rate of complex formation is zero). This allows separation of variables and integration. Eq. 16 is the result; it describes the relationship between a concentration at a given time, $B_{(t)}$, and B, the concentration at equilibrium:

$$\ln \frac{B}{(B - B_{(t)})} = (k_{+1} \cdot [L]_T - k_{-1}) \cdot t = k' \cdot t \qquad \text{(eq. 16)}$$

Eq. 17 describes the relationship between an apparent velocity constant k' and $[L]_T$ and can be deduced from eq. 16. k' is obtained from the slope of the straight line, which results from plotting the logarithmic expression in eq. 16 against time t.

$$k' = k_{+1} \cdot [L]_T - k_{-1} \qquad \text{(eq. 17)}$$

Eq. 17 suggests two possible experimental approaches for the evaluation of k_{+1}. First, the constant k_{-1} might be determined by an independent experiment measuring the off-rate (see next section). Subsequently, k_{+1} is derived from k'. Second, the association experiment might be repeated with different total concentrations of the ligand (remember the assumptions made for the evaluation of k_{+1}. F has to remain constant during the experiment). k' is then plotted against $[L]_T$. k_{+1} is derived from the slope of the resulting straight line, whereas k_{-1} can be obtained from the intercept on the y-axis.

The linearity in all these plots is based on the assumption, that the binding sites, which are subject to this analysis, are homogenous and do not interact with one another. If there is either any heterogeneity of binding sites or in any sense cooperativity of binding, this approach will fail.

3.2.2. The off-rate

When determining k_{-1}, one has to consider only the dissociation process. The receptor and the tracer ligand are mixed and incubated until equilibrium is established. The bound ligand concentration equals the equilibrium concentration B. After equilibration, the association process is blocked either by the dilution with the incubation buffer to a volume exceeding at least 100 times the initial assay volume or by strongly reducing the radioligand's specific activity by the addition of an excess of the unlabelled ligand. The dissociation of the radioligand is measured as the loss of bound radioactivity $B_{(t)}$ over time.

Eq. 18 describes the rate of the dissociation process:

$$\frac{dB}{dt} = -k_{-1} \cdot B_{(t)} \qquad \text{(eq. 18)}$$

Integration of this equation results in eq. 19:

$$\ln \frac{B_{(t)}}{B} = -k_{-1} \cdot t \qquad \text{(eq. 19)}$$

By plotting $\ln (B_{(t)}/B)$ against the time t, k_{-1} is derived from the slope of the resulting straight line. Again, B represents the bound ligand's concentration at equilibrium at time t = 0.

The determination of k_{-1} by dilution as well as by competition with an excess of the nonradioactive ligand should give the same result. Obtaining a different result from these two methods indicates cooperativity. If the dissociation measured by dilution is faster, this might be caused by positive cooperativity. A faster dissociation measured by competitive inhibition may indicate negative cooperativity.

3.3. Inhibition experiments

Inhibition experiments, also called displacement experiments, investigate the binding of nonlabelled ligands by displacing a radioligand from its binding site. In this section, only competitive inhibition will be considered: the inhibitor as well as the radioligand bind to the same site. These experiments result in the determination of an IC_{50} or a K_i, the dissociation constant of the receptor-inhibitor complex.

IC_{50} is the competing ligand's total concentration $[I]_T$, which forces the displacement of 50% of the bound radioligand's concentration, obtained in the absence of inhibitor. This concentration depends on both the radioligand's total concentration and the K_i. According to Cheng and Prusoff [28], a relationship between these two parameters is described by the following equation:

$$IC_{50} = K_i \cdot \left(1 + \frac{F}{K_d}\right) \qquad (eq.\ 20)$$

This equation is only valid if the bound ligand's concentration B is very low compared to $[L]_T$ and if B_{max} is much lower than K_d. B should never exceed about 10% of $[L]_T$. Under these conditions, F approximately equals $[L]_T$. It is very important that this condition is fulfilled. A higher depletion of the free radioligand (at least above 20%) as well as an incomplete equilibration would otherwise lead to artificially flattened binding curves.

In practice, a constant total concentration of the radioligand, $[L]_T$, is incubated with an increasing total concentration of the inhibitor, $[I]_T$. The radioligand's total concentration is usually on the order of K_d. The radioligand displaced from the receptor-ligand complex by the competition of the inhibitor for the binding sites is measured, i.e. the difference of radioligand binding with and without inhibitor.

The value of IC_{50} can be obtained by the following methods:
a) $P' = B_{([I])}/B_{([I] = 0)}$ is plotted against the logarithm of the inhibitor's total concentration $[I]_T$ (see Fig. 8a). P' is the fractional occupation of the binding sites at an inhibitor concentration $B_{[I]}$. The reference point for P' is the initial concentration $B_{([I] = 0)}$ in absence of the inhibitor. This concentration is normally different from B_{max}. The resulting plot shows a so-called displacement curve. The IC_{50} is derived from this plot at the inhibitor's concentration, at which 50% of the initially bound radioligand remains bound.
b) A more precise determination of IC_{50} is obtained by a logit-log transformation of the binding data.

$$\text{logit}(P') = \log\left(\frac{P'}{1 - P'}\right) \qquad (eq.\ 21)$$

The logit (P') is plotted against the logarithm of the inhibitor concentration (see Fig. 8b). This results in a straight line, which equals the logarithm of IC_{50} at logit (P') = 0.

A special case of displacement experiments is the inhibition of the radioligand's binding by increasing concentrations of its nonlabelled counterpart. This way, one of the criteria of specific binding, reversibility of ligand binding, is demonstrated.

Displacement experiments are very useful in indirectly characterizing ligands which have a low affinity for a binding site. Also, the heterogeneity of a receptor preparation can be demonstrated according to the characteristic binding profile of different receptors or receptor subtypes. A less selective radioligand with a high affinity, mostly an antagonist, is displaced by a unlabelled ligand, which binds with less affinity but more selectivity to one of the species of binding sites. An example for this approach is given by Birdsall and Hulme [11]. The binding of [^{3}H]-propylbenzi-

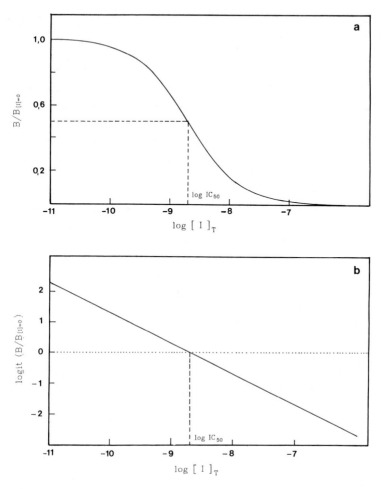

Fig. 8. Result of a fictive inhibition experiment. a) Displacement of the radioligand by increasing concentrations of the competing ligand I. b) Corresponding linearization by a logit-log plot. The log IC_{50} is indicated. The curves are calculated assuming a K_d of the radioligand of $1 \cdot 10^{-8}$ M, a K_i of the competing ligand of $1 \cdot 10^{-8}$ M, and a total concentration of the radioligand $[L]_T$ of $1 \cdot 10^{-8}$ M. To prevent depletion of the free radioligand in absence of the competing ligand, the concentration of the bound radioligand should not exceed 10% of the free ligand concentration.

lylcholine, a nonselective, high-affinity antagonist, to muscarinic binding sites in the rat medulla pons was inhibited by carbachol, an agonist that binds with a lower affinity. The data allowed the differentiation of three subpopulations of binding sites. These sites are termed SH, H, and L. They drastically differ in their affinity for carbachol but not for the tritiated antagonist. A similar heterogeneity of agonist binding has also been found for other neurotransmitter receptors such as adrenoceptors, dopamine, or opiate receptors.

Different affinity for agonists is also seen for G-protein-coupled receptors due to the status of the G-proteins. Treatment of such receptor preparations with guanine nucleotides leads to a loss of agonist high-affinity binding. In many cases, due to a lack of suitable radioactively labelled agonists, the different receptor conformations induced by the functional status of associated G-proteins can only be detected indirectly by displacement of a radioactively labelled antagonist with increasing agonist concentrations. In the presence of guanine nucleotides, e.g. the nonhydrolyzable guanylyl imidodiphosphate (GppNHp), the high-affinity binding is abolished. Depending on the GppNHp concentration the IC_{50} of inhibition of the antagonist binding by the agonists becomes higher. The inhibition curve (according to Fig. 8a) shifts to the right. An example is the inhibition of binding of the antagonist [^{3}H]-methylscopolamine to solubilized muscarinic receptors from rat heart by oxotremorine-M, a muscarinic agonist. This is described by Poyner [29]. In the presence of 100 μM GppNHp, the inhibition curve drastically shifts to the right indicating a change in the dissociation constant K_d of about one order of magnitude for at least a part of the binding sites (from about $5 \cdot 10^{-9}$ M to circa $8 \cdot 10^{-8}$ M).

Heterogeneity of a receptor preparation might become evident by measuring B and F at different inhibitor concentrations and plotting the data as direct binding curves (B against F) or according to the transformation of Eadie and Hofstee (B against the quotient B/F). The Eadie-Hofstee transformation is similar to the Scatchard plot (see section 3.1.1., eq. 9). Its basic form in absence of an inhibitor is shown in eq. 22:

$$B = - K_d \cdot \frac{B}{F} + B_{max} \qquad (eq.\ 22)$$

3.4. Computer-based versus graphical evaluation

Generally, it seems preferable to analyze the untransformed binding data by a numerical approach [17–20,30,31]. Several computer programs are available, which analyze data using non-linear least-squares fit based on the Marquardt algorithm. Application of these programs provides the best analysis of binding data based on a specific mathematical model. Among several models, the multiple-sites model is very useful. It describes the total concentration of the bound ligand as the sum of binding to several specific binding sites ($B_{max\ i}$, $K_{d\ i}$) including a nonspecific component, which is through the constant C linearly related to the concentration of the free ligand. A form of this model is shown in eq. 23:

$$B = C \cdot F + \sum_i \frac{B_{max\ i} \cdot F}{K_{d\ i} + F} \qquad (eq.\ 23)$$

$$B_\Sigma = B_{ns} + \sum_i B_{s\ i}$$

With complex binding situations, the use of the numerical approach for analysis is advisable. Up to three different species of binding sites might be discriminated using this approach.

Nevertheless, the quality of data also limits the applicability of numerical analysis. A computerized analysis will never give good results from bad data. Therefore, one should examine very carefully whether the quality of the data will support the assumption of a complex model. Bürgisser [19] demonstrated (using Monte-Carlo Studies) the difficulties in choosing the appropriate model for binding data. One third of randomly chosen binding studies from the literature, where a curvilinear Scatchard plot was interpreted as showing two different species of binding sites, turned out to be unreliable. Hence, if the quality of the data does not guarantee the reliability of a complex model, a simpler model should be chosen. Furthermore, one has to remember that the fit is achieved for a previously chosen mathematical model. Therefore, the result is the best fit only for this model. For instance, a program will probably give the best evaluation of K_d and B_{max} for linearly correlated data if the fit was based on the Scatchard equation. Hence, one must try more than one model and then decide which of these models will result in the best fit.

4. Experimental techniques

4.1. The separation of free and bound ligands

As discussed above, a critical point in the analysis of a binding experiment is the reliable determination of the bound as well as the free ligand concentration. For this purpose, the separation of both states has to be achieved in a quick and reproducible way, which should also allow minimization of the nonspecific binding. The critical parameter in this context is the off-rate. The choice of the separation technique mainly depends on it. A ligand, which dissociates very slowly from the binding site (i.e. has a high affinity), will not leave the binding site even if the surrounding environment is virtually washed free of the unbound ligand. Therefore, its separation will not be as critical as for a ligand having a lower affinity. This allows the achievement of a much more stringent technique, probably resulting in a better signal as well as a better reproducibility.

The following sections show some experimental techniques, which are commonly used to achieve a reliable separation of free and bound ligand. These techniques are also summarized in Table VI.

4.1.1. Filtration
Separation by filtration is used for particulate receptor preparations. This technique needs a dissociation constant K_d lower than 10^{-8} M to 10^{-9} M depending on the temperature of the washing solution (see below). The critical step is the subsequent

TABLE VI
Comparison of methods for separation of free and bound ligands

Method	Complete separation	Simplicity of operation	Time of separation	Specific to nonspecific ratio	Reproducibility of results
Rapid filtration	good	good	sec	high	good
Centrifugation	fair	fair	min	fair	fair
Dialysis	poor	poor	day	low	poor
Gel filtration	fair	fair	min	high	fair
Precipitation	fair	fair	min	fair	fair
Absorption	fair	fair	min	fair	poor

Reproduced from Wang et al. [33] with permission of IRL Press.

washing of the filters. One has to be careful to wash out neither the filter-bound receptor nor the receptor-bound ligand. On the other hand, washing the filters is necessary for better removal of the free ligand and reduces the amount of nonspecific binding. The critical parameter is, as stated above, the off-rate. During washing, the receptor-ligand complex is in a large volume of ligand-free buffer for up to a half minute. During this time, only very little of the bound ligand should dissociate from the receptor. Because a low temperature slows down the dissociation process, ice-cold washing solutions should be used. The extent of washing has to be optimized. The aim is low nonspecific binding and almost no loss of the bound ligand. A loss of the bound ligand may possibly not be recognized. Therefore, as a control the free ligand concentration for some samples should be determined independently by centrifugation (see below).

In these assays glass fiber filters (Whatman GF/X) and cellulose acetate filters are commonly used. The filters should be stored in washing buffer until they are used. Just before filtration of the incubation mixture, the filter should again be wetted with ice-cold washing buffer. The incubation mixture is filtered by suction and the filter washed several times with an appropriate volume of ice-cold washing buffer. The radioactive filtrate is retained in the reservoir of the filtration apparature. Finally, the filter-bound radioactivity is determined. The use of a filtration manifold allows the collection of data points in a short time by applying a vacuum to many filtration slots.

A modification of this method may be applied to solubilized receptors. Most receptors are negatively charged at neutral pH, which is due to their rather acidic isoelectric points. Therefore, these receptors might be retained on anion exchange filters. Commonly used are either glass fiber filters, which were incubated for about 2 hours in a 0.3% aqueous polyethyleneimine solution, or commercially available anion exchange filters, e.g. Whatman DEAE 81 filters.

4.1.2. Centrifugation
Another technique is the separation of the receptor-ligand complex by centrifugation. The centrifugational forces necessary to sediment the bound ligand should exceed 14 000 × g. This sort of separation technique is applicable to particulate preparations of a receptor. A dissociation constant K_d in the range of 10^{-6} M to 10^{-9} M is required. While separation by filtration is finished in less than one minute, separation by centrifugation takes several minutes. Centrifugation time depends on the composition of the incubation buffer. Diluted buffers of low ionic strength need more time for centrifugation, whereas divalent cations in the solution often may reduce the centrifugation time, which is necessary for sedimenting the receptor-ligand complex.

An advantage of this method is the persistent equilibrium during centrifugation. A loss of bound ligand only may occur due to washing of the pellet. Therefore, the time of centrifugation normally does not influence the result. Unfortunately, this method often is limited by showing a high level of nonspecific binding. According to Hulme [25], for a reliable experiment the nonspecific binding should not exceed specific binding at 10% occupation of the specific binding sites.

In practice, the incubation takes place in microcentrifugation tubes until equilibrium is established. Then, the receptor-ligand complex is sedimented by centrifugation for several minutes. Owing to the temperature dependence of the binding equilibrium, centrifugation takes place at the same temperature chosen for the incubation. After centrifugation an aliquot of the supernatant is removed to determine the concentration of the unbound radioligand. If possible, the pellets are washed with ice-cold washing buffer or deionized water in order to remove remaining free ligand. The radioactivity of the pellet is measured. For ^{3}H-labelled ligands, the pellets have to be solubilized before counting.

4.1.3. Gel filtration
For binding experiments with solubilized receptors, separation by nonequilibrium gel filtration is used. This method allows the separation within minutes. The dissociation constant should be lower than about 10^{-8} M. Disposable polypropylene columns, which contain about 2 ml of Sephadex G50, are commonly used for separation. After equilibration with the detergent-containing incubation buffer, a small volume of the incubation mixture is applied to the column. Elution is carried out with the ice-cold incubation buffer. A total of 1 ml of the eluate is collected in a scintillation vial. Collecting starts with the application of the incubation mixture. Again, all separation steps are carried out at a low temperature. As is necessary for filtration assays, the free ligand concentration F of some samples should be checked independently. Equilibrium dialysis (see below) must be used for this purpose.

Another approach which is similar to equilibrium dialysis (see section 4.1.4.) is described by Hummel and Dreyer [33]. A gel filtration column as well as a sample of the receptor preparation are equilibrated with the same total radioligand concentration $[L]_T$. The sample is applied to the column and eluated by the equilibration solution containing the radioligand concentration $[L]_T$. The radioactivity profile of elution is detected. The receptor-ligand complex elutes at the excluded volume. This is seen as a radioactive peak indicating the sum of free and bound ligand concentration. The free ligand concentration F equals $[L]_T$. As an internal control the radioactivity is decreased at a point after elution of the receptor-ligand complex. The total of reduction should equal the increase in radioactivity due to binding. This method can be used for soluble or solubilized proteins.

The radioligand used should have a high affinity for the receptor ($K_d < 10^{-8}$ M) for two reasons: 1. the binding equilibrium has to be established during gel filtration and 2. high concentrations of the receptor are required in order to give a measurable signal for the bound ligand compared to the free ligand. Furthermore, since the column must remain equilibrated by the radioligand during the complete elution, this method is very expensive and bears an enhanced danger of radioactive contamination. Finally, only a small number of samples can be tested due to the complicated procedure. Therefore, this method is not used very often for binding studies.

4.1.4. Dialysis

Dialysis may be used for ligands with a dissociation constant up to 10^{-6} M. The principle of this method is the incubation of a receptor with its ligand in a volume, which is separated into two compartments by a dialysis membrane. This membrane is permeable only for the free ligand. At equilibrium, the receptor's compartment contains a ligand concentration that is the sum of free and bound ligand, whereas the receptor-free compartment shows only the equilibrium concentration of the free ligand.

Apparatuses are available for this purpose, which allow parallel dialysis of several samples. However, this method is not used very often. Disadvantages are a long incubation time, which takes at least several hours, and, as a result, rather small numbers of samples that can be analyzed at the same time. Furthermore, the determination of the free and the bound ligand concentration often is rather poor and lacks a good reproducibility.

4.1.5. Other techniques

Two additional methods for separating have been described elsewhere. The first one is the precipitation of solubilized receptors by polyethyleneglycol. This technique allows a fair separation within minutes. A second method is the removal of the free ligand by adsorption to activated charcoal. After equilibration, a slurry of activated charcoal is added. After a short incubation period, the charcoal is removed by low-speed centrifugation, which does not sediment the receptor-ligand complex. This technique also allows a fair separation within minutes, but suffers in some cases from a lack of reproducibility.

4.2. Remarks on the assay setup

In this section, some principle considerations for designing a binding assay shall be outlined. The topics mentioned in this section were described partially before. Important parameters that must be controlled are the buffer composition, which has to ensure the stability of the receptor and the ligand, the temperature during the incubation period, the temperature during the separation process, and the incubation time for experiments, demanding equilibration. As discussed before, in optimizing a binding assay the aim is to minimize nonspecific binding as well as to maximize specific binding of the ligand. Both, the receptor and the ligand have to remain stable until the separation of free and bound ligand is finished.

The composition of the *incubation buffer* must ensure a physiological pH. A moderate ionic strength should be chosen. Divalent cations might be added in order to stabilize the receptor proteins and the membranes of a preparation. Inhibitors of proteases should be added. Especially for peptide ligands, the proteolytic degradation might be a factor which should carefully be examined.

Loss of ligands by *adsorption* to the walls of the test tubes must also be considered. Hydrophobic peptides are especially prone to this phenomenon. This loss may be reduced by adding low concentrations of nonionic detergents or by rinsing the tube walls with 'inert' proteins like albumin. Up to 1 mg/ml albumin can be added to the incubation mixture. This also helps to *stabilize receptor proteins at a high dilution*. Another advantage is a more nonspecific protection of the receptor against proteolysis.

Incubation usually will take place at *physiological temperatures* of 25–37°C, whereas the *separation* of free and bound ligands should be performed at a *temperature between 0°C and 4°C* in order to prevent a loss of the bound ligand by dissociation. Exceptions are the separation by centrifugation, equilibrium gel filtration according to Hummel and Dreyer [33], and equilibrium dialysis. For these techniques the temperature should remain constant during incubation and separation. Under certain circumstances, lower temperatures for the incubation may be chosen. One reason could be the stabilization of a diluted receptor; another reason might be to reduce the rate of association for a kinetic experiment. For nonequilibrium separation techniques, such as filtration assays or nonequilibrium gel filtration, the correct determination of the free ligand concentration of a number of samples should be checked independently by centrifugation or equilibrium dialysis, respectively.

The *quality of the ligand* must be examined before the experiment according to the requirements mentioned in section 2.3.2.

The *minimal time* necessary *for complete equilibration* must be determined in a kinetic experiment. This is the minimal incubation time for all other types of binding experiments.

Another important point is the *definition of B_{max}*. According to Hulme [25], the concentration of binding sites should range between 0.1–0.5 times the value of the K_d.

Before the concentration of binding sites B_{max} is fixed for a binding experiment, a further check is recommended. It is important to test the dependence of the bound ligand concentration on the volume of the receptor preparation used in the assay. For this purpose, a high concentration of the ligand is incubated with increasing amounts of the receptor preparation. The chosen receptor concentration should be in the linear segment of the resulting plot. A deviation from linearity indicates that components of the receptor preparation interfere with ligand binding. This, for example, could be due to endogenous ligands or ligand-degrading activities, which generally should both be eliminated from a preparation.

References

1. Closse, A., Frick, W., Dravid, A., Bolliger, G., Hauser, D., Sauter, A. and Tobler, H.J. (1984) Naunyn-Schmiedeberg's Arch. Pharmacol. 327, 95–101.
2. Green, J.P. (1990) Trends Pharmacol. Sci. 11, 13–16.
3. Kenakin, T.P. (1989) Trends Pharmacol. Sci. 10, 18–22.
4. Hulme, E.C. (Ed.) (1990) Receptor Biochemistry – A Practical Approach, IRL Press, Oxford.
5. Hulme, E.C. (Ed.) (1990) Receptor-Effector Coupling – A Practical Approach, IRL Press, Oxford.
6. Hulme, E.C. (Ed.) (1992) Receptor-Ligand Interactions – A Practical Approach, IRL Press, Oxford.
7. Klotz, I.M. (1983) Trends. Pharmacol. Sci. 4, 253–255.
8. Levitzki, A. (1976) In: Receptors and Recognition 2, Series A (Cuatrecasas, P. and Greaves, M.F., Eds.) pp. 199–229, Chapman & Hall, London.
9. Reid, E., Cook, G.M.W. and Morré, D.J. (Eds.) (1984) Investigation of Membrane-located Receptors, Methodological Surveys in Biochemistry and Analysis 13, Plenum Press, New York, London.
10. Repke, H. and Liebmann, C. (1987) Membranrezeptoren und ihre Effektorsysteme, Verlag Chemie, Weinheim.
11. Birdsall, N.J.M. and Hulme, E.C. (1983) Trends Pharmacol. Sci. 4, 459–463.
12. Ehle, B., Lemoine, H. and Kaumann, A.J. (1985) Naunyn Schmiedeberg's Arch. Pharmacol. 331, 52–59.
13. Cuatrecasas, P. and Hollenberg, M.D. (1975) Biochem. Biophys. Res. Commun. 62, 31–41.
14. Hollemans, H.J.G. and Bertina, R.M. (1975) Clin. Chem. 21, 1969–1773.
15. Faguet, G.B. (1986) J. Cell. Biochem. 31, 243–250.
16. Munson, P.J. and Rodbard, D. (1983) Science 220, 979–981.
17. Leatherbarrow, R.J. (1990) Trends. Biochem. Sci. 15, 455–458.
18. Thakur, A.K., Jaffe, M.L. and Rodbard, D. (1980) Anal. Biochem. 107, 279–295.
19. Bürgisser, E. (1984) Trends Pharmacol. Sci. 5, 142–144.
20. Duggleby, R.G. (1991) Trends Biochem. Sci. 16, 51–52.
21. Feldman, H.A. (1983) J. Biol. Chem. 258, 12865–12867.
22. Ketelslegers, J.-M., Pirens, G., Maghuin-Rogister, G., Hennen, G. and Frère, J.-M. (1984) Biochem. Pharmacol. 33, 707–710.
23. Mendel, C.M. and Mendel, D.B. (1985) Biochem. J. 228, 269–272.
24. Peters, F. and Pingoud, V.A. (1982) Biochem. Biophys. Acta 714, 442–447.
25. Hulme, E.C. (1990) In: Receptor-Effector Coupling – A Practical Approach, Appendix I (Hulme, E.C., Ed.) pp. 203–215, IRL Press, Oxford.
26. Eldefrawi, M.E. and Eldefrawi, A.T. (1973) Biochem. Pharmacol. 22, 3145–3150.

27. Strange, P.G. and Williamson, R.A. (1990) In: Receptor Biochemistry – A Practical Approach, (Hulme, E.C., Ed.) Chapter 3, pp. 79–97, IRL Press, Oxford.
28. Cheng, Y.-C. and Prusoff, W.H. (1973) Biochem. Pharmacol. 22, 3099–3108.
29. Poyner, D. (1990) In: Receptor-Effector Coupling – A Practical Approach, (Hulme, E.C., Ed.) Chapter 2, pp. 31–58, IRL Press, Oxford.
30. Efron, B. and Tibshirani, R. (1991) Science 253, 390–395.
31. Munson, P.J. and Rodbard, D. (1980) Anal. Biochem. 107, 220–239.
32. Hummel, J.P. and Dreyer, W.J. (1962) Biochem. Biophys. Acta 63, 530–532.
33. Wang, J.-X., Yamamura, H.I., Wang, W. and Roeske, W.R. (1992) In: Receptor-Ligand Interactions – A Practical Approach, (Hulme, E.C., Ed.) Chapter 6, pp. 213–234, IRL Press, Oxford.

F. Hucho (Ed.) *Neurotransmitter Receptors*
© 1993 Elsevier Science Publishers B.V. All rights reserved.

CHAPTER 3

Receptor regulation

FRANÇOIS NANTEL and MICHEL BOUVIER

Department of Biochemistry and Groupe de Recherche sur le Système Nerveux Autonome, Université de Montréal, Montréal, Qc, H3C 3J7, Canada

1. Introduction

Over the past two decades, numerous studies have provided a wealth of information on the biochemical events leading to the transfer of information across biological membranes during the neurotransmission process. In particular, the applications of molecular biology techniques have revealed the identity and structure of many of the proteins which are involved in the signal transduction events.

One of the most fascinating features of these signalling systems is their high degree of plasticity. In fact, regulatory processes acting at various levels of the signalling pathways have been shown to modulate their responsiveness. These processes permit to the cell to adapt to its environment and play a major role in the sorting and the integration of the information detected at the receptor level. Several independent mechanisms have been shown to regulate receptor reactivity in response to their own activity and to the activation of other receptor signalling pathways. Alterations of these normal regulation processes probably underlie certain pathological conditions related to hyper- or hyposensitivity to hormonal or neurotransmitter stimulation.

One important regulatory mechanism of receptor function is known as agonist-induced desensitization. It is characterized by the fact that, upon stimulation, the intensity of a response mediated by the receptor wanes over time despite the continuous presence of a stimuli of constant intensity. The phenomenon of desensitization has practical clinical implications as it tends to limit the efficacy of many therapeutic agents which have membrane receptors for acting sites.

Historically, it was thought that the appearance of desensitization resulted solely from a decrease in receptor number following sustained stimulation. However, it soon became evident that desensitization was much more complex. In fact, the effi-

---

Abbreviations: β_2AR, β_2-adrenoceptor; PKA, cAMP-dependent protein kinase; PKC, phospholipid- and calcium-dependent protein kinase; βARK, β-adrenoceptor kinase; cAMP, cyclic adenosine monophosphate; IRSK, insulin receptor serine kinase; GABA, γ-aminobutyric acid; NAChR, nicotinic acetylcholine receptor

cacy of transduction pathways has been shown, in many cases, to be dynamically regulated in the absence of change in receptor density. Moreover, in many instances, the desensitization occurs long before any down-regulation of the receptor can be detected. Covalent modifications, such as phosphorylation of the receptors, have been shown to contribute to the desensitization by interfering with the ability of the receptor to recognize ligands or transduce its stimuli.

Most of the receptors belong to one of the following superfamilies: ligand-gated ion-channel receptors (ex: the acetylcholine nicotinic receptor), receptors baring intrinsic enzymatic activity (ex: the insulin receptor); and the G-protein-coupled receptors (ex: the β_2-adrenoceptor, β_2AR). Regulatory processes modulating the reactivity of receptors belonging to each of those three groups have been described. In the present chapter, we will review several aspects of the molecular mechanisms which play a role in receptor regulation. Although regulation of members of the three superfamilies will be discussed, special attention will be given to the G-protein-coupled receptors which are the most thoroughly studied in that respect.

2. Receptor regulation

2.1. G-protein-coupled receptors

As the initial point of interaction of many hormones and neurotransmitters, the G-protein-coupled receptors are uniquely positioned to regulate the responsiveness of a given signalling pathway. In fact, these receptors are subjected to various forms of regulation which play important roles in the complex integration of transmembrane signalling. All the G-protein-coupled receptors share similar structural features. In particular, a characteristic common to all of the cloned G protein-coupled receptors is their proposed membrane topology consisting of a hydrophobic core of seven transmembrane α helices, an extracellular amino-terminal segment, and a cytoplasmic C-terminal tail (Fig. 1; see [1] for review). Similarly, many of the mechanisms involved in their regulation appear to be common to all the members of this receptor family. In many respects, β_2AR may be considered to be prototypical of the G protein-coupled receptors and has been used as a study model for this receptor family. β_2AR is coupled via a stimulatory G-protein (G_s) to the stimulation of adenylyl cyclase and therefore to an increased production of the second messenger cAMP. It is believed that at least three distinct processes, at the level of the receptor alone, are involved in the desensitization phenomenon: uncoupling, sequestration, and down-regulation [2]. The uncoupling refers to a rapid decrease in the receptor-mediated activation of adenylyl cyclase with no change in receptor number. This has been correlated to a decreased ability of agonists to stabilize the guanyl nucleotide-sensitive high affinity form of the receptor necessary for enzyme activation. This high affinity state is believed to consist of a ternary complex composed of the hormone, the recep-

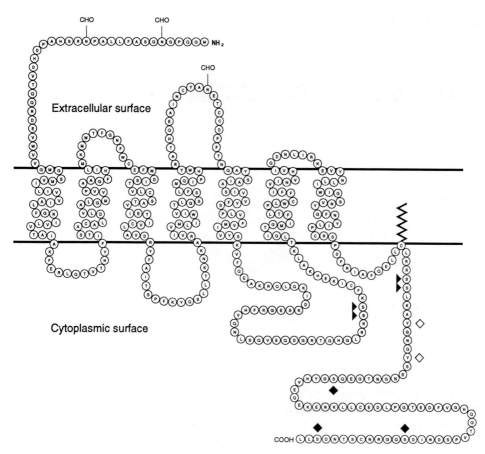

Fig. 1. Schematic representation of the β_2-adrenoceptor with potential N-glycosylation sites (CHO), tyrosine residues implicated in the downregulation process (◇), and potential phosphorylation sites by βARK (◆), PKA, and PKC (▶).

tor, and G_s. Thus, a decreased ability of the receptor to interact with G_s, probably resulting from the phosphorylation of the receptor (as will be discussed later) appears responsible for functional uncoupling. The sequestration consists of a rapid (minutes) cellular redistribution of β_2AR from the cell surface to a poorly defined 'light membrane' fraction. The sequestration of the receptor can be assessed in binding experiments such as those described in Figure 2A. Indeed, receptors that are sequestered retain their ability to bind hydrophobic ligands such as [^{125}I]pindolol. Thus, the level of sequestration is determined by comparing the ability of a hydrophobic (propranolol) and a hydrophilic (CGP 12177) ligand to compete with [^{125}I]pindolol in whole cell binding experiments [3]. It is believed that sequestration is rapidly reversible and allows the recycling of functional receptors back to the cell surface [4]. The

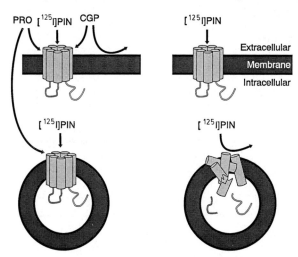

Fig. 2. Schematic representation of sequestration (A) or downregulation (B) experiments on β_2AR. [^{125}I]PIN, [^{125}I]pindolol; PRO, propranolol; CGP, CGP-12177.

down-regulation, which follows long-term agonist stimulation (hours), is characterized by a loss of binding sites for the ligand [^{125}I]pindolol (Fig. 2B). This phenomenon appears to be controlled by mechanisms acting both at the level of gene expression and at the level of intracellular receptor degradation following its internalization. In fact, the treatment of various cell types with β-agonists has been shown to promote a cAMP-dependent destabilization of the β_2AR mRNA, leading to a decrease in the steady-state level of this message [5–7]. However, this process appears to be only one of several mechanisms involved in the down-regulation process. An increased degradation rate of the receptor, upon long term stimulation, is also believed to contribute to the down-regulation [8,9]. Molecular determinants in the receptor structure, which could play a role in the down-regulation of β_2AR, have recently been identified. These include two tyrosine residues, located in the carboxyl terminal of β_2AR (Fig. 1); the mutation of which significantly delays and decreases the amplitude of agonist-induced down-regulation [10]. Interestingly, the mutation of tyrosine residues located in the carboxyl tail of receptors, such as the receptor for low density lipoproteins and mannose-6-phosphate, also produces significant decreases in their rate of internalization [11,12]. These tyrosine residues might dictate the interactions of the receptor with cytosolic proteins involved in internalization and/or degradation processes.

The phosphorylation and dephosphorylation of proteins constitute one of the most extensively studied post-translational covalent modifications known to have regulatory significance. Phosphorylation of β_2AR is believed to play an important role in the expression of agonist-induced desensitization. Indeed, increased phosphorylation of the receptor (up to 2–3 mol phosphate/mol receptor) has been shown to accompany the development of desensitization (see [1] for review). This phosphorylation event is believed to regulate the ability of β_2AR to interact with G_s and has little effect on sequestration and down-regulation processes. Studies conducted in vitro have demonstrated that the β_2-adrenoceptor can be phosphorylated by at least three distinct protein kinases, the cAMP-dependent protein kinase (PKA), the phospholipid- and calcium-dependent protein kinase (PKC), and the β-adrenoceptor kinase (βARK) ([13–15]; Figs. 1 and 3). Whereas PKA and PKC are kinases with broad substrate spectrums, βARK demonstrates a much more restricted specificity. It was demonstrated that βARK phosphorylates only the agonist-occupied form of the receptor [16] and that phosphorylation of serine and threonine residues in the C-terminus tail of β_2AR play a major role in the rapid desensitization which follows stimulation [17]. Three serines have been proposed to represent specially good substrates for βARK (Fig. 1). Phosphorylation of β_2AR by βARK allows its interaction with another protein, termed β-arrestin. The formation of the β_2AR/β-arrestin complex is believed to interfere with the coupling of the receptor to G_s ([18]; Fig. 3). By its ability to phosphorylate only the agonist-occupied receptor, βARK has attracted much attention in regard to its potential role in agonist-induced desensitization.

The amino acid sequence of the β_2-adrenergic receptor also reveals two putative

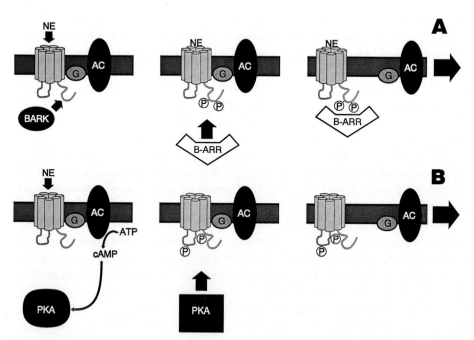

Fig. 3. Suggested mechanism of action of the phosphorylation of β_2AR by βARK (A) or PKA (B). NE, norepinephrine; β-ARR, β-arrestin; AC, adenylyl cyclase, P; phosphorylation.

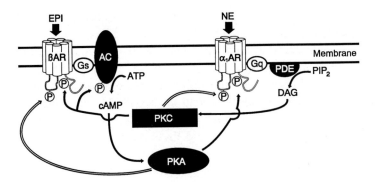

Fig. 4. Illustration of cross-talk regulation between the adrenergic-coupled adenylyl cyclase pathway and phosphatidylinosityl pathways. NE, norepinephrine; EPI, epinephrine; PDE, phospholipase C; PIP_2, phosphatidylinositol diphosphate; DAG, diacylglycerol; P, phosphorylation.

PKA phosphorylation sites (Figs. 1 and 3B). Low levels of agonist (1–10 nM isoproterenol) stimulation were found to preferentially induce phosphorylation of the receptor at PKA sites, whereas higher levels of agonist (1–10 μM isoproterenol) stimulation promoted phosphorylation on both the PKA and βARK sites [19]. Moreover, it was also shown that the phosphorylation induced by βARK is extremely rapid (seconds), while that induced by PKA occurs within minutes of stimulation [20]. These observations are compatible with the idea [21] that the desensitization induced by βARK would be crucial for receptors located in the synaptic cleft where the levels of transmitters are high and the responses must be rapid (Fig. 3A). Peripheral receptors that are subjected to much lower doses of catecholamines would mostly be desensitized through the PKA-dependent pathway (Fig. 3B). In contrast to the receptor phosphorylated by βARK, the PKA-phosphorylated receptor does not display a high affinity for β-arrestin [22] and the mechanism by which this phosphorylation decreases the interaction of the receptor with G_s remains to be elucidated.

In addition to being phosphorylated, the $β_2AR$ is subjected to an additional post-translational modification of cysteine 341 by a the fatty acid palmitate ([23]; Fig. 1). This modification is believed to anchor the C-terminus tail to the plasma membrane and create a fourth intracellular loop. Palmitoylation has been proposed as a potential regulatory mechanism for other transmembrane receptors such as the transferrin receptor, the acetylcholine nicotinic receptor, and rhodopsin. A mutated, nonpalmitoylated, $β_2AR$ (in which cysteine 341 is replaced by glycine) displays a markedly reduced ability to form the guanyl nucleotide-sensitive, high affinity agonist binding state and to activate adenylyl cyclase [23]. These results suggest that palmitoylation plays a crucial role in determining the ability of the receptor to couple to the stimulatory pathway of adenylyl cyclase. More recently, agonist-modulated palmitoylation of $β_2AR$ has been observed to support a potential role for this modification in dynamic regulation [24].

Because of the wide variety of hormones and neurotransmitters acting through G-protein-coupled receptors and given their similarities, it was proposed that functional interactions between them might contribute to the elaboration of coherent cellular responses. In this respect, cross-talk regulation between receptors linked to the phosphatidylinositide (PI) turnover pathway and the β-adrenergic-stimulated adenylyl cyclase activity has attracted considerable attention [25]. It was suggested that phosphorylation of β_2AR by PKC (Fig. 1) contributes to this regulation. Indeed, β_2AR can act as a substrate for PKC [13] and it has been demonstrated that such phosphorylation, upon PKC activation in whole cells, leads to the uncoupling of the receptor from the adenylyl cyclase stimulatory pathway [26]. It is therefore plausible that stimulation of receptors such as the α_1AR, which leads to the production of IP$_3$ and diacylglycerol and activates PKC, could promote the phosphorylation and desensitization of β_2AR. Conversely, elevation of cAMP level, brought about by the activation of β_2AR, can lead to the phosphorylation of α_1AR (Fig. 4) and hence contribute to the regulation of its reactivity [25]. In addition to the receptor, phosphorylation of other components of the signalling pathways (G-proteins and effectors) has been suggested to contribute to the regulation of the signalling efficacy. The catalytic subunit of adenylyl cyclase itself has been shown to be a substrate for both PKA and PKC [27,28]. Such phosphorylation has been proposed to have an effect on the responsiveness of the enzyme.

2.2. Tyrosine kinase receptors

The tyrosine kinase receptor family includes the receptors for insulin, insulin-like growth factor, platelet-derived growth factor, epidermal growth factor, and others. They are known to play a central role in cellular metabolism, growth, and differentiation. Stimulation of these receptors leads to the phosphorylation of a wide variety of substrates, including the receptor itself, on tyrosine residues. The activation of the receptor's tyrosine kinase activity is essential to its biological activity and represents the initial step in a complex cascade of events. The most extensively studied receptor of this group, the insulin receptor, will be used as an archetype of this receptor family.

The insulin receptor is a tetrameric oligomer consisting of two extracellular α subunits, each of them containing an insulin-binding site, and two transmembranous β subunits harboring the tyrosine kinase activity. Binding of insulin to its receptor rapidly induces its autophosphorylation on tyrosine residues located in the β subunits [29]. This phosphorylation further activates the tyrosine kinase activity of the receptor, thus favoring the phosphorylation of other substrates such as the insulin receptor substrate I [21]. In addition to this rapid positive-feedback loop acting on the insulin receptor enzymatic activity, a series of regulatory processes have been shown to modulate the cellular effects of insulin. These include the regulation of receptor number, affinities, and kinase activities.

Prolonged exposure to insulin has been shown to cause a down-regulation of cell

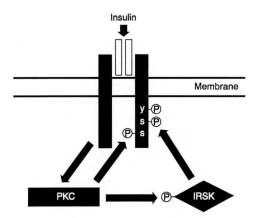

Fig. 5. Schematic representation of the hypothetical roles of PKC and IRSK in insulin-receptor regulation. P, phosphorylation.

surface receptors (see [30] for review). In 3T3-L1 and 3T3-C2 cells, insulin down-regulates its receptor by increasing the rate of receptor degradation. In fact, in most cell lines studied, down-regulation of the insulin receptor results from its endocytosis via either smooth or clathrin-coated vesicles and lysosomal degradation. However, in cultured human lymphocytes (IM9 cells), insulin treatment has been shown to decrease the rate of synthesis of the receptor, thus contributing to its down-regulation. These observations may suggest that tissue-specific factors may contribute to regulate the expression of the gene and inactivation of the receptor.

Other processes, such as covalent modifications of the receptor protein, also regulate cell sensitivity to insulin stimulation. As discussed above, the autophosphorylation of tyrosine residues, located in the β subunits of the receptor, has been shown to stimulate its tyrosine kinase activity. However, phosphorylation of the receptor on serine and threonine residues in the C-terminus tail of the β subunits have been reported to decrease the enzymatic activity and to lower the affinity of the receptor for insulin [31]. Such phosphorylation can be induced by phorbol esters and agents that elevate intracellular levels of cAMP or directly by purified PKA and PKC in vitro. This may suggest that the activation of other signalling pathways, which lead to the activation of these serine and threonine protein kinases, could regulate the cellular response to insulin.

The insulin receptor was also found to be phosphorylated by a more specific kinase, the insulin receptor serine kinase (IRSK), that copurifies with the insulin receptor. Phosphorylation sites for this kinase are serine residues different from those phosphorylated by PKA or PKC. IRSK is activated through phosphorylation by PKC. Figure 5 illustrates a hypothetical mechanism of action for IRSK [30]. According to this model, stimulation of the insulin receptor leads to the activation of PKC which subsequently phosphorylates and activates IRSK. Both kinases then phosphorylate the insulin receptor at different serine residues, thereby reducing the tyrosine kinase activity of the receptor and contributing to the termination of the signal.

2.3. Ligand-gated ion-channel receptors

Ligand-gated ion channels include receptors for acetylcholine, glutamate, glycine, and GABA. Binding of neurotransmitters to the receptor causes the opening of an integral ion channel and permits the flux of ion through the plasma membrane. The nicotinic acetylcholine receptor (NAChR), present in the electric organ of *Torpedo*, has been the most extensively studied receptor of this family.

NAChR is a pentameric transmembrane oligomer consisting of four different polypeptides with a stoichiometry of $\alpha_2\beta\gamma\delta$. The acetylcholine-binding sites have been located on the α-subunits. Desensitization of NAChR is a phenomenon in which the channel becomes unresponsive to further activation when it is exposed to the neurotransmitter for a relatively long period of time (tens of milliseconds). Under those circumstances, the NAChR channel remains closed even while being occupied by molecules of acetylcholine [32]. Interestingly, the desensitized receptor displays a higher affinity for the agonist than does the resting receptor [33]. The exact mechanism leading to this desensitization state remains poorly understood. However, phosphorylation of the receptor has been suggested to play an important role this process.

All four subunits are known to be phosphorylated at serine residues with the α-, β-, γ-, and δ-polypeptides having respectively 1, 1, 2, and 5 phosphorylation sites. Experiments using phorbol ester or purified PKC have shown NAChR to be phosphorylated on the α- and δ-subunits, while PKA phosphorylates the γ- and δ-subunits. Phosphorylation by both these kinases has been proposed to increase the rate of desensitization by several fold without affecting the initial rate of ion flux [32].

More recently, NAChR was also found to be phosphorylated on tyrosine residues located in the β-, γ-, and δ-subunits by an endogenous tyrosine protein kinase. Tyrosine phosphorylation was found to have no effect on the single channel open time or conductance but, like serine phosphorylation, it causes an eight to ten fold increase in the rate of rapid desensitization. Interestingly, the effects induced by phosphorylation of each subunit were found to be additive [34].

Some hypotheses have been formulated on the regulation of NAChR by acetylcholine and other neurotransmitters. In cultured cells, calcitonin gene-related peptide, which is coreleased with acetylcholine at the neuromuscular junction, and noradrenaline, acting through β_2AR, may induce NAChR desensitization through a cAMP-dependent receptor phosphorylation process [35,36]. Rapid desensitization following acetylcholine stimulation has also been proposed to result, in part, from an indirect activation of PKC which could feedback on NAchR by phosphorylating the α- and δ-subunits and thus increasing the rate of desensitization [33].

3. Conclusion

Although each receptor has specific regulatory mechanisms, some general processes appear to be broadly encountered. Most noticeable is the role played by the phosphorylation of the receptor proteins. In all the cases presented here, this post-translational modification appears as a crucial event involved in the onset or modulation of rapid agonist-promoted desensitization. Another fairly common feature is the occurrence of receptor internalization and degradation. The existence of common molecular denominators involved in these processes for many receptors will most likely be identified in the near future.

References

1. Dohlman, H.G., Thorner, J., Caron, M.G. and Lefkowitz, R.J.(1991) Annu. Rev. Biochem. 60, 653–688.
2. Benovic, J.L., Bouvier, M., Caron, M.G. and Lefkowitz, R.J. (1988) Annu. Rev. Cell. Biol. 4, 405–427.
3. Hertel, C., Muller, P., Portenier, M. and Staehelin, M. (1983) Biochem. J. 216, 669–674.
4. Sibley, D.R., Strasser, R.H., Benovic, J.L., Daniel, K. and Lefkowitz, R.J. (1986) Proc. Natl. Acad. Sci. USA. 83, 9408–9412.
5. Bouvier, M., Collins, S., O'Dowd, B.F., et al. (1989) J. Biol. Chem. 264, 16786–16792.
6. Collins, S., Bouvier, M., Bolanowski, M.A., Caron, M.G. and Lefkowitz, R.J. (1989) Proc. Natl. Acad. Sci. USA. 86, 4853–4857.
7. Hadcock, J.R. and Malbon, C.C. (1988) Proc. Natl. Acad. Sci. USA. 85, 5021–5025.
8. Rebois, R.V. and Patel, J. (1985) J.Biol.Chem. 260, 8026–8031.
9. Morishima, I., Thompson, W.S., Robison, G.A. and Strada, S.J. (1980) Mol. Pharmacol. 18, 370–378.
10. Valiquette, M., Bonin, H., Hnatowich, M., Caron, M.G., Lefkowitz, R.J. and Bouvier, M. (1990) Proc. Natl. Acad. Sci. USA. 87, 5089–5093.
11. Canfield, W.M., Johnson, K.F., Ye, R.D., Gregory, W. and Kornfeld, S. (1991) J. Biol. Chem. 266, 5682–5688.
12. Bansal, A. and Gierasch, L.M. (1991) Cell 67, 1195–1201.
13. Bouvier, M., Leeb-Lundberg, L.M.F., Benovic, J.L., Caron, M.G. and Lefkowitz, R.J. (1987) J. Biol. Chem. 262, 3106–3113.
14. Benovic, J.L., Pike, L.J., Cerione, R.A., et al. (1985) J. Biol. Chem. 260, 7094–7101.
15. Benovic, J.L., Strasser, R.H., Caron, M.G. and Lefkowitz, R.J. (1986) Proc. Natl. Acad. Sci. USA. 83, 2797–2801.
16. Benovic, J.L., Mayor, F., Jr., Staniszewski, C., Lefkowitz, R.J. and Caron, M.G. (1987) J. Biol. Chem. 262, 9026–9032.
17. Bouvier, M., Hausdorff, W.P., De Blasi, A., et al. (1988) Nature 333, 370–373.
18. Benovic, J.L., Kühn, H., Weyand, I., Codina, J., Caron, M.G. and Lefkowitz, R.J. (1987) Proc. Natl. Acad. Sci. USA. 84, 8879–8882.
19. Hausdorff, W.P., Bouvier, M., O'Dowd, B.F., Irons, G.P., Caron, M.G. and Lefkowitz, R.J. (1989) J. Biol. Chem. 264, 12657–12665.
20. Lohse, M.J., Benovic, J.L., Caron, M.G. and Lefkowitz, R.J. (1990) J. Biol. Chem. 265, 3202–3209.
21. Hausdorff, W.P., Caron, M.G. and Lefkowitz, R.J. (1990) FASEB J. 4, 2881–2889.
22. Pitcher, J., Lohse, M.J., Codina, J., Caron, M.G. and Lefkowitz, R.J. (1992) Biochemistry 31, 3193–3197.

23. O'Dowd, B.F., Hnatowich, M., Caron, M.G., Lefkowitz, R.J. and Bouvier, M. (1989) J. Biol. Chem. 264, 7564–7569.
24. Mouillac, B., Caron, M., Bonin, H., Dennis, M. and Bouvier, M. (1992) J. Biol. Chem. 267, 21733–21737.
25. Bouvier, M. (1990) Ann. N.Y. Acad. Sci. 594, 120–129.
26. Bouvier, M., Guilbault, N. and Bonin, H. (1991) FEBS Lett. 279, 243–248.
27. Yoshimasa, T., Sibley, D.R., Bouvier, M., Lefkowitz, R.J. and Caron, M.G. (1987) Nature 327, 67–70.
28. Yoshimasa, T., Bouvier, M., Benovic, J.L., Amlaiky, N., Lefkowitz, R.J. and Caron, M.G. (1988) In: Molecular Biology of Brain and Endocrine Peptidergic Systems (McKerns, K.W. and Chretien, M., Eds.) pp. 123–139, Plenum Publishing Co., New York.
29. Zick, Y. (1989) CRC Crit. Rev. Biochem. 24, 217–269.
30. Knutson, V.P. (1991) FASEB J. 5, 2130–2138.
31. Rosen, O.M. (1987) Science 237, 1452–1457.
32. Huganir, R.L. and Greengard, P. (1990) Neuron 5, 555–567.
33. Neubig, R.R., Boyd, N.D. and Cohen, J.B. (1982) Biochemistry 21, 3460–3467.
34. Hopfield, J.F., Tank, D.W., Greengard, P. and Huganir, R.L. (1988) Nature 336, 677–680.
35. Mulle, C., Benoit, P., Pinset, C., Roa, M. and Changeux, J.-P. (1988) Proc. Natl. Acad. Sci. USA. 85, 5728–5732.
36. Klein, W.L., Sullivan, J., Skorupa, A. and Aguilar, J.S. (1989) FASEB J. 3, 2132–2140.

II. Prototype Receptors

F. Hucho (Ed.) *Neurotransmitter Receptors*
© 1993 Elsevier Science Publishers B.V. All rights reserved.

CHAPTER 4

The nicotinic acetylcholine receptor

FERDINAND HUCHO

Institut für Biochemie, Freie Universität Berlin, Thielallee 63, D-14195 Berlin, Germany

1. Introduction

One could describe the history of neuropharmacology and 'receptorology' over the last one and a half century as the history of the discovery and investigation of acetylcholine receptors: among the first neurotoxin effects localized was the inhibitory action of the arrow poison curare, a substance acting in our language as a cholinergic antagonist. Claude Bernard [1] showed around the middle of last century that it paralyzed not by inactivating muscle contraction but by blocking the communication between nerve and muscle at the (again in our nomenclature) neuromuscular endplate, a cholinergic synapse. At about the same time E. Du Bois-Reymond [2] in Berlin presented evidence for a chemical communication between nerve and muscle. Langley, [3] around the turn of the century, coined the term 'receptor' (originally he called it 'receptive substance') for the substance 'receiving' nicotine or curare at the muscle. He also discovered the principal difference between the action of the agonist nicotine and the antagonist curare. About the same time the concept, and later on also the term, 'synapse' was coined by the muscle physiologist Sherrington [4]. Acetylcholine was the first neurotransmitter to be identified (by Loewi and Sir Henry Dale). Its quantal release and its effects on the postsynaptic membrane were shown by Bernhard Katz [5]. The history as outlined here briefly is a history of methods allowing the investigation of the biological phenomenon nerve impulse transmission with increasing resolution. The final steps in this direction were accomplished within the last two decades when the nicotinic acetylcholine receptor (AChR) was the first to be identified on a molecular basis and purified to homogeneity. Leading names in this field are Jean-Pierre Changeux [6], Arthur Karlin [7], Michael Raftery [8] and many others. These scientists rendered AChR accessible by investigating its molecular structure in relation to its mechanism of functioning. Biochemical analysis was greatly stimulated when at the same time electrophysiology reached molecular, and in connection with recombinant DNA techniques almost atomic, resolution through the development of the patch-clamp methods by Neher and Sakmann [9] and through the introduction of cloning and heterologous expression methods to neurobiology, primarily by Shosaku Numa [10]. At all these stages the nicotinic cholinergic synapse and, especially in the most recent phase, AChR proved to be an almost ideal model for elucidating general principles of signal transmission between cells and also signal transduction through plasma membranes. Because of these principal advantages AChR is today the neurotransmitter receptor we know best.

2. Function and occurrence

There are two types of AChR: a predominantly excitatory receptor in the peripheral nervous system and a second type in autonomic ganglia and the central nervous system. They are usually referred to as the peripheral (or muscle-like) and the neuronal receptors, respectively. The peripheral receptor is not even located in neuronal cells, but rather in the postsynaptic membrane of the neuromuscular synapse, the 'endplate', where it mediates the stimulation of muscle contraction by a nerve impulse. The location of neuronal receptors is complex and will be described in Section 4.3. Their function and mode of action is largely unknown and their biochemistry is very different from that of the peripheral receptors and not well elucidated.

All AChR are members of the superfamily of 'ligand-gated ion-channels receptors'. Their physiological role is to facilitate a flow of cations (Na^+, K^+, and also, to a certain degree, divalent ions such as Ca^{2+}) through a pore in the postsynaptic membrane which opens upon arrival of the agonist acetylcholine that has been released from the nerve terminal. As with all members of this superfamily, the pore is integrated within the receptor molecule and interacts with the agonist-binding site in the same manner as active and regulatory sites in allosteric enzymes [11,12]. Accordingly, AChR exists in several functional (and conformational) states, among which the resting (R) and the active (channel open) state A are the most obvious ones. The channel is opened by sequential binding [13] of two agonist molecules:

$$R + L \rightarrow RL + L \rightarrow R2L \rightarrow A2L$$

A third state has been observed under certain artificial conditions, e.g., during application of acetylcholine when the acetylcholinesterase is blocked. Prolonged presence of an agonist at the receptor seems to shift it to a 'desensitized state' [14]. In this conformation the ion channel is closed, although the agonist is bound with an increased affinity. Various kinetic models for the interconversion between these states have been proposed, circular and linear, and several further 'intermediate' states have been postulated [15].

Our picture of the peripheral receptors is much more advanced, mainly for two reasons. Since electrophysiologists have used the endplate for a long time as an experimental model for synapses in general, many of the principles of synaptic transmission have been elucidated, thus yielding a wealth of information about the AChR molecule. The other reason for the advanced state of research is an almost ideal model tissue, the electric organ of fish such as the electric eel *Electrophorus electricus* and the various species of the electric ray *Torpedo*. These tissues, introduced more than half a century ago by David Nachmansohn [16], are extremely favorable for receptor biochemistry because they are densely innervated by exclusively cholinergic neurons and therefore are a very rich source of AChR. About 10 mg receptor protein can be easily obtained from 100 g electric tissue. Only about a fiftyfold purification is required to prepare homogeneous AChR from *Torpedo* electric organs. Fortunately, it has been discovered that these receptors are closely related to the evolutionarily much younger species of higher vertebrates, including man.

3. *Pharmacology and toxicology*

A third reason for the advanced picture of the peripheral AChR may be seen in its elaborate pharmacology and toxicology [17–19], which was the basis for the identification, characterization and, finally, isolation of this receptor. Substances interacting specifically with AChR are subdivided into three classes. (a) *Agonists*, such as the physiological neurotransmitter acetylcholine or the exogenous activator nicotine, are molecules binding to the receptor thereby triggering the opening of its ion channel. (b) *Competitive antagonists* such as the arrow poison D-tubocurarine also bind with similar affinity and probably to the same binding site. Unlike the agonists, their binding has no effect other than blocking agonist binding. (c) *Noncompetitive antagonists* make up the third class of ligands [20,21]. They do not compete with the agonist molecule for a common binding site. However, they still block the agonist's action, probably by binding more directly to the ion channel, thereby inhibiting the flux of cations. Table I shows some typical representatives of the three classes of molecules. The compounds included are of a confusing structural variety. Neither a general pattern nor chemical characteristics are obvious. Furthermore, at present it seems completely impossible to predict from looking at a compound's formula whether a drug will behave as an agonist, or a competitive, or non-competitive antagonist. This basic pharmacological problem remains unsolved even after one hundred years of receptor research.

The so-called α-neurotoxins, peptide neurotoxins from elapid snakes [22], have been especially useful in AChR research [23,24]. They form a family of small proteins with highly homologous primary structures [19,22]. The most prominent representative of this class, the α-bungarotoxin from the krait *Bungarus multicinctus* (M_r 8000), binds virtually irreversibly (with $K_d = 10^{-11}$) as a competitive antagonist and is used after radioiodination in binding assays. Since tertiary structures from X-ray [25,26] and NMR analysis [27] are available for several of these toxins, they may be further used for mapping the antagonist-binding sites of the receptor protein. The structure of α-bungarotoxin, as shown in Figure 1, has several characteristic features. Like the other snake venom neurotoxins which have been analyzed, the polypeptide chain is

TABLE I
The pharmacology of the nicotinic acetylcholine receptor. A selection of typical representatives of the three classes of cholinergic effectors

Agonists	Antagonists	
	competitive	non-competitive
Acetylcholine	d-Tubocurarine	Histrionicotoxin
Carbamoylcholine	Flaxedil	Triton X–100 (n=9·10)
Decamethonium	Hexamethonium	Phencyclidine
Suberyldicholine	α-Bungarotoxin	TPMP$^+$
		Lidocain
		Trimethisoquin
		Chlorpromazine

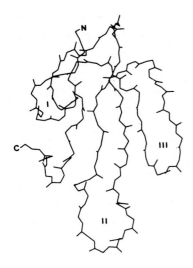

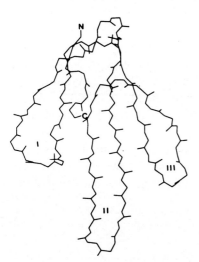

Fig. 1. α-Neurotoxins. Structures as obtained by X-ray and NMR analysis [25–27]. Top: α-Bungarotoxin, a so-called 'long' α-neurotoxin, containing 72 amino acids. Bottom: Erabutoxin b, a 'short' neurotoxin (62 amino acids), isolated from the venom of the sea snake *Laticauda semifasciata* [25]. N, the amino terminal; C, the carboxy terminal end of the polypeptide chain; I, II, and III, the loop structures characteristic for all α-neurotoxins (Reproduced from Hucho [19] with permission of Springer and from Endo and Tamiya [22] by courtesy of the Editors of *Pharmacol. Ther.*)

folded into a core stabilized by four disulfide bonds from which three loops protrude interacting through a hydrogen-bonded three-stranded β-pleated sheet. The amino acid side chains essential for toxicity seem to be located on loops II and III.

4. Biochemistry

4.1. AChR from Torpedo electric tissue

4.1.1. The molecule, quaternary structure, electron microscopy

In 1972 this receptor was shown for the first time to be an entity distinct from the other acetylcholine-binding synaptic protein, the acetylcholinesterase [28]: An affinity chromatography designed by Jean-Pierre Changeux and his coworkers at the Institut Pasteur in Paris separated an acetylcholine-hydrolyzing from a ^{3}H-bungarotoxin-binding protein. Subsequently, with receptor-rich tissues, binding assays [24,29] and appropriate detergents at hand for extracting the receptor from its location in the electrocyte, the receptor was purified to homogeneity from electric eel and electric ray by various laboratories around the world [29–32]. These are some of its biochemical characteristics [33]:

The AChR is an integral membrane glycoprotein. Its relative molecular mass M_r is about 290,000. This figure includes about 20,000 Dalton carbohydrate. The AChR has a heteropentameric quaternary structure: it is composed of two identical polypeptide chains (α apparent $M_r = 40,000$, as determined by SDS polyacrylamide gel electrophoresis) and three chains differing from these (β: $M_r = 48,000$; γ: $M_r = 60,000$; δ: $M_r = 68,000$). The resulting quaternary structure therefore is $\alpha_2\beta\gamma\delta$. Torpedo AChR (but not receptors from other sources) forms dimers of two pentameric complexes covalently linked by a disulfide bond between the δ-subunits. No functional significance is assigned to this dimerization. All five polypeptide chains span the membrane and all are glycoproteins [34], the carbohydrate antennae being of the

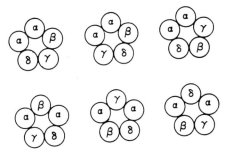

Fig. 2. The six possible circular arrangements of the subunits of an $\alpha_2\beta\gamma\delta$-heteropentamer. The $\alpha\beta\alpha\gamma\delta$ arrangement (read clockwise) or its mirror image is presently favored for AChR [51], although other arrangements are discussed [7]. The quaternary structure is important, among others, for the role of individual subunits in the allosteric regulation of the receptor.

high mannose type [35,36]. All polypeptide chains are phosphorylated [34–37], some in several positions of the primary structure (see below). The β-chain is acylated by a fatty acid. The AChR is a calcium-binding protein [38]. More than 60 calcium ions have been found [39], some which are only loosely bound and are released upon agonist binding, and others which are present even after separating the polypeptide chains by SDS-polyacrylamide gel electrophoresis . No functional role for these ions has been shown.

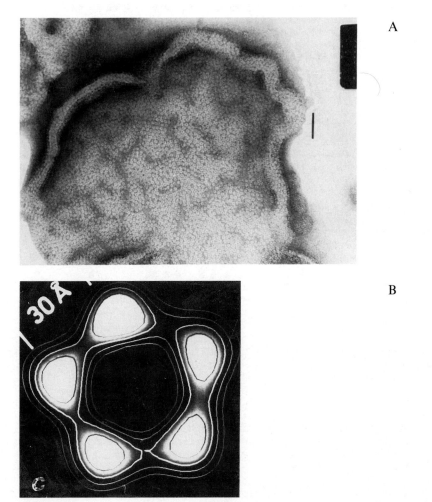

Fig. 3. Electron microscopy of *Torpedo* AChR. A: receptor-rich membrane negatively stained, B: images obtained by image processing. (Reproduced from Kunath et al. [42] by courtesy of the Editors of *Electron Microsc. Rev.*; see also [43].)

TABLE II

Amino acid sequences of the precursors of the four acetylcholine receptor subunits from *Torpedo californica*; the alignment was done according to Higgins and Sharp using the CLUSTAL computer program (K-tuple value = 1, gap penalty = 5, open gap and unit gap cost = 10, Dayhoff Mutation Data Matrix MDM-78). Squares (■) indicate amino acid residues conserved in all four polypeptide chains; over the whole consensus length, 18.2% of the amino acids are identical.

```
                ■   Signal sequence        1 ■ ■ ■     ■      ■
α-subunit   MILCSYWHVGLVLLLFSCCGLVLG SEHETRLVANLL--ENYNKVIRPVEHHTHFVDITV     033
β-subunit   MENVRRMALGLVVMMALALSGVGA SVMEDTLLSVLF--ETYNPKVRPAQTVGDKVTVRV     033
γ-subunit   MV------LTLLLIICLALEV-RS ENEEGRLIEKLL--GDYDKRIIPAKTLDHIIDVTL     033
δ-subunit   MGNI---HFVYLLISCLYYSGCSG VNEEERLINDLLIVNKYNKHVRPVKHNNEVVNIAL     035
            ■■ ■    ■         ■■   ■■■■■   ■  ■ ■        ■ ■■ ■■ ■■
α-subunit   GLQLIQLISVDEVNQIVETNVRLRQQWIDVRLRWNPADYGGIKKIRLPSDDVWLPDLVLY       093
β-subunit   GLTLTNLLILNEKIEEMTTNVFLNLAWTDYRLQWDPAAYEGIKDLRIPSSDVWQPDIVLM       093
γ-subunit   KLTLTNLISLNEKEEALTTNVWIEIQWNDYRLSWNTSEYEGIDLVRIPSELLWLPDVVLE       093
δ-subunit   SLTLSNLISLKETDETLTSNVWMDHAWYDHRLTWNASEYSDISILRLPPELVWIPDIVLQ       095
            ■■ ■■     ■    ■   ■■■■ ■■■■  ■■■■ ■■■                      ■
α-subunit   NNADGDFAIVHMTKLLLDYTGKIMWTPPAIFKSYCEIIVTHFPFDQQNCTMKLGIWTYDG       153
β-subunit   NNNDGSFEITLHVNVLVQHTGAVSWQPSAIYRSSCTIKVMYFPFDWQNCTMVFKSYTYDT       153
γ-subunit   NNVDGQFEVAYYANVLVYNDGSMYWLPPAIYRSTCPIAVTYFPFDWQNCSLVFRSQTYNA       153
δ-subunit   NNNDGQYHVAYFCNVLVRPNGYVTWLPPAIFRSSCPINVLYFPFDWQNCSLKFTALNYDA       155
                                             ■■■■          ■
α-subunit   TKVSI----SPESDR--P------DLSTFMESGEWVMKDYRGWKHWVYYTCCPDTPYLDI       201
β-subunit   SEVTLQHALDAKGER--EVKEIVINKDAFTENGQWSIEHKPSRKNWR----SDDPSYEDV       207
γ-subunit   HEVNLQLSAEEGEA----VEWIHIDPEDFTENGEWTIRHRPAKKNYNWQLTKDDTDFQEI       209
δ-subunit   NEITMDLMTDTIDGKDYPIEWIIIDPEAFTENGEWEIIHKPAKKNIYPDKFPNGTNYQDV       215
            ■   ■  ■■      ■■■■■■■■■    ■■    ■  ■       ■■ ■■    ■■
α-subunit   TYHFIMQRIPLYFVVNVIIPCLLFSFLTGLVFYLPTDSGEKM-TLSISVLLSLTVFLLVI       260
β-subunit   TFYLIIQRKPLFYIVYTIIPCILISILAILVFYLPPDAGEKM-SLSISALLAVTVFLLLL       266
γ-subunit   IFFLIIQRKPLFYIINIIAPCVLISSLVVLVYFLPAQAGGQKCTLSISVLLAQTIFLFLI       269
δ-subunit   TFYLIIRRKPLFYVINFITPCVLISFLASLAFYLPAESGEKMST-AISVLLAQAVFLLLT       274
            ■■ ■■  ■■■              ■                       ■■    ■           ■
α-subunit   VELIPSTSSAVPLIGKYMLFTMIFVISSIIITVVVINTHHRSPSTHTMPQWVRKIFIDTI       320
β-subunit   ADKVPETSLSVPIIIRYLMFIMILVAFSVILSVVVLNLHHRSPNTHTMPNWIRQIFIETL       326
γ-subunit   AQKVPETSLNVPLIGKYLIFVMFVSMLIVMNCVIVLNVSLRTPNTHSLSEKIKHLFLGFL       329
δ-subunit   SQRLPETALAVPLIGKYLMFIMSLVTGVIVNCGIVLNFHFRTPSTHVLSTRVKQIFLEKL       334
            ■
α-subunit   PNVMFF--------STMKRASKEKQE---------NKIF----ADDIDISDISGK-QVTG       358
β-subunit   PPFLWIQRPVTTPSPDSKPTIISRAN---------DEYFIRKPAGDFVCPVDNARVAVQP       377
γ-subunit   PKYLGMQLEPSEETPEKPQP---RRRSSFGIMIKAEEYILKKPRSELMFEEQKDRHGLKR       386
δ-subunit   PRILHMSRADESEQPDWQNDLKLRRSSSVGYISKAQEYFNIKSRSELMFEKQSERHGL--       392
                                        ■        ■          ■
α-subunit   -EVIFQT-----------PLIKNPDVKSAIEGVKYIAEHMKSDEESSNAAEEWKYVAMV       405
β-subunit   -ERLFSEMKW--HLNGLTQPVTLPQDLKEAVEAIKYIAEQLESASEFDDLKKDWQYVAMV       434
γ-subunit   VNKMTSDIDIGTTVDLYKDLANFAPEIKSCVEACNFIAKSTKEQNDSGSENENWVLIGKV       446
δ-subunit   VPRVTPRIGFGNNNENIAASDQLHDEIKSGIDSTNYIVKQIKEKNAYDEEVGNWNLVGQT       452
              ■             ■■  ■
α-subunit   IDHILLCVFMLICIIGTVSVFAGRLIELSQEG----------------              461
β-subunit   ADRLFLYVFFVICSIGTFSIFLDASHNVPPDNPFA--------------              493
γ-subunit   IDKACFWIALLLFSIGTLAIFLTGHFNQVPEFPFPGDPRKYVP------              506
δ-subunit   IDRLSMFIITPVMVLGTIFIFVMGNFNHPPAKPFEGDPFDYSSDHPRCA              522
```

There are six possible circular arrangements (plus six mirror images) for the five subunits $\alpha_2\beta\gamma\delta$, in a rosette (Fig. 2), among which $\alpha\beta\alpha\gamma\delta$ is the most likely one for several reasons. In addition to the five membrane-spanning subunits, a peripheral protein of 43,000 Dalton, called the 43 k or v_1 protein, is stoichiometrically associated with the receptor complex at its cytosolic surface. Its function is thought to be anchoring the receptor to the cytoskeleton [12].

AChR is the first [41] and hitherto only neurotransmitter receptor to be made visible in great detail by electron microscopy [42]. Receptor-rich membranes isolated from homogenates of *Torpedo* electric organs by sucrose density gradient centrifugation exhibit after negative staining 'doughnut-like' structures, rosettes with a diameter of about 75 Å and a central pit of about 30 Å (Fig. 3). Computer-aided image processing resolved these rosettes into rings of five maxima surrounding an electron-dense center, probably representing the five receptor subunits [43]. The center seems to be the entrance to the transmembrane ion channel. Side views [44] of the molecule showed that approximately 50% of the protein protrude from the extracellular side of the membrane, while only about 15% extend from the intracellular surface to the cytoplasm. The remaining 35% are buried within the membrane.

Even higher resolution images were obtained with electron diffraction analysis of quick-frozen receptor-rich membranes artificially arranged in two dimensional crystalline arrays [44]. At about 25 Å resolution, and in the absence of negatively staining heavy metal ions, the receptor presented itself as a molecule about 120 Å long, with a diameter of about 80 Å. A central pore, with a diameter of 25–30 Å in its upper part, extends through the entire length. Close to the cytosolic surface of the membrane, this pore has a constriction too narrow to be resolved by electron microscopy. This constriction may play a role in selecting which ions are allowed to permeate (see below, Section 4.1.4).

4.1.2. Primary structure, evolutionary aspects
The amino acid sequences of the precursors of the four receptor subunits [10,45,46] (and of the 43 k protein [47]) were deduced from cDNAs, cloned and sequenced (mainly by Numa and his coworkers in Kyoto) (Table II). They confirmed the homology and evolutionary relationship of the four subunits as suggested on the basis

TABLE III
Molecular properties of the *Torpedo* AChR subunits

Subunits	Length of signal peptide	Apparent M_r (SDS PAGE)	Mature chain	Calculated M_r
α	24	40,000	437	50,116
β	24	48,000	469	53,681
γ	17	60,000	489	56,279
δ	24	65,000	501	57,565

TABLE IV

A selection of AChR α-subunits. Squares indicate conserved residues. The sequences are taken from [10, 46, 97, 98]

```
                          Signal sequence                          1■     ■    ■    ■
Torpedo α1     MILCSYWHVGLVLLLFSCC--GLVL-----------------G SEHETRLVANLLENY           015
Human α1       M------EPWPLLLLFSLCSAGLVL-----------------G SEHETRLVAKLFKDY
Human α3       MAL---------------AVSLPLACRARLLLLLLSLLPVARA SEAEHRLFERLFEDY
Chick BgtBP1   MGLRA-----LMLWLLAAA--GLVRES-------------LQ GEFQRKLYKELLKNY
                          ■■■              ■ ■■ ■■      ■■ ■   ■    ■       ■  ■■
Torpedo α1     NKVIRPVEHHTHFVDITVGLQLIQLISVDEVNQIVETNVRLRQQWIDVRLRWNPADYGGI   075
Human α1       SSVVRPVEDHRQVVEVTVGLQLIQLINVDEVNQIVTTNVRLKQQWVDYNLKWNPDDYGGV
Human α3       NEIIRPVANVSDPVIIHFEVSMSQLVKVDEVNQIMETNLWLKQIWNDYKLKWNPSGYGGA
Chick BgBP1    NPLERPVANDSQPLTVYFTLSLMQIMDVDEKNQVLTTNIWLQMYWTDHYLQWNVSEYPGV
                ■  ■■■  ■■■         ■            ■       ■■ ■■■■ ■■■■    ■■
Torpedo α1     KKIRLPSDDVWLPDLVLYNNADGDFAIVHMTKLLLDYTGKIMWTPPAIFKSYCEIIVTHF   135
Human α1       KKIHIPSEKIWRPDLVLYNNADGDFAIVKFTKVLLQYTGHITWTPPAIFKSYCEIIVTHF
Human α3       EFMRVPAQKIWKPDIVLYNNAVGDFQVTTKTKALLKYTGEVTWIPPAIFKSSCKIDVTYF
Chick BgtBP1   KNVRFPDGLIWKPDILLYNSADERFDATFHTNVLVNSSGHCQYLPPGIFKSSCYIDVRWF
                ■■■ ■ ■■■■■                              ■■■        ■    ■ ■■
Torpedo α1     PFDQQNCTMKLGIWTYDGTKVSISPESDRPDLSTFMESGEWVMKDYRGWKHWVYYTCCPD   195
Human α1       PFDEQNCSMKLGTWTYDGSVVAINPESDQPDLSNFMESGEWVIKESRGWKHSVTYSCCPD
Human α3       PFDYQNCTMKFGSWSYDKAKIDLVLIGSSMNLKDYWESGEWAIIKAPGYKHDIKYNCCEE
Chick BgtBP1   PFDVQKCNLKFGSWTYGGWSLDLQWQEADISGYISN--GEWDLVGIPGKRTESFYECCKE
                ■ ■■■       ■ ■         ■■■ ■■■■■■ ■■■■■ ■ ■  ■   ■■■  ■■
Torpedo α1     TPYLDITYHFIMQRIPLYFVVNVIIPCLLFSFLTGLVFYLPTDSGEKMTLSISVLLSLTV   255
Human α1       TPYLDITYHFVMQRLPLYFIVNVIIPCLLFSFLTGLVFYLPTDSGEKMTLSISVLLSLTV
Human α3       I-YPDITYSLYSRRLPLFYTINLIIPCLLISFLTVLVFYLPSDCGEKVTLCISVLLSLTV
Chick BgtBP1   -PYPDITFTVTMRRRTLYYGLNLLIPCVLISALALLVFLLPADSGEKISLGITVLLSLTV
                ■ ■   ■■■    ■■ ■ ■■■       ■■■■■■■■■■■■ ■■■ ■■ ■  ■■■ ■ ■■
Torpedo α1     FLLVIVELIPSTSSAVPLIGKYMLFTMIFVISSIIITVVVINTHHRSPSTHTMPQWVRKI   315
Human α1       FLLVIVELIPSTSSAVPLIGKYMLFTMVFVIASIIITVIVINTHHRSPSTHVMPNWVRKV
Human α3       FLLVITETIPSTSLVIPLIGEYLLFTMIFVTLSIVITVFVLNVHYRTPTTHTMPSWVKTV
Chick BgtBP1   FMLLVAEIMPATSDSVPLIAGYFASTMIIVGLSVVVTVIVLQYHHHDPDGGKMPKWTRVI
Torpedo α1     FIDTIPNVMFFSTMK---------------------------------------       330
Human α1       FIDTIPNIMFFSTMK---------------------------------------
Human α3       FLNLLPRVMFMTRPTSNEGNAQKPRPLYGAELSNLNCFSRAESKGCKEGYPCQDGMCGYC
Chick BgtBP1   LLNWCA---WFLRMK---------------------------------------
                             ---------------■                             ■
Torpedo α1     --------------RASKEKQENKIFADDIDISDISGKQVTGEVIFQTPLIKNPDVKSA   375
Human α1       --------------RPSREKQDKKIFTEDIDISDISGKPGPPPMGFHSPLIKHPEVKSA
Human α3       HHRRIKISNFSANLTRSSSSESVDAVVS----LSALS-----------------PEIKEA
Chick BgtBP1   --------------RPGEDK-----------------------------------VRPA
Torpedo α1     ------------------------------------------------------------
Human α1       ------------------------------------------------------------
Human α3       ------------------------------------------------------------
Chick BgtBP1   CQHKQRRCSLSSMEMNTVSGQQCSNGNMLYIGFRGLDGVHCTPTTDSGVICGRMTCSPTE
                               ■■■                   ■  ■■ ■■■ ■■ ■
Torpedo α1     ------------------IEGVKYIAEHMKSDEESSNAAEEWKYVAMVIDHILLCVFM   415
Human α1       ------------------IEGIKYIAETMKSDQESNNAAAEWKYVAMVMDHILLGVFM
Human α3       ------------------IQSVKYIAENMKAQNEAKEIQDDWKYVAMVIDRIFLWVFT
Chick BgtBP1   EENLLHSGHPSEGDPDLAKILEEVRYIANRFRDQDEEEAICNEWKFAASVVDRLCLMAFS
                ■ ■
Torpedo α1     LICIIGTVSV-------------FAGRLIELSQEG 437 COOH
Human α1       LVCIIGTLAV-------------FAGRLIELNQQG
Human α3       LVCILGTAGL------FLQPLMAREDA
Chick BgtBP1   FVTIICTIGILMSAPNFVEAVSKDFA
```

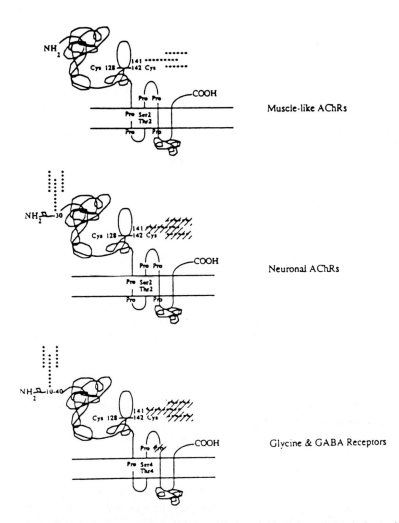

Fig. 4. Models of the muscle-like AChRs, the neuronal AChRs, and the glycine and GABA$_A$ receptors. Only those features which are common to all of the members of each class of channels are shown in the figure. From hydropathic profiles, all of the channels contain four domains which are long enough and hydrophobic enough to span the lipid bilayer. Cysteine residues are present in all of the channels which are separated by 13 amino acids and are shown as linked residues numbered Cys-128 and -142. The series of stars represent potential sites of asparagine-linked glycosylation. Hatched marks (glycosylation or proline residues) indicate that some but not all of the members of this group of channels possess such sites or residues. The four to five serine and threonine residues present in M2 of the muscle-like and neuronal AChRs are denoted Ser$_2$/Thr$_2$, and the seven to eight residues present in M2 of the GABA and glycine receptors are denoted Ser$_4$/Thr$_4$. (Figure and legend reproduced with kind permission from Toni Claudio [51].)

of previous partial protein sequences [48]. The length of the polypeptide chains as calculated from the cDNA sequences and the calculated molecular weights are shown in Table III.

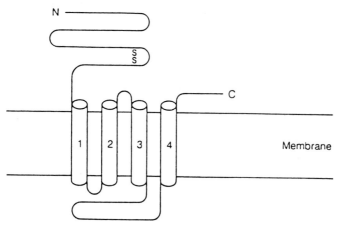

Fig. 5. The 4-helix model of AChR and the other ligand-gated ion-channel receptors of this superfamily.

The amino acid sequences as shown in Figure 2 indicate the close evolutionary relationship between the four subunits and suggest that their genes originated from a common ancestral gene. The α, β, γ, and δ subunits show 19% identities and 54% homologies (identities plus conservative substitutions). Furthermore, their sequences are fairly well conserved during evolution (Table IV).

Even more striking than the homology between the AChR subunits is the similarity with glycine [49] and $GABA_A$ [50] receptor subunits. In addition to significant sequence homologies, there are several features that these receptors have in common [51,52] (Fig. 4). Their ligand-binding subunits all have four strongly hydrophobic amino acid sequences which are thought to form membrane-spanning α-helices [53] (see below). They have a disulfide bridge forming a loop of 13 amino acids. There are proline residues located in conspicuous positions, e.g. in the middle of the putative transmembrane helix M1 and in the sequence connecting helices M2 and M3. These and other common features justify calling them a 'superfamily of ligand gated ion channels' [50]. One may speculate that they all originate early in evolution from a common ancestral receptor.

One special feature of the AChR α-subunit primary structure must not be forgotten, because it divides the evergrowing number of cloned receptor subunits and their corresponding genes in 'α' and 'non-α' subunits. α-Subunits, whose function is binding of the ligand acetylcholine (see Section 4.1.4), possess tandem cysteine residues [54] in positions α-Cys 192 and α-Cys 193 (*Torpedo*) forming a disulfide bridge. The functional significance of this is discussed below. A second disulfide bridge is found in α- and in all other AChR subunits. It forms a loop between α-Cys 128 and α-Cys 142 (all positions given refer to the Torpedo sequence).

The AChR primary structures contain several consensus sequences for post-translational modifications. Among the various Asn-X-Ser/Thr motifs, the potential glycosylation sites, only α-Asn 141 has been shown to be glycosylated [36]. The location of the other oligosaccharide residues is unknown. The phosphorylation sites will be discussed in connection with the functional topography of the receptor molecule (Section 4.1.4).

4.1.3. Secondary structure, transmembrane folding
FTIR spectroscopy provided evidence that *Torpedo* AChR contains about 32% α-helical and 36% β-structure [55]. By means of secondary structure algorithms, it has been predicted that all AChR subunits contain four membrane-spanning α-helices, M1–M4 [56]. Since the existence of signal peptides was deduced from the cDNA sequences, the N-terminus of the polypeptide chains must be located on the extracellular side of the membrane. A four-helix model (Fig. 5) therefore means that the C-terminus is also extracellular. Biochemical evidence supports this theoretical model, while other models of the transmembrane architecture, having three [57], five [53], and even seven [58] membrane-spanning helices have also been proposed. These models are largely based on mapping experiments using monoclonal antibodies [51]. The four-helix model is favored by most researchers in the field, but one should keep in mind that this and the alternative models are based mainly on 'predictions', i.e. on more or less intelligent guessing. Final proof must await X-ray analysis of crystalline receptor.

4.1.4. Functional topography
The RTE model of receptors (see Chapter 1) suggests that, like the other receptors of its class, AChR consist of three functional domains: the signal-receiving (= transmitter binding) moiety R, the effector component E (= the ion channel), and the transducing domain T which mediates the transduction of the signal from R to the effector E. Receptor research attempts to elucidate the molecular structure of these three domains which, according to our definition of Type I receptors, are integral parts of the AChR heteropentamer.

R, the binding site for agonists and competitive antagonists, has been localized by affinity labeling on the two α-subunits [59]. The tandem cysteine residues Cys 192/Cys 193 are part of the binding site. Blocking only one of these with e.g. the affinity reagent ^{3}H-4-(N-maleimido)benzyltrimethylammonium (MBTA) after reduction or removing it by site-directed mutagenesis abolishes binding of acetylcholine (but not of the competitive antagonist α-bungarotoxin) [60].

The sequence around the tandem cysteines (which is in *Torpedo californica* AChR -RGWKHWVYYTCCPDTPYLDITYH-) is highly conserved. It was proposed to form part of the α-neurotoxin-binding site [61], because synthetic peptides with this sequence bind the toxin with significant affinity [62] and antibodies raised against this peptide compete for the ligand-binding site on the receptor [63]. Other subunits of the receptor seem to participate in toxin binding as well, which explains its very high affinity ($K_d = 10^{-11}$) [64–66].

Several other amino acids have been shown by various means to contribute to ligand-binding sites (Tyr 190 [67], Tyr 93 [68], and Trp 149 [69]). They are not positioned on a continuous stretch of the α-subunit's primary structure, indicating that the ligand-binding pocket is formed by several, possibly distant (in terms of the primary sequence), loops of the polypeptide chain [20].

T, the signal-transducing moiety, is the molecular basis of the allosteric properties of AChR. From the very beginning of its investigation, AChR was described as an allosteric protein [11,12,20]. The two acetylcholine-binding sites interact cooperatively (yielding a sigmoidal dose-response curve in electrophysiological experiments with the neuromuscular endplate or with electrocytes from the electric eel). Furthermore, the agonist-binding sites interact in a positively cooperative manner with the ion channel. Various noncompetitive antagonists (see below) shown to bind to amino acid side chains in the ion channel increase the affinity of the agonist for its binding site which is probably at least 40 Å away.

Finally E, the effector moiety, is the ion channel. As mentioned before, it is a cation channel; anions do not pass. It is less selective with respect to the type of cation. K^+ permeates only slightly better than Na^+, monovalent cations are preferred but divalent ions permeate also. The order of permeability for monovalent ions as determined from the reversal potentials with frog muscle is $Cs^+ > Rb^+ > K^+ > Na^+ > Li^+$ and for divalent cations $Mg^{2+} > Ca^{2+} > Ba^{2+} > Sr^{2+}$ [71]. Various organic cations have been used to determine the diameter of the pore, which was found to be on the order of 0.64 mm [72]. This is larger than a fully dehydrated cation: sodium and potassium ions permeate in a partially hydrated form. The channel itself is hydrated and has been described as a 'water-filled pore'.

Other properties of the ion channel are as follows [73]:
a) H^+ blocks the channel. The pK value of the dissociating amino acid side chain is 4.8, possibly a carboxylic group.
b) Opening and closing of the channel is an 'all or nothing' event. There are no partially opened channels. The transition between these states occurs within 20 μs, the time resolution of the patch-clamp technique. Multiple and discrete conductance states ('substates') of individual channels have been observed.
c) The mean open time of the channel depends on the temperature, the membrane potential, and the agonist. It is 0.6 ms with carbamoylcholine, 2.4 ms with acetylcholine, and 5.6 ms with suberyldicholine (11–15°C, 60–80 mV).
d) The mean open time depends on the location of the channel in the cell membrane: extrasynaptic receptor channels have open times three to five times longer than endplate channels.
e) The mean open time changes during ontogeny. With synapse maturation it decreases, a process not simply correlated with receptor density.
f) The conductance of the open channel is 15–30 pS. About 10^4 ions permeate per ms.
g) Myasthenia gravis, an autoimmune disease directed against the endplate and its

receptors, alters the number but not the properties of remaining functional receptors.

However, what is the proof for our statement that the channel is an integral part of the receptor protein? The evidence is twofold: AChR purified by affinity chromatography using an immobilized ligand directed against the binding site can be reconstituted into an artificial lipid membrane yielding a completely functional 'ligand-gated ion channel'. Furthermore, cloned mRNA coding for the precursors of the four different receptor subunits can be expressed e.g. in *Xenopus* oocytes, again yielding functional ligand gated ion channels. By the way, all four types of subunits must be co-expressed to obtain functional channels with the peripheral AChR (The δ-subunit can be replaced to a certain extent by others.) We shall see that this is different with the neuronal receptors. To express the α-bungarotoxin binding site only, two subunits are necessary, α and δ or δ and γ [74].

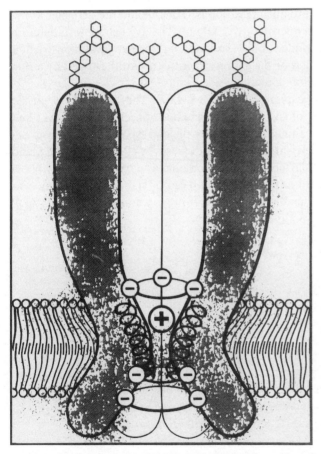

Fig. 6. The helix-M2 model of AChR [76]. The model combines data obtained from biochemical [75], molecular biological [77] and electron microscopical [44] investigations. For explanation see text.

The purpose of the ligand gated ion channel is to lower the energy barrier preventing charged particles (sodium or potassium ions) from passing the lipid bilayer of the plasma membrane. Photoaffinity labeling of the channel with radioactive organic cations has given insight into the molecular architecture of the transmembrane part of the AChR [75]. The wall of the pore is formed by the helices M2, probably contributed by all five receptor subunits. They all seem to be oriented in such a way that their helix M2 is located near the lumen of the channel. This model, called the helix-M2 model (Fig. 6) [76], has been confirmed by an elegant experiment combining recombinant DNA techniques with patch-clamp electrophysiology [77]. The electric properties of the AChR channel from the electric tissue of the *Torpedo* ray are different from the corresponding channel in the calf. Exchanging only the helix M2 of the *Torpedo* receptor's δ-subunit for the corresponding amino acid sequence from calf AChR (by site-directed mutagenesis) gives this chimera the single channel conductance of the calf pore. The dimensions of the helix M2 pore are 0.64 nm at the narrowest site. The diameter deduced from the photolabeling cation is 1.15 nm, and the diameter at the entrance visualized by electron microscopy is 3.0 nm. The five helices M2 therefore seem to form a funnel. As mentioned above (Section 4.1.1) the narrowest constriction may represent the selectivity filter or the gating structure (or both), formed by a ring of negatively charged and rotating glutamate side chains [78].

Finally, there are several phosphorylation sites [34,37]. They are located on the supposedly intracellular loops of the receptor subunits connecting helices M3 and M4. As shown experimentally and as suggested by the corresponding consensus sequences, the AChR is phosphorylated by protein kinases A (PKA), C (PKC), and tyrosine kinases. The receptor heteropentamer contains a maximum of nine phosphate groups. They are believed to be involved in receptor stabilization during synapse development [79,81], rather than in short-term regulation of its activity (e.g. desensitization) [82].

4.2. Muscle AChR

4.2.1. Biochemistry
Torpedo and higher vertebrate muscle AChR are very similar, this justifies investigating the former as a model for the latter. Muscle receptors are also heteropentamers; but there is a significant difference: the γ-subunit is present only in the fetal AChR. In adult bovine and rat receptors it was found that the γ-subunit is replaced by a novel (though homologous) subunit called ε [83,84]. This developmental exchange correlates with changes in single channel properties observed during synaptogenesis. The homologies of the corresponding subunits in *Torpedo* and bovine AChR are α: 79%; β: 57%; γ: 54%; δ: 59%; ε: 57% (compared with *Torpedo* γ).

4.2.2. Development [85]
Early in embryogenesis, before muscle innervation, acetylcholine receptors are dif-

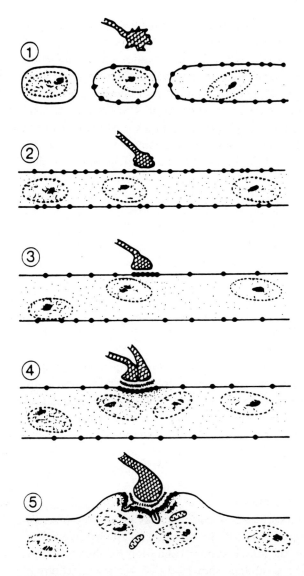

Fig. 7. Expression of AChR during formation of the neuromuscular junction (black dots: AChR). 1) Fusion of myoblasts into myotubes; AChR biosynthesis is enhanced. 2) The exploratory motor axon approaches. 3) The growth cone contacts the myotube and a subneural cluster of AChR forms. 4) Several motor nerve endings converge on the subneural cluster of AChR. 5) One motor nerve ending becomes stabilized; subneural folds develop and interactions with the cytoskeleton become apparent (Figure and legend reproduced with kind permission from J.-P. Changeux [85].)

fusely distributed over the muscle surface (about 20 receptors per μm^2). After innervation they cluster densely (about 20,000 per μm^2) at the site of the ingrowing axon

(Fig. 7). The metabolic stability increases in parallel, the half-life of the subsynaptic receptors being about 14 days and the extrasynaptic ones 7 hours. The number of extrasynaptic receptors increases considerably after denervation (a phenomenon called supersensitivity) and decreases again after innervation (or after electrical stimulation of the muscle). The conclusion from this is that receptor density and stability is regulated by some trophic factor from the ingrowing neuron. Besides neurotrophic factors, components of the basal lamina (e.g. agrin [86]) and cytoskeletal elements have been identified as being involved in forming and maintaining receptor clusters.

Muscle activity primarily affects the diffuse extrasynaptic AChR population: electrical stimulation decreases their density. Blocking neuromuscular transmission by blocking the receptors with α-bungarotoxin or inhibiting presynaptic action potentials by blocking sodium channels with tetrodotoxin increases their number 50–100-fold [85]. One especially interesting factor should be mentioned, a 43,000 Dalton protein called ARIA (**A**cetylcholine **R**eceptor **I**nducing **A**ctivity). Its physiological role is not yet clear and its primary structure has turned out to be significantly homologous with the scrapie prion protein [87].

4.3. Neuronal AChR [88]

4.3.1. Biochemistry and molecular biology
Neuronal AChR, present in pheochromocytoma (PC12) cells, autonomic ganglia, and the central nervous system (CNS), are very different from their peripheral counterparts. Although activated by the agonists acetylcholine and nicotine as well, they do not bind the antagonist neurotoxin α-bungarotoxin (for one exception, see below). Instead, the so-called neuronal or κ-toxin, present in the venom of the same elapid snakes and structurally closely related to the α-neurotoxins, binds and blocks these receptors [89]. In addition, there are 'α-bungarotoxin-binding proteins' in the brain that will be discussed below.

Low-stringency screening of cDNA libraries obtained from neuronal tissue with oligonucleotide probes derived from sequences of peripheral AChR produced several AChR subunits chracterized as neuronal AChR. Six neuronal α-subunits (named α_2–α_7, α_1 in this nomenclature refers to the muscle subunit) and four 'β-subunits' (β_2–β_5) were found in various animals. The α-subunits showed the characteristic feature of all α-subunits of this superfamily, the tandem cysteine residues. The β-subunits do not possess the tandem cysteine. Since they have nothing particular in common with β_1, the muscle β-subunit, they are often refered to as 'non-α'-subunits. cDNA sequences related to peripheral γ-, ε-, and δ-subunits were not detected. The quaternary structures of neuronal receptors are still unknown. Homo- and heterooligomeric receptors may be present. Expressing various combinations of α- and non-α-subunits in *Xenopus* oocytes and measuring their properties by patch-clamp electrophysiology resulted in an impressive picture: apparently, both the types of α- and β-subunit give the resulting AChR a distinct set of electrophysiological and pharma-

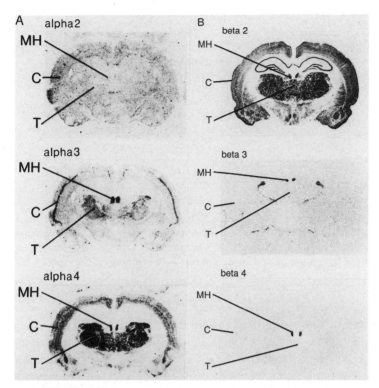

Fig. 8. Specific expression of neuronal AChR subunits in different areas of the rat brain. A. α-Subunit transcript distribution (as detected by in-situ hybridization). B. β-Subunit transcript distribution. The comparison shows that the subunits are expressed to a different extent in different brain areas. As a result, AChRs with different subunit compositions (quaternary structures) are formed. The functional significance of these differences are not yet clear. (Reproduced with kind permission from Evan S. Deneris [88].)

cological properties. With regard to toxin sensitivity, three classes of receptor could be produced: those sensitive to α-, those to κ-neurotoxin, and those insensitive to both. Even a homo-oligomeric functional AChR composed of only α7 was obtained [90]. This neuronal species was effectively blocked by the classical 'peripheral' snake venom toxin α-bungarotoxin. All these subunits show sequence homologies of about 40–70% among themselves and with peripheral AChR subunits.

In addition, there are α-bungarotoxin binding proteins in the CNS that do seem not to be linked to functional acetylcholine receptors [91,92]. Their function is completely unknown. The situation is even more confusing since similar α-bungarotoxin binding proteins have been detected in non-neuronal and nonmuscular tissue and cells [93].

4.3.2. Localization of neuronal receptor subunits in the brain
In-situ hybridization of gene transcripts with radioactive oligonucleotide probes and

TABLE V
Toxin sensitivities of neuronal nicotinic receptor subunit combinations expressed in oocytes

	α-Bgt (0.1 μM)	n-Bgt (0.1 μM)	NSTX (2 nM)	LTX-1 (10 μM)	G1A (10 μM)	M1 (10 μM)
α2β2	0	0	+++	++	0	0
α3β2	0	+++ (0.01 μM)	+++	++	0	0
α4β2	0	+	+++	+++	0	0
α4nα1*	0 (0.3 μM)	0 (0.5 μM)				
α3β4*	0	0				
α7*	+++					

*Avian subunit combinations.
0, no blockade at indicated concentration; +, ++, partial blockade at indicated concentration; +++, nearly complete or complete blockade at indicated concentration; α-Bgt, α-bungarotoxin; n-Bgt, neuronal bungarotoxin; NSTX, neosurugatoxin; LTX-1, lophotoxin; G1A, α-conotoxin G1A; M1, α-conotoxin M1. Taken from Deneris et al. [88] by courtesy of the Editors of Trends in Pharmacological Sciences.

autoradiography of brain slices were used to find out where the subunits are expressed in the brain. Figure 8 and Table V indicate the variety of receptors that are possible by combining different subunits in the respective brain areas. Immunoaffinity methods used for purifying nicotine-binding proteins from chicken and rat brain and biochemical analysis of the homogenous protein showed that these receptors were composed predominantly of α_4 and β_2-subunits.

4.4. AChR from other sources

Most of the data described in this Chapter refer to AChR from *Torpedo*, higher vertebrates, and birds (chicken). It should be mentioned briefly that AChR from a variety of other animals have been either isolated biochemically or characterized by recombinant DNA techniques. Among these the nicotinic AChR from insects is of special interest [94]. In general, they are more closely related to vertebrate neuronal than to muscle-like receptors. A special case seems to be the AChR from locusts. Apparently one type of subunit [95] alone, reconstituted in an artificial lipid-membrane system, is able to form functional receptor channels. Several α-like (ALS: putative *Drosophila* α-like subunit) and non-α-like (ARD: putative *Drosophila* non-α-like subunit) subunits have been identified by cloning and sequencing cDNA from *Drosophila* [94]. They turned out to form seemingly homo-oligomeric ligand-gated ion channels when expressed alone in the usual expression systems, although with rather unusual pharmacologies. Especially SAD, the 'functional *Drosophila* α-subunit' (also called Dα2), requires agonist concentrations in the 10 mM range and is not blocked by nicotinic antagonists like d-tubocurarine and α-bungarotoxin. Another interesting AChR has been partially cloned and sequenced from an Israeli poisonous snake [96]. This AChR does not bind and is not blocked by α-bungarotoxin. Its amino acid sequence especially in the neighborhood of the tandem cysteine residues is significantly different.

References

1. Bernard M.C. (1857) Leçon sur les Effets des Substances Toxiques et Medicamenteuses, p. 238–306 Baillière, Paris.
2. DuBois-Reymond, E. (1877) Gesammelte Abh. Allg. Muskel Nervenphys. 2, 700.
3. Langley, J.N. (1907–1908) J. Physiol. 36, 347–384.
4. Eccles, J.C. and Gibson, W.C. (197) Sherrington, His Life and Thought, Springer, Berlin, Heidelberg, New York.
5. Katz, B. (1969) The Release of Neural Transmitter Substances, Liverpool University Press, Liverpool.
6. Changeux, J.-P. (1981) Harvey Lect. 75, 85–254.
7. Karlin, A. (1991) Harvey Lect. 85, 71–107.
8. Conti-Tronconi, B.M. and Raftery, M.A. (1982) Annu. Rev. Biochem. 51, 491–530.
9. Neher, E. and Sakmann, B. (1976) Nature 160, 799–802.
10. Noda, M. Takahashi, H., Tanabe, T., Toyosato, M., Furutani, Y., Hirose, T., Asai, M., Inayama, S., and Numa, S. (1982) Nature 299, 793–797.
11. Karlin, A. (1967) J. Theor. Biol. 16, 306–320.
12. Changeux, J.-P., Devillers-Thiéry, A. and Chemouilli, P. (1984) Science 225, 1335–1345.
13. Maelicke, A. (1984) Angew. Chem. Int. 23, 195–221.
14. Katz, B. and Thesleff, S. (1957) J. Physiol. 138, 63–80.
15. Sakmann, B. and Neher, E. (Eds) (1983) Single Channel Recording, Plenum Press, New York.
16. Nachmansohn, D. (1959) Chemical and Molecular Basis of Nerve Activity, p. 235, Academic Press, New York.
17. Taylor, P. (1990) In: The Pharmacological Basis of Therapeutics (Goodman Gilman, A., Rall, T.W., Nies, AS. and Taylor, P., Eds) 8th edn, p. 122, Pergamon, New York.
18. Taylor, P. (1990) In: The Pharmacological Basis of Therapeutics, (Goodman Gilman, A., Rall, T.W., Nies, AS., Taylor, P., 8th edn, p. 166, Pergamon, New York.
19. Hucho, F. (1992) In: Handbook of Experimental Pharmacology, Vol. 102 – Selective Neurotoxicity (Herken, H. and Hucho, F., Eds) pp. 577–610, Springer, Berlin.
20. Changeux, J.-P. (1990) Fidia Research Foundation Neuroscience Award Lectures 4, 21–168.
21. Swanson, K.L. and Albuquerque, E.X. (1992) In: Handbook of Experimental Pharmacology, Vol. 102 – Selective Neurotoxicity (Herken, H. and Hucho, F., Eds) pp. 577–610, Springer, Berlin.
22. Endo, T. and Tamiya N. (1987) Pharmacol. Ther. 34, 403–451.
23. Menez, A., Morgat, J.L., Fromageot, P., Ronseray, A.M. Boquet, P. and Changeux, J.-P. (1971 FEBS Lett. 17, 333–338.
24. Duguid, J.R. and Raftery, M.A. (1973) Biochemistry 12, 3593–3597.
25. Low, B.W., Preston, H.S., Sato, A., Rosen, L.S., Searl, J.E., Rudko, A.D. and Richardson, J.R. (1976) Proc. Natl Acad. Sci. USA 73, 2991–2994.
26. Walkinshaw, M.D., Sanger, W. and Maelicke, A. (1980) Proc. Natl Acad. Sci. USA 77, 2400–2404.
27. Endo, T., Inagaki, F., Hayashi, K. and Miyazawa, T. (1981) Eur. J. Biochem. 122, 541–547.
28. Olsen, R., Meunier, J.C. and Changeux, J.-P. (1972) FEBS Lett. 28, 96–100.
29. Meunier, J.C., Sealock, R., Olsen, R. and Changeux, J.-P. (1974) Eur. J. Biochem. 45, 371–394.
30. Schmidt, T.J. and Raftery, M.A. (1972) Biochem. Biophys. Res. Commun. 49, 572.
31. Karlin, A. and Cowburn, D.A. (1973) Proc. Natl Acad. Sci USA 70, 3636–3640.

32. Eldefrawi, M.E. and Eldefrawi, A.T. (1973) Arch. Biochem. Biophys. 159, 362.
33. Hucho, F. (1986) Eur. J. Biochem. 158, 211–226.
34. Vandlen, R.L., Qu, W.C.-S., Eisenach, J.C. and Raftery, M.A. (1979) Biochemistry 18, 1845–1854.
35. Nomoto, H., Takahashi, N., Nagaki, Y., Endo, S., Arata, Y. and Hayashi, K. (1986) Eur. J. Biochem. 157, 133–142.
36. Poulter, L., Earnest, J.P., Stroud, R.M. and Burlingame, A.L. (1989) Proc. Natl Acad. Sci. USA 86, 6645–6649.
37. Huganir, R.L. and Greengard, P. (1987) TINS 8, 472–477.
38. Rübsamen, H., Eldefrawi, A.T., Eldefrawi, M.E. and Hess, G.P. (1978) Biochemistry 17, 3818–3825.
39. Chang, H.W. and Neumann, E. (1976) Proc. Natl Acad. Sci. USA 73, 3364–3368.
40. Karlin, A., Hiltzman, E., Yodh, N., Lobel, P., Wall, J. and Hainfeld, J. (1983) J. Biol. Chem. 258, 6678–6681.
41. Nickel, E. and Potter, L.T. (1973) Brain Res. 57, 508–517.
42. Kunath, W., Giersig, M. and Hucho, F. (1989) Electron Microsc. Rev. 2, 349–366.
43. Brisson, A. and Unwin, P.N.T. (1985) Nature 315, 474–477.
44. Toyoshima, C. and Unwin, N. (1988) Nature 336, 214–250.
45. Noda, M., Takahashi, H., Tanabe, T., Toyosato, M., Kikyotani, S., Furutani, Y., Horose, T., Takashima, H., Inayama, S., Miyata, T. and Numa, S. (1983) Nature 302, 528–532.
46. Noda, M., Takahashi, H., Tanabe, T., Toyosato, M., Kikyotani, S., Horose, T., Asai, M., Takashima, H., Inayama, S., Miyata, T. and Numa, S. (1983) Nature 301, 251–255.
47. Frail, D.E., Mudd, J., Shah, V., Carr, C., Cohen, J. and Merlie, J.P. (1987) Proc. Natl Acad. Sci. USA 84, 6302–6306.
48. Raftery, M.A., Hunkapiller, M., Strader, C.D. and Hood, L.E. (1980) Science 208, 1454–1457.
49. Grenningloh, G., Riehnitz, A., Schmitt, B., Methfessel, C., Zensen, M., Bevreuther, K., Gundelfinger, E.D. and Betz, H. (198) Nature 328, 215–220.
50. Schofield, P.R., Darlison, M.G., Fujita, N., Burt, D.R., Stephenson, F.A., Rodriguez, H., Rhee, L.M., Ramachandran, J., Reale, V., Glencorse, T.A., Seeburg, P.H. and Barnard, E.A. (1987) Nature 324, 221–227.
51. Claudio, T. (1989) In: Frontiers in Molecular Biology: Molecular Neurobiology Volume, (Glover, D.M. and Hames, B.C., Eds) pp. 63–142, IRL, Oxford Univ. Press, Oxford.
52. Betz, H. (1990) Neuron 5, 383–392.
53. Finer-Moore, J. and Stroud, R.M. (1984) Proc. Natl Acad. Sci USA 81, 155–159.
54. Kao, P.N. and Karlin, A. (1986) J. Biol. Chem. 261, 8085–8088.
55. Naumann, D., Schultz, C., Görne-Tschelnokow, U. and Hucho, F. (1992) Biochemistry 32, 3162–3168.
56. Claudio, T., Ballivet, M., Patrick, J. and Heinemann, S. (1983) Proc. Natl Acad. Sci USA 80, 1111–1115.
57. Maelicke, A. (1988) In: Handbook of Experimental Pharmacology, (Whittaker, V.P., Ed) Vol. 86, pp. 267–313, Springer, Berlin, Heidelberg, New York.
58. Criado, M., Hochschwender, S., Sarin, V., Fox, J.L. and Lindstrom, J. (1985) Proc. Natl Acad. Sci. USA 82, 2004–2008.
59. Weill, C.L., McNamee, M.G. and Karlin, A. (1974) Biochem. Biophys. Res. Commun. 61, 997–1003.
60. Mishina, M., Tobimatsu, T., Imoto, K., Noda, M., Takahashi, T., Wuma, S., Methfessel, C. and Sakmann, B. (1985) Nature 313, 364–369.
61. Wilson, P.T., Gershoni, J.M., Hawrot, E. Lentz, T.L. (1984) Proc. Natl Acad. Sci. USA 81, 2553–2557.
62. Neumann, D., Barchan, D., Safran, A., Gershoni, J.M. and Fuchs, S. (1986) Proc. Natl Acad. Sci. USA 83, 3008–3011.
63. Conti-Tronconi, B.M., Tang, F., Diethelm, B.M., Spencer, S.R., Reinhardt-Maelicke, S. and Maelicke, A. (1990) Biochemistry 29, 6221–6230.

64. Hucho, F. (1979) FEBS Lett. 103, 27–32.
65. Witzemann, V., Muchmore, D. and Raftery M.A. (1979) Biochemistry 18, 5511–5518.
66. Kreienkamp, H.-J., Utkin, Y.N., Weise, C., Machold, J., Tsetlin, V.I. and Hucho, F. (1992) Biochemistry 31, 8235–8244.
67. Abramson, S.N., Li Y., Culver, P. and Taylor, P. (1989) J. Biol. Chem. 253, 12666–12672.
68. Galzi, J.L., Revah, F., Black, D., Goeldner, M., Hirth, C. and Changeux, J.-P. (1990) J. Biol. Chem. 265, 10430–10437.
69. Dennis, M., Giraudat, J., Kotzyba-Hibert, F, Goeldner, M., Hirth, C., Chang, J.Y., Lazure, C., Chretién, M. and Changeux, J.-P. (1988) Biochemistry 27, 2346–2357.
70. Herz, J.M., Johnson, D.A. and Taylor, P. (1989) J. Biol. Chem. 264, 12439–12448.
71. Adams, D.J., Dwyer, T.M. and Hille, B. (1980) J. Gen. Physiol. 75, 493–510.
72. Huang, L.Y.M., Catterall, W.A. and Ehrenstein, G. (1978) J. Gen. Physiol. 71, 397–410.
73. Hille, B. (1992) Ionic Channels of Excitable Membranes. 2nd edn. Sinauer Press, Massachusetts.
74. Blount, P. and Merlie, J.P. (1989) Neuron 3, 349–357.
75. Hucho, F, Oberthür, W. and Lottspeich, F. (1986) FEBS Lett 205, 137–142.
76. Hucho, F. and Hilgenfeld, R. (1989) FEBS Lett. 257, 17–23.
77. Imoto, K., Methfessel, C., Sakmann, B., Mishina, M., Konno, T., Nakai, J., Bujo, H., Mori, Y., Fukuda, K. and Numa, S. (1988) Nature 324, 670–674.
78. Imoto, K., Busch, C., Sakmann, B., Mishina, M., Konno, T., Nakai, J., Bujo, H., Mori, Y., Fukuda, K. and Numa, S. (1988) Nature 335, 645–648.
79. Saitoh, T. and Changeux, J.-P. (1980) Eur. J. Biochem. 105, 51–62.
80. Ross, A.F., Rapuano, M., Schmidt, J.H. and Prives, J.M. (1987) J. Biol. Chem. 262, 14640–14647.
81. Green, W.N., Ross, A.F. and Claudio, T. (1991) Proc. Natl Acad. Sci USA 88, 854–858.
82. Huganir, R.L., Delcour, A.H., Greengard, P. and Hess, G.P. (1986) Nature 321, 744–776.
83. Takai, T., Noda, M., Mishina, M., Shimizu, S., Furutani, Y., Kayano, T., Ikeda, T., Kubo, T., Takahashi, H., Takahashi, T., Kuno, M. and Numa, S. (1985) Nature 315, 761–764.
84. Witzemann, V., Stein, E., Barg, B., Konno, T., Koenen, M., Kues, W., Criado, M., Hofmann, M. and Sakmann, B. (1990) Eur. J. Biochem. 194, 437–448.
85. Laufer, R. and Changeux, J.-P. (1989) Mol. Neurobiol. 3, 1–53.
86. Reist, N.E., Werle, M.J. and McMahan, U.J. (1992) Neuron 8, 865–868.
87. Harris, D.A., Falls, D.L., Johnson, F.A. and Fischbach, G.D. (1991) Proc. Natl Acad. Sci. USA 88, 7664–7668.
88. Deneris, E.S., Connolly, J., Rogers, S.W. and Duvoisin, R. (1991) TIPS 12, 34–40.
89. Chiappinelli, V.A. (1983) Brain Res. 277, 9–21.
90. Couturier, S., Bertrand, D., Matter, J.-M., Hernandez, M.-C., Bertrand, S., Millar, N., Valera, S., Barkas, T. and Ballivet, M. (1990) Neuron 5, 847–856.
91. Quick, M. and Geertsen, S. (1988) Can. J. Physiol. Pharmacol. 66, 971–979.
92. Clementi, F., Cabrini, D., Gotti, C. and Sher, E. (1986) J. Neurochem. 47, 291–297.
93. Chini, B., Clementi, F., Hukovic, N. and Sher, E. (1992) Proc. Natl. Acad. Sci. USA 89, 1572–1576.
94. Gundelfinger, E.D. (1992) TINS 15, 206–211.
95. Hanke, W. and Breer, H. (1986) Nature 321, 171–174.
96. Neumann, D., Barchan, D., Horowitz, M., Kochva, E. and Fuchs, S. (1989) Proc. Natl Acad. Sci. USA 86, 7255–7259.
97. Fornasari, D., Chini, B., Tarroni, P. and Clementi, F. (1990) Neurosci. Lett. 111, 351–356.
98. Schoepfer, R., Conroy, W., Gore, M. and Lindstrom, J. (1990) Neuron 5, 35–48.

F. Hucho (Ed.) *Neurotransmitter Receptors*
© 1993 Elsevier Science Publishers B.V. All rights reserved.

CHAPTER 5

The β-adrenoceptors

MARTIN J. LOHSE[1], RUTH H. STRASSER[2] and ERNST J.M. HELMREICH[3]

[1]*Laboratorium für Molekulare Biologie, Genzentrum der Universität München,* [2]*Medizinische Klinik der Universität Heidelberg,* [3]*Physiologische Chemie I, Theodor-Boveri-Institut für Biowissenschaften (Biozentrum) der Universität Würzburg, Germany*

1. Introduction

Signal transmission from β-adrenoceptors to stimulatory G-protein (G_s) and adenylyl cyclase and signal transmission in the course of visual excitation in retinal rod outer segment membranes are the best-studied examples of G-protein-coupled receptor actions. As a fact, the β-adrenoceptor has become a paradigm for all hormone receptors where signals are transmitted and amplified by G-proteins. In this review an attempt was made to cover important aspects of β-adrenoceptor function, structure (E.J.M.H.) and regulation, including desensitization mechanisms (M.J.L.), both under physiological and pathophysiological conditions and pharmacological and clinical aspects (R.H.S.). The broad range of this survey and the vast amount of data which had to be considered made it inevitable that only certain aspects could be dealt with. We are aware that the selection of the topics is somewhat arbitrary. The reasons for this bias are to be found in the research interests of the authors. Although, no attempt was made to cover the field in a comprehensive fashion, the literature rele-

---

Abbreviations: βAR, β-adrenergic receptor, β-adrenoceptor; Protein kinases [E.C. 2.7.1.37]; βARK, specific β-adrenoceptor kinase. PKA, Protein kinase A, cyclic AMP-dependent protein kinase; PKC, Protein kinase C, Diacylglycerol activated protein kinase; Adenylyl cyclase [E.C. 4.6.1.1.]; Phospholipase C [E.C. 3.1.4.3]; M_1-, M_2- and M_3-receptors, muscarinic acetylcholine receptors 1, 2 and 3, respectively. TSH, Thyroid stimulating hormone. G-proteins, heterotrimeric GTP-binding proteins composed of α-, β- and γ-subunits; Gs, stimulatory G-protein; Gi, inhibitory G-protein; Go, brain-derived G-protein; Gt, transducin, rhodopsin-linked G-protein. GAP, GTPase-activating protein; GTP[γS], guanosine 5'-0-(3-thiotriphosphate); Gpp[NH]p, guanosine imidodiphosphate; CGP-12177 is a β-adrenergic antagonist and the Ciba Geigy product 12177, 4-[3-(*t*-butyl-amino)-2-hydroxypropoxy]benzimidazol-2-on. Iodocyanopindolol and carazolol,1-(9H-carbazol-4-yloxy)-3-[(1-methylethyl)-amino]-2-propanol are β- adrenergic blockers. Isoproterenol is a β-adrenergic agonist. HPLC; High performance liquid chromatography. CHO-cells; Chinese hamster ovary cells. EGF, epidermal growth factor.

vant to the topics under discussion was considered through 1991. Hormones and their actions have been dealt with before [1].

2. Pharmacological characterization

2.1. β_1-, β_2-, β_3-adrenoceptors

Based on the diversity of response produced in different tissues by catecholamines, Ahlquist in 1948 [2] presented a conceptual framework whereby receptors for adrenergic compounds were subdivided into two classes, α- and β-adrenergic receptors, which were differentiated on the basis of the relative potencies of different catecholamine compounds in eliciting a given physiological response. He observed, for example, that epinephrine was more potent than norepinephrine, which in turn was much more potent than isoproterenol in bringing about vasoconstriction in smooth muscle. In contrast, other responses such as positive inotropic effects in heart or vasodilation were elicited more readily by isoproterenol than by epinephrine or norepinephrine. Thus the former effects were postulated to be mediated by α-adreno- and the latter by β-adrenoceptors. By further refining the scale of relative potencies of β-adrenergic agents according to their lipolytic, cardiac, bronchodilatory, and vasodilatory effects, Lands et al. [3] distinguished two subtypes of β-adrenergic receptors, termed β_1- and β_2-adrenoceptors. According to this subclassification, norepinephrine and epinephrine are nearly equipotent agonists for β_1-adrenoceptors, whereas epinephrine is much more potent than norepinephrine at β_2-adrenergic receptors.

A heterogeneous population of β-adrenergic receptors which can be differentiated pharmacologically may coexist in the same tissue. The first evidence of such coexistence was obtained by Ablad et al. [4] who demonstrated that a β_1-selective antagonist is more potent in inhibiting the positive inotropic effect in the heart mediated by norepinephrine than by epinephrine. Heterogeneity of the receptor population in human heart has been demonstrated by radioligand-binding assays with selective β-antagonists [5–7]. Radioligand-binding techniques and Northern blot analyses [8] have also made it possible to demonstrate that a single cell is capable of expressing both β_1- and β_2-receptor subtypes. In the heart, both β_1- and β_2-adrenergic receptors contribute to the positive inotropic and positive chronotropic effects of β-agonists [9]. Different locations may be responsible for different effects. For example, β_2-adrenergic receptors in human atria constitute about 30–40% of all β-adrenergic receptors, but only about 20% in ventricles. Both, β_1- and β_2-receptor subtypes, if tested separately, can mediate positive chronotropy in atria and positive inotropy mainly in ventricles. Consequently, a β_1-agonist elicits mainly inotropic activity in ventricles, whereas a β_2-agonist causes more chronotropic effects through its action in atria, where β_2-receptors are predominantly localized [10,11]. Using synthetic β-agonists and β-antagonists, an additional subtype of β-adrenergic receptors has been identified in white adipocytes. Originally, this receptor was classified as a $β_1$-adrenergic receptor with equal potency for epinephrine and norepinephrine [3], but epinephrine-induced lipolysis and cAMP accumulation in adipocytes were blocked equally well with both $β_1$- and $β_2$-specific antagonists, when applied at high concentrations. The atypically low affinities of $β_1$- and $β_2$-selective antagonists for β-adrenergic receptors in adipocytes [12] were combined with low degree of stereoselectivity [13–15]. Finally, the application of new synthetic β-agonists which selectively activate brown and white adipocytes clearly indicated that an atypical β-adrenergic receptor is expressed in these cells, which is mainly responsible for lipolysis, also in human adipocytes [16–20]. Recently, an atypical β-adrenergic receptor referred to as $β_3$-adrenoceptor was cloned, sequenced, and expressed in CHO cells by Emorine et al. [21]. While this $β_3$-adrenergic receptor shares many features with the β-adrenergic receptors of adipocytes, it also differs in some aspects, such as affinity for the β-antagonist [^{125}I]iodocyanopindolol [21,22]. The low affinity of the $β_3$-adrenergic receptor compared with that of classical β-adrenergic receptors for the endogenous naturally occurring catecholamines raises the question whether it is of physiological significance. For example, this receptor might come into play at high levels of circulating catecholamines, which accompany pathophysiological situations such as prolonged shock or are found under physiological conditions in the synaptic cleft of sympathetic neurons. Moreover, it should be noted that the $β_3$-adrenergic receptor lacks phosphorylation sites for the cAMP-dependent (A) protein kinase and for the β-adrenergic receptor kinase (βARK) [21]. This could enable this type of receptor to escape desensitization even when exposed to high concentrations of catecholamines. Actually, different tissue expression and affinity and differences in regulatory response apply to other β-receptor subtypes as well and play an important role in pathophysiological conditions, such as hyperthyroidism. (see: Section 5.2.1.).

Pharmacological information of this kind is undoubtedly of great value for pharmaceutical research and for the clinician who can take advantage of the selective expression and specific properties of receptor subtypes by using highly discriminating antagonists in his therapy. However, the uncertainties in the relationship between chemical structures and activities of adrenoceptor ligands caution against relying too much on pharmacological classifications, based on selective ligand-binding properties of the receptor (see Boege et al. [23]). One should remember that differential effects of β-agonists may not be solely determined by distinct receptor subtypes, but might reflect events at some postreceptor step in the signal transmission chain. Moreover, recent evidence indicates that one and the same receptor activated by one type of agonist may have multiple functions (see for examples, Ashkenazi et al. [24], Rooney et al. [25]). Thus, the functional correlates of a pharmacological differentiation into distinct subtypes of β-adrenergic receptors are not yet sufficiently clarified. In the next section, an attempt is made to analyze the structure and function relationships of β-adrenoceptors on a molecular level.

3. Function and structures

3.1. Function

3.1.1. Activation of G-proteins

Two models are currently under discussion to explain how β-receptors activate GTP-binding proteins (cf: Birnbaumer et al. [26]). One model visualizes the role of the receptor as the catalyst of the exchange of GDP bound to the α-subunit of the heterotrimeric $\alpha\beta\gamma$-G-protein complex by GTP. This model was originally introduced by Cassel and Selinger [27,28] who discovered the hormonal stimulation of GTPase activity of G-proteins. The model implies that once GTP is bound and the active species of G-proteins has been formed, the receptor is no longer required. The other model incorporates subunit dissociation as an essential feature of G-protein activation (cf: [29]). It assumes that GDP-GTP exchange requires formation of a receptor-G-protein complex, an assumption which is also implicit in the Cassel-Selinger model.

An indirect proof for the interaction between β-receptors and G-proteins was first provided by the high affinity state of the receptor for agonists formed in the presence of G-proteins. In membranes (but *without* addition of GTP, GDP, GMP, or analogs) β-receptors exhibit high affinity for the agonist [30]. On addition of guanine nucleotides or analogs, a shift from the high to a low affinity state was observed [31]. The low affinity state is also formed in the course of receptor purification, that is in the absence of G-proteins [32]. Consequently, on reconstitution in phospholipid vesicles of isolated low affinity receptors and G_s, the high affinity form of the receptor was reformed [33]. Moreover, several experiments have shown that nonactivated G-proteins are mostly complexed with GDP [34] and that agonist-receptor complexes promote GDP-GTP exchange [28,35]. However, the dissociation model postulates furthermore that on binding GTP the activated $G\alpha\beta\gamma$-complex dissociates in the GTP-bound α-subunit and the $\beta\gamma$-complex. This postulate is supported by the finding that purified G-proteins activated by the binding of nonhydrolyzable GTP analogs, such as Gpp(NH)p, GTPγS and concentrations of Mg^{2+} ions in the mM range or $[AlF_4]^-$-Mg^{2+} and GDP which mimics GTP as shown by Pfister et al. [36], dissociate into α-subunits and $\beta\gamma$-subunits [37–41]. Moreover, there is evidence to suggest that $G_{s\alpha}$-subunits are associated with purified adenylyl cyclase [42–44]. $G_{s\alpha}$- subunits were also shown to bind to dihydropyridine-sensitive skeletal muscle Ca^{2+}- channels in a reconstituted membrane system [45,46]. Asano et al. [47] and Brandt and Ross [48] have inferred from the rates of GTP[γS] binding in the presence of β-receptors reconstituted in liposomes, that the activated receptors facilitate the release of GDP and the binding of GTP to receptor-bound G_s and promote G_s activation. However, there still remain many open questions before a convincing molecular activation model of G-proteins can be formulated. For example, the role of the $\beta\gamma$- subunits is not clear: $\beta\gamma$-dimers have been shown to stimulate at low concentrations and inhibit at high concentrations the receptor-mediated activation of inhibitory and stimulatory G-proteins in membranes and in reconstituted vesicles [49–52]. A molar excess of $\beta\gamma$ also delays the rate of activation of purified G_s by GTP[γS] [53]. Birnbaumer et al. [26] have suggested that for this to occur $G_{s\alpha}$ must dissociate from $\beta\gamma$ on activation in order to be able to activate the effector. Thus, the inhibition by $\beta\gamma$-subunits is thought to result from reformation of the inactive G-$\alpha\beta\gamma$ holocomplex. But, excess $\beta\gamma$-subunits might also prevent release of α-subunits from a receptor-$\beta\gamma$ complex (see scheme) or might reflect inhibition of adenylyl cyclase [54]. Kurstjens et al. [55] have shown that bovine brain $\beta\gamma$-subunits reconstituted in liposomes even promote the binding of α_o-subunits to the non activated β-adrenoceptor from turkey erythrocytes. Although, G_o and a β-receptor are unlikely to be physiological coupling partners, this interaction resulted in stimulation of basal GTPase activity, suggesting that the basal activity in the hormonally nonactivated state might be due to a precoupled G-protein-β-receptor complex. These findings are reminiscent of observations by Cerione et al. [56] who have reported on stimulation of GTPase activity on coreconstitution of G_i with β-adrenoceptor in lipid vesicles. Although, it has been suggested before, that a receptor-G-protein complex may not dissociate, at least not until the activated G-protein meets its target [57–59], there is evidence [42–46] that an activated α-GTP-subunit binds to an appropriate effector and activates it. For a more recent review, see Birnbaumer [352]. Whether $\beta\gamma$-subunits interact directly with target enzymes such as adenylyl cyclase and other effector systems remains an open question [60,61]. A controversy concerning the role of $\beta\gamma$-subunits of G_{i3} (or G_k) in the regulation of muscarinic acetylcholine receptor-gated K^+-channels has received much attention [62–65]. Tang and Gilman [54] have recently reported that $\beta\gamma$-subunits are capable of modulating the activity of certain adenylyl cyclases in the presence of $G_{s\alpha}$ which is the α-subunit of the G-protein that activates adenylyl cyclase. One form of adenylyl cyclase was inhibited by $\beta\gamma$, whereas others were either activated or not affected by it.

The following scheme is a modification of a presently widely accepted model for the activation of G-proteins and incorporates interactions between $\beta\gamma$-subunits, β-receptors, and adenylyl cyclases. It shows the activation and shutoff of the β-adrenoceptor-stimulated adenylyl cyclase system.

$$R + G_{s\alpha\beta\gamma} \cdot GDP \underset{}{\overset{[\beta\gamma]}{\rightleftharpoons}} R \cdot G_{s\alpha\beta\gamma} \cdot GDP;$$

$$R \cdot G_{s\alpha\beta\gamma} \cdot GDP \underset{}{\overset{\substack{\text{Hormone} \\ + \\ \text{GTP/Mg}^{2+}}}{\rightleftharpoons}} R^* \cdot G^*_{\alpha\beta\gamma} \cdot GTP \underset{}{\overset{[\beta\gamma]}{\rightleftharpoons}} R^* \cdot \beta\gamma + G^*_{s\alpha} \cdot GTP;$$

$$G^*_{s\alpha} \cdot GTP \cdot C \underset{}{\overset{[\beta\gamma]}{\rightleftharpoons}} G^*_{s\alpha} \cdot GTP \cdot C^* \cdot \beta\gamma \underset{}{\overset{[\beta\gamma]}{\rightleftharpoons}} C \cdot \beta\gamma \rightleftharpoons C + \beta\gamma + G_{s\alpha} \cdot GDP + P_i;$$

$$G_{s\alpha} \cdot GDP \underset{}{\overset{[\beta\gamma]}{\rightleftharpoons}} G_{s\alpha\beta\gamma} \cdot GDP;$$

R and R* are non-activated and hormone-activated β-adrenoceptors; G and G* are non-activated and activated stimulatory G_s-proteins; C and C* are non-activated and activated adenylyl cyclase, respectively; and [$\beta\gamma$] signifies the steps affected by addition of $\beta\gamma$-subunits [52,54,55,66].

An attempt was made recently using fluorescence resonance energy transfer to estimate the affinities of α_o- and $\beta\gamma$-subunits for each other and of $\beta\gamma$-subunits for the turkey-erythrocyte β-receptor [41]. The affinities of isolated, bovine brain α_o- and $\beta\gamma$-subunits for each other were about 5–10 times stronger, both in lipid vesicles and in detergent solutions, than the affinities of the non-activated and the activated β-receptor for $\beta\gamma$. The high affinity interactions between α_o- and $\beta\gamma$-subunits were nearly completely abolished in the presence of GTP[γS] and mM concentrations of Mg^{2+} ions, whereas the interactions between the β-receptor and the $\beta\gamma$-subunits were not affected. It seems therefore that the promotion of β-receptor-mediated activation of GTPase activity of G_s by $\beta\gamma$-subunits in reconstituted proteoliposomes is a consequence of an increase in the rate of coupling of the receptor with the α subunit of the G-protein in the presence of $\beta\gamma$-subunits [52]. An excess of $\beta\gamma$-subunits might help maintain a β-receptor-$\beta\gamma$ complex in competition with the G-protein complex. To sum up: These recent findings make it attractive to consider the possibility that β- and γ-subunits of G-proteins may have slightly different functions. Whereas, the $\beta\gamma$-subunits can bind both to their corresponding α-subunits and to receptors and effectors, the γ-subunits which are isoprenylated and anchored to the membrane [353] may, in addition, act like a swivel to transfer α- and β-subunits from receptor to effector, thus facilitating the interactions with both these membrane-spanning proteins. In order to prove this proposition, one would need quantitative data for the affinities of all the mutually interacting partners, and in addition, one would need to know the actual concentrations of the various donor-acceptor couples in the cell. Further evidence for a role of $\beta\gamma$-subunits in anchoring α-subunits to receptors has recently been presented [360,361]. Moreover, it was shown that differences exist between β- and γ-subunits with respect to the specificity of receptor-G-protein-coupling [362,363].

3.1.2. The catalytic function of the β-receptor
Levitzki and Helmreich [67] discussed possible modes of receptor-G-protein-effector coupling. Some of these ideas are still worth considering. The proposed precoupled model included components permanently coupled to each other like regulatory and catalytic domains in aspartate transcarbamoylase, whereas the model where receptor and holoenzyme (G-protein-catalyst complex) dissociate when the receptor becomes activated resembled cAMP-dependent protein kinase where regulatory and catalytic parts dissociate on activation. There are variations of this theme. In the floating receptor model, first suggested by Cuatrecasas et al. [68], equilibria for hormone-activated and non-activated receptors and for free and receptor-G-protein-coupled catalysts were introduced. The collision-coupling model by Levitzki and colleagues [57,58] became the most popular model of that sort. It predicts that the rate constant

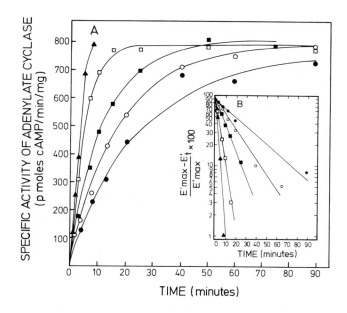

Fig. 1. Determination by receptor number of the rate of activation of adenylylcyclase (G_s) by guanine nucleotide. Turkey erythrocyte membranes with 1.12 pmol β-adrenoceptors per mg (▲) were alkylated, leading to loss of 42 (□), 70 (■), 87 (○), and 91% (●) of receptors. The corresponding fractional rates of activation of G_s by Gpp(NH)p, 0.374 min^{-1} in nonalkylated membranes, were reduced by 42, 52.4, 87.3, and 92%, respectively. This finding indicates that receptors act catalytically rather than stoichiometrically. A stoichiometric mode of action predicts that as receptors are lost, fewer G_s molecules are activated. (Reproduced from Tolkovsky and Levitzki [57] by courtesy of the authors and the Editors of *Biochemistry*.)

of enzyme (adenylyl cyclase) activation is proportional to the activated receptor concentration, whereas the maximal number of catalytic units that are eventually activated is independent of it. As shown in the Figure 1, reproduced from [57], this prediction was supported by experiments where the rate of activation of adenylyl cyclase by G_s, irreversibly activated with the nonhydrolyzable analog Gpp(NH)p, was proportional to the number of isoproterenol-activated receptor molecules. These experiments were carried out with turkey erythrocyte membranes where the number of receptors was stepwise reduced by irreversible inactivation with an alkylating receptor ligand. These experiments support a catalytic function of the receptor, since as the number of receptors was lowered the rate of activation declined correspondingly, but the remaining intact receptors were still capable of activating all the target molecules (G_s and adenylyl cyclase) in the membrane. Levitzki and colleagues [57,58] have emphasized the point that the committed step in the collision-coupling model is the encounter (k_3) of the hormone-receptor-complex with its coupling partner G-protein (and eventually catalyst) since $k_5 \gg k_4$.

$$H + R \xrightleftharpoons[K_H]{} H \cdot R + G_s \xrightleftharpoons[k_4]{} [HR \cdot G_s \xrightleftharpoons[]{k_3} HR \cdot G^*]$$

$$\xrightarrow{k_5} HR + G^* \cdot GTP \xrightarrow{k_6} G_s \cdot GDP + P_i;$$

Subsequent experiments by Citri and Schramm [69] showed also that the primary function of the hormone-activated β-adrenoceptor is the catalytic activation of the G-protein, G_s, but they did not support the idea that the encounter of HR with G_s (k_3) is the committed step. An attempt was made to directly verify collision coupling by studying the lateral mobility of β-adrenoceptor in the membranes of secondary liver cells in culture using the fluorescence recovery after photobleaching [FRAP] technique [70]. The conclusion was that the majority of the β-adrenoceptors is immobilized in the membrane over a temperature range of 4–37°C. However, the possibility was considered that the G-protein and the catalytic units may be sequestered in the same local regions in the membrane as the β-receptors, so that lateral diffusion over micrometer distances (measured by FRAP) may not be required and lateral or rotational diffusion over submicrometer distances may suffice for activation. Lohse et al. [354,355] have recently re-investigated the mechanism of receptor activation and coupling in human platelet membranes using the technique of Tolkovsky et al. [57,58] of irreversibly blocking receptors, in this case A2 adenosine receptors, with a photoaffinity label. Based on these experiments, a model was proposed [355] which incorporates not only the catalytic activation of receptors emphasized by Tolkovsky et al. [57,58], but also a restricted membrane domain where the collision between receptor and effector is thought to occur. Therefore, the question of the role of mobility of the components of the β-adrenergic signal-transmission chain remains open (see also [71]).

3.1.3. Amplification
There is little doubt that receptors can act as catalysts in G-protein activation and that one receptor molecule can activate more than one G-protein. This is also supported by experiments with purified components incorporated into phospholipid vesicles [72–74]. Asano et al. [47] calculated that one receptor promoted the binding of GTP[γS] to about 10 molecules of G_s. However, much more is known about response amplification in the visual system. The fact that rhodopsin, in contrast to β-adrenoceptors, carries an intrinsic signal which sensitively records activation is mainly responsible for that. Liebman et al. [75] (see also Stryer [76]) have reasoned that a single bleached rhodopsin molecule could collide with more than a thousand transducin molecules per second and if each rhodopsin-transducin collision was productive, about 1000 activated transducin · GTP molecules would be formed per second per each activated rhodopsin molecule. This estimate rests on the property of the activat-

ing rhodopsin species, Meta II-rhodopsin, which like an activated β-receptor has a rather long (20–30 s) lifetime. For the next step, it is assumed that about 10–20 activated transducin·GTP molecules are required to activate one molecule of phosphodiesterase resulting in a gain in the activity of the phosphodiesterase by about 100-fold, which might be the same or only slightly greater than the gain of adenylylcyclase activity. Hence, the much greater gain in the efficacy of visual signal transmission in comparison with adenylyl cyclase activation is mainly due to the much greater turnover of cGMP-phosphodiesterase, which hydrolyzes 1000–2000 cyclic GMP molecules per second per each molecule of activated phosphodiesterase compared to only about 20–40 molecules of cAMP produced per second by the activated adenylyl cyclase [42,43,72,77].

The regulation of the rhodopsin-regulated cGMP-phosphodiesterase activity is better understood than that of β-receptor-activated adenylyl cyclase. In the case of visual signal transduction, the kinetics of activation of cGMP-phosphodiesterase is compatible with the physiological response to light [78,79]. As in the case with hormonal signal transduction, the turnoff involves three different moieties: the activated receptor (R*); the G_α^* (T_α^*)-activated species of the G-protein (G_α^* (T_α^*) · GTP); and the activated form of the effector which is the $\alpha\beta$-form of phosphodiesterase in visual signal transfer. Rhodopsin, like β-adrenoceptor, is assumed to be deactivated by phosphorylation and interaction with arrestin [80]. This is discussed in Section 4.1.1. T_α · GTP is thought to remove the inhibitory γ-subunit from the phosphodiesterase and to release it again when T_α^* · GTP is deactivated by its own GTPase, which itself is activated by the interaction with the effector molecule. The released inhibitory γ-subunit is now free to recombine with the $\alpha\beta$-phosphodiesterase forming the inactive $\alpha\beta\gamma$-complex.

Related with deactivation but distinctly different is the recovery process which is not discussed here. For information on this topic, see references [80–83]. What is discussed briefly, is the fact that the rates of transducin GTPase, like that of all heterotrimeric G-proteins, are very slow; these GTPases hydrolyze only a few molecules of GTP per min [84–88]. Based on these rates, the T_α^* · GTP complex would be rather long-lived with a lifetime of tens of seconds and correspondingly phosphodiesterases would be active much too long which would be incompatible with the physiological light response. In order to find an explanation for this discrepancy, Vuong and Chabre [89,90] have carried out time-resolved microcalorimetric measurements of GTP hydrolysis with flash-activated rod outer-segment suspensions, which led them to conclude, in accordance with Arshavsky et al. [91], that the slow GTP hydrolysis is greatly enhanced once T_α^* · GTP associates with the effector. Thus, cGMP-phosphodiesterase, is believed to act like a GTPase activating protein, GAP, which was shown to stimulate the GTPase of ras-p21 (cf: [92]). The GTPase activity of G_s in the β-receptor-adenylyl cyclase system might be regulated in a similar but inhibitory fashion by interaction with phosducin (cf: Bauer et al. [93]).

3.2. Structures

There are at least three β-receptor subtypes, β_1, β_2, β_3, in mammalian cells which all activate G_s and adenylyl cyclase. In nonmammalian systems β-adrenoceptors have been cloned and sequenced from frog and turkey erythrocytes. The first 55,000-fold purification of a β-adrenoceptor was achieved with frog erythrocyte membranes solubilized with digitonin [94]. The next purification reported was that of β_1-adrenoceptor from turkey erythrocytes again by affinity chromatography but combined with HPLC [95]. Other laboratories have followed [96]. The purified β-adrenergic receptor from turkey erythrocytes was found to consist of two equally active components, a larger one with an apparent M_r of about 50 kDa and a smaller one with Mr of about 40 kDa, which could be separated electrophoretically [95,97].

The list of adrenoceptors which have been cloned and with the primary sequence determined includes the human β_1-adrenoceptor [98], the β_1-adrenoceptor from rats [99], the hamster β_2-receptor [100], the human β_2-receptor [101], the mouse β_2-receptor [102], and the rat β_2-receptor [103,104]. A gene for a human β_3-adrenoceptor was isolated and sequenced and expressed in Chinese hamster ovary cells [21]. In differentiated adipocytes a 2.3 kilobase mRNA was detected which hybridized with a human β_3-adrenoceptor probe. Therefore, the predominant β-adrenoceptor population in 3T3 adipocytes, which is also found in mouse white adipose tissue, appears to be the β_3-receptor [22].

Henderson and Unwin [105] originally visualized seven transmembrane α-helices in bacteriorhodopsin by low-energy electron diffraction at 0.6 nm resolution. Henderson et al. [106] have more recently presented a refined structure at 0.35 nm resolution. Accordingly, for β-adrenoceptors and rhodopsin and a great number of other G-protein-linked receptors, a structure with seven highly conserved transmembranous segments, each with 20–28 hydrophobic amino acids linked by three external and three (or four) cytoplasmic loops has been proposed (see Fig. 2). Recently, a projection structure of rhodopsin was obtained at 9Å resolution; this structure is in general agreement with the previously proposed models [364]. The transmembrane sequences of all receptors with seven transmembrane helices are usually quite similar, whereas the N- and C-terminal regions and the large cytoplasmic loop connecting transmembrane segments V and VI are different. Therefore, it is necessary to see which functions are common to all receptors of this type and which are different and to find out how they are related to structural variations. For such studies receptor mutants and chimeric receptors were constructed and expressed. The results of these experiments [107] have recently been discussed [108].

The overall sequence similarity among adrenoceptors is about 49% and about 70% when only the transmembrane β-helical segments are considered. Each member of this receptor class binds to specific ligands with different affinities. Moreover, β-adrenoceptors are coupled to G_s whereby adenylyl cyclase is activated, while α_2-receptors are coupled to G_i where activation results in inhibition of adenylyl cyclase.

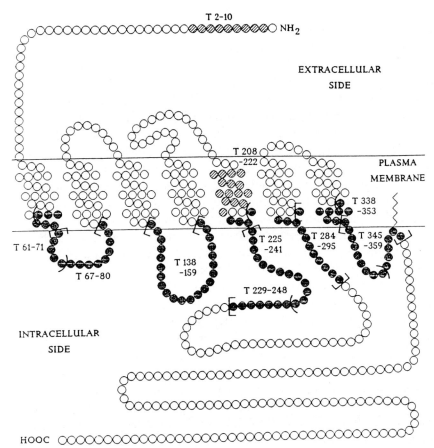

Fig. 2. Location of synthetic peptides deduced from the structure of the turkey β-adrenoceptor. The model illustrating an additional intracellular loop anchored at Cys 358 is adapted from Yarden et al. [351] and O'Dowd et al. [142]. Filled circles within round or square brackets delineate the sequence of amino acid residues from which the synthetic peptides were derived. The sequences of synthetic peptides serving as controls are marked by hatched circles. (Reproduced from Muench et al. [127] by courtesy of the authors and the Editors of *Eur. J. Biochem.*).

Incidentally, the same mechanisms are believed to be operative in the activation of G_i as in the case of G_s; therefore, inhibitory signal transfer is not considered separately. When transmembrane segment VII of the α_2-adrenoceptor was replaced with that from the β_2-adrenoceptor, the chimeric-$\alpha_2\beta_2$-receptor had binding preference for β-adrenergic ligands. From that, it was concluded that the major specificity determinants for ligand binding are located in that transmembrane domain. When transmembrane helices V and VI and the loop connecting them were substituted with the corresponding regions from the β_2-adrenoceptor, the chimera bound α_2-specific ligands but coupled to G_s and stimulated adenylyl cyclase like a β_2-adrenoceptor. It therefore seems that the selection for G-proteins takes place in a domain consisting of transmembrane segments V and VI and the interconnecting loop.

The β-receptor from turkey erythrocytes (Fig. 2) does not quite fit into the scheme of the other β-adrenoceptors [109,110]. Wong et al. [111] have replaced the whole loop connecting transmembrane segments V and VI and the N-terminal 11 amino acid long part of the loop from the M_1 muscarinic acetylcholine receptor with the corresponding sequences from the turkey β-adrenoceptor. This chimeric M_1-β_1-receptor responded to muscarinic agonists with activation of adenylyl cyclase in addition to stimulation of phospholipase C, which is characteristic for the M_1 receptor. Originally, this was thought to be a peculiar property of the chimeric receptor, but in the meantime evidence has came to light indicating that the native β-adrenoceptor from turkey erythrocytes is also capable of activating independently both adenylyl cyclase and phospholipase C. These independent activations are apparently mediated by two different G-proteins; one of which is G_s, while the other is not identified but is a cholera toxin-insensitive G-protein [25]. The larger (P 50) of the two forms of turkey erythrocyte β-adrenoceptor is converted in membranes in a time- and temperature-dependent reaction to the smaller (P 40) form [95,112–115]. The limited proteolysis removes a fragment from the 50 kDa polypeptide to which are joined N-linked complex carbohydrate chains [115,116]. The proteolytic reaction was found to be attenuated in intact erythrocytes and it was speculated that the large (millimolar) amounts of reduced glutathione and glutathione reductase present in erythrocytes [117] are responsible for inhibition of a proteinase which cleaves the 50 kDa form to the 40 kDa one [116]. The deglycosylated truncated P 40 lacks an N-terminal epitope [118], and because P 40 retains Arg^{29} [119], it follows that proteolysis must have occurred between residues 15 and 28. However, there is also proteolytic processing of the unusually long C-terminal domain of the turkey β-receptor. Truncation of this region causes the receptor to be expressed at higher levels and renders it more susceptible to agonist-induced endocytosis and downregulation [120,121]. Thus, both the P 40 and the P 50 forms are proteolytically processed in the C-terminal region; however, the P 40 form is cleaved additionally in the N-terminal region [116,118]. It seems that cleavage of the C-terminal domain makes the truncated receptor more susceptible to proteolysis in the N-terminal domain. Luxembourg et al. [122] have speculated that only C-terminally truncated receptors have access to a compartment where proteolytic cleavage of the N-terminal region and the conversion of P 50 to P 40 occurs.

3.2.1. Functional correlations with structural elements
Experiments with chimeric receptors and deletion mutants and with receptors modified by site-directed mutagenesis have so far implicated cytoplasmic domains, the interconnecting transmembrane segments V and VI in G-protein coupling [123,124]. The β_1-adrenoceptor-G_s interaction in turkey erythrocyte membranes [125–127] and

the rhodopsin-transducin interactions [128] were probed with synthetic peptides. Peptides from the second loop and from the N-terminal regions of the third and fourth cytoplasmic loops, respectively, were found to interfere in a cooperative fashion with hormone-stimulated β-receptor G_s-activation. These results are summarized in Table I from Muench et al. [127]. In the case of the turkey β-adrenoceptor, it is the second intracellular loop represented by the peptide T138-T159 which interferes most effectively with adenylyl cyclase stimulation, respectively G_s-coupling (see also [124]). Considering results from substitution mutants, the major interaction appears to occur with the N-terminal part of the second loop [123,129]. From peptide competition experiments it becomes apparent that the N-terminal part of the third intracellular loop, represented by the peptides T225-T241 and T229-T248, is also participating in the interaction, but only the C-terminal part of the seventh loop, T345-T359 appears

TABLE I
Effects of synthetic peptides on hormone-mediated activation of adenylyl cyclase in turkey erythrocyte membranes

Peptide sequence[a]	Amino acid composition[b]	Adenylyl cyclase activity at 100 μM peptide (%) of initial	EC_{50} (μM)
Intracellular βAR peptides			
T61–71	LVIAAIGRTQR	80	> 200
T67–80	GTRQRLQTLTNNLFI	99	> 200
T138–159	DRYLAITSPFRYQSLMTRARAK	10	15
T225–241	FVYLRVYREAKEQIRKI	25	45
T229–248	RVYREAKEQIRKIDRCEGGRF	15	35
T234–248	AKEQIRKIDRCEGRF	78	> 200
T284–295	REHKALKTLGII	130	> 200
T338–353	FNPIIYCRSPDFRKAF	100	> 200
T345–359	RSPDFRKAFKKRLLCF	4	20
Combination of intracellular AR peptides			
T229–248 + T138–159		1	3
T229–248 + T345–359		2	4
T138–159 + T345–359		0	4
T229–248 + T345–359 + T138–159		0	1.5
Extracellular and transmembrane AR peptides			
T2–10	GDGWLPPDC	94	n.d.
T208–222	AIASSIISFYIPLLI	82	n.d.

The initial adenylyl cyclase activity (100%) was measured in the absence of synthetic peptides. [a]Position and [b]amino acid composition of peptides correspond to the turkey erythrocyte β-AR sequence [351]. (Adopted from Muench et al. [127] by courtesy of the authors and the Editors of *Eur. J. Biochem.*)

to be directly involved (see Table I). This region may be a coupling site common to all G-protein-linked receptors (see also [130]). Peptide T284-T295 representing the C-terminal part of the positively charged large third intracellular loop deserves attention. This peptide was capable of eliciting G_s-mediated adenylyl cyclase activation, circumventing the receptor [125]. It is of interest in that context that Hausdorff et al. [131] constructed a $β_2$-adrenoceptor mutant by deletion of the conserved sequence KEHKALK which corresponds to the sequence REHKALK (T284-T291) of the turkey $β$-adrenoceptor. This mutant receptor showed impaired G_s-dependent adenylyl cyclase activation, but was still capable inducing the high affinity state of the receptor for agonists characteristically binding G_s. Thus, the three domains at the cytoplasmic site of the $β$-adrenoceptor which are represented by the peptides, T138-T159, T225-T241, T229-T248 (and T284-T291) seem to participate in receptor-G_s coupling. The fact that these peptides are nonoverlapping but are functioning synergistically raises the question to which of the subunits of the $αβγ$-heterotrimeric complex of G_s they might bind. The receptor recognition sites for G_s seem to be different from the corresponding sites on the adenylyl cyclase. More information of this kind is needed, however, in order to rationalize the structural basis for selectivity of $β$-receptor-G-protein coupling.

The hydrophobic domains of the amphipathic transmembrane helices of the adrenoceptors have been implicated in ligand binding. Studies using $α_2β_2$-adrenoceptor chimeras indicated distinct binding sites discriminating among ligands [107,132,133]. However, no particular transmembrane domain could be singled out and made solely responsible for ligand binding. Using site-directed mutagenesis, the residues Ser 204 and Ser 207 in the transmembrane segment V of the $β_2$-adrenoceptor were implicated in binding of catecholamines and receptor activation [134]. These two serine residues are conserved among all receptors with seven transmembrane sequences binding catecholamines. Another, conserved amino acid residue involved in agonist binding is the acidic Asp 113 in transmembrane segment III of the $β_2$-adrenoceptor, although in this case when replacing Asp 113 by Asn the affinity for antagonists was likewise diminished [135]. As was suggested by Strader et al. [130] Ser 204 and Ser 207 might be involved in hydrogen bonding to the adjacent ring hydroxyl groups of the catechol moiety, whereas Asp 113 could form an ion pair with the protonated amine. Asp 104 on helix II is the only polar residue in the transmembrane helices which is conserved in all receptors of this kind. In the case of the Lutropin receptor it was suggested that this acidic residue is the counter-ion to the Na^+ cation [356] which modulated the ligand-binding affinity and the coupling efficiency of these receptors [357].

Various $β$-adrenergic agonists differ in affinity to $β$-adrenoceptors, but increased affinity is not necessarily associated with increased efficacy [136]. Moreover, one and the same ligand can act as an antagonist, partial agonist or full agonist [137,138]. For example, in cells expressing a $β_2$-adrenoceptor mutant with exchange of Asp 113 by Glu, classical antagonists were found to act as partial agonists [139]. Furthermore, antagonists for the β_1- and β_2-adrenoceptors act as partial agonists for β_3-receptors; also all three β-adrenoceptors have an Asp in the transmembrane region III [21]. Heithier et al. [140] have probed the environment of the ligand-binding site of the turkey β_1-adrenoceptor with a more hydrophilic fluorescent antagonist, a CGP 12177 analog, and Tota and Strader [141] have characterized the binding site of a hamster β_2-adrenergic receptor expressed in baculovirus-infected *Spodoptera frugiperda* cells with the more hydrophobic antagonist carazolol as fluorescent probe. The results obtained with the two fluorescent ligands, differing with respect to their relative hydrophilic-hydrophobic character, were quite different, whereas the hydrophobic probe sensed a hydrophobic environment for the binding site, a more hydrophilic environment was sensed by the hydrophilic probe. This again emphasizes the importance of ligand structure in the selection of receptor-binding sites.

An important structural property common to receptors featuring seven transmembranous segments is the presence of several conserved cysteine residues. Cys 341 in the carboxyterminal part near transmembrane segment VII is palmitoylated in β_2-adrenoceptor [142] and in rhodopsin [143]. Its replacement results in a partial loss of G-protein-coupling activity. At least two cysteinyl-disulfide bonds seem to be required in the β_2-adrenoceptor for ligand binding: Cys 106 and 184 and Cys 190 and 191 [144–146]. It should be noted that these cysteins are not located within the transmembranous domains thought to constitute the ligand binding site, but in the extracellular loops. Therefore their role in ligand binding must be indirect (cf: 108]). There is no evidence that sulfhydryl groups are involved in receptor aggregation. We have studied ligand-induced aggregation of β-adrenoceptor in turkey erythrocyte membranes [23] using a cross-linking reagent highly effective in cross-linking EGF dimers. But in contrast to EGF, the β_1-adrenoceptor did not form dimers or oligomers on hormone binding and activation.

Each adrenoceptor which was sequenced was found to be glycosylated [147–150]. The first β-receptor gene which was cloned and sequenced was the gene encoding the hamster β_2-adrenoceptor [100]. The mammalian β_2-receptors consist of a single polypeptide chain of M_r 64,000–75,000 containing N-linked high mannose and complex type carbohydrate chains [151–154]. The glycosylation pattern of the β_2-adrenoceptor in A 431 cells [155] and in S 49 mouse lymphoma cells [156] was like that of hamster lung β_2-adrenoceptors. However, the functional consequences of deglycosylation of β_2-adrenoceptor are controversially discussed. Whereas George et al. [156] reported unchanged coupling with G_s and adenylyl cyclase, Boege et al. [155] found a coupling defect with a deglycosylated receptor from tunicamycin-treated A 431 cells.

4. Regulation of β-adrenoceptor function

Early theories of receptor-mediated signal transduction tended to assume that such systems were static with unchanging properties in a given cell or tissue. In fact, this assumption was the basis of the receptor theory developed initially by Clark et al. [157] and modified later by Stephenson [158] and Furchgott [159]. However, it soon became apparent that receptor-generated signals are not constant, but highly modulated such that the response to a given stimulus may be temporarily changed by a number of factors such as repetitive or prolonged stimulation by the specific agonist and the intervention of other factors. Various processes have been described over the past ten years that can affect the response elicited by stimulation of β-adrenergic receptors. Many, but not all of these appear to act at the level of the receptors, and will be discussed in Section 4.1. These mechanisms are most conveniently divided into those which alter the function of the receptors without affecting their number and those which change the number of receptors present in a given cell or tissue. The latter will be discussed in the Section 4.2. on receptor expression.

4.1. Desensitization

Changes in receptor function result in a loss of its activity which is often referred to as desensitization. In other words, only a virgin unmodified receptor is capable of producing a maximal signal. β-Adrenoceptor desensitization can be defined in terms of a reduction of the receptor's capacity to activate its corresponding G protein, G_s, and, hence, of adenylyl cyclase. Two forms of desensitization can be distinguished: homologous desensitization which affects only those receptors that are stimulated by the same agonist initiating the process and heterologous desensitization which is a more generalized process that can be caused by other unrelated agents. On a mechanistic level, functional and spatial uncoupling of receptors and G_s can occur. Spatial uncoupling appears to involve the removal of the receptors from the cell surface; this process is also called sequestration. Functional uncoupling refers to a process where receptors and G_s remain in the same membranous locations. This process seems to be closely linked to phosphorylation of the receptors. Desensitization of β-adrenergic receptors can be followed in various ways. One way is to measure activation of adenylyl cyclase by β-receptor agonists in membranes. In this case desensitization is characterized by agonist concentration-response curves that are shifted to higher concentrations with reduced maximal response [160] (see Fig. 3). Such experiments, however, do not usually allow for the differentiation between changes in receptor function and changes occurring at the postreceptor level even if post-receptor functions are tested separately, for example by $[AlF_4]^-$ or forskolin-stimulated activities. Alterations in receptor function can be assessed more specifically when receptors in a donor cell are fused with G_s and adenylyl cyclase present in a recipient cell which does not contain receptors [161–163] or when desensitized receptors are isolated and purified and then fused with a recipient cell [164]. Probably, the best way to determine receptor function is to co-reconstitute purified receptors with G_s into phospholipid vesicles and measure their capacity to activate the GTPase activity of G_s

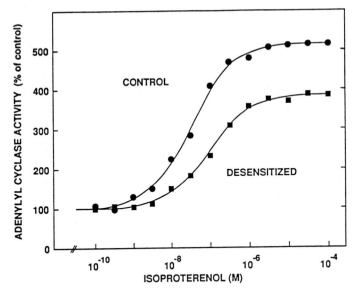

Fig. 3. Desensitization of β-adrenoceptors. β_2-adrenoceptors were desensitized in human A 431 cells by incubation with 10 μM (-)isoproterenol for 10 min. Desensitized (■); Control (non-desensitized cells) (●). Stimulation of adenylyl cyclase activity by the β-receptor agonist (-)isoproterenol was measured in membranes. Note that compared to controls, the curve for the desensitized cells is shifted to the right and has a lower maximum.

[33,72,73,165]. Using such reconstituted systems it has recently been possible to reproduce both heterologous and homologous desensitization with purified proteins [166,167].

4.1.1. Phosphorylation

Functional uncoupling of β-adrenergic receptors from G_s can be achieved by phosphorylation of the receptors by any one of three different protein kinases: the specific β-adrenergic receptor kinase (βARK), of which two isoforms have been cloned [168,169]; protein kinase A [170]; and protein kinase C [171]. Phosphorylation of β-adrenergic receptors by protein kinases A and C is by itself sufficient to decrease receptor function [167], whereas phosphorylation by βARK only makes possible the binding of a cytosolic protein, namely β-arrestin. In this case it is the binding of β-arrestin which leads to the uncoupled state of the receptor [166,172]. The latter mechanism is analogous to that of the inactivation of the 'light'-receptor rhodopsin by rhodopsin kinase, a homologue of βARK [173], and by arrestin [80,174]. Phosphorylation by βARK triggers the homologous form of rapid desensitization of the receptors [175–177]. The high degree of receptor specificity occurs because βARK phosphorylates only the activated, agonist-occupied form of the receptors [178]. The reason for this specificity might be multisite phosphorylation which allows the discrimination of various conformational states. Apparently, βARK recognizes not only the C-terminus of the receptors, where it phosphorylates multiple serine and threonine residues [179], but also additional sites, whose accessibilities may differ between the active and the inactive conformation. Three lines of evidence support this hypothesis: (a) a peptide corresponding to the first intracellular loop of the β_2-adrenergic receptor interferes with the action of βARK, which it may do by binding to an allosteric switch site [179,180]; (b) there is no well-defined consensus sequence for βARK-phosphorylation sites; and (c) synthetic peptides corresponding to C-terminal regions of the β_2-adrenergic receptor which are phosphorylated by βARK are about 1000 times less effective substrates for βARK than the agonist-activated receptor [180,181]. Apart from agonist-occupied G-protein-coupled receptors, there are no other known biological substrates for βARK. However, the specificity of βARK for different activated G-protein-coupled receptors is not known. βARK has been shown to phosphorylate β_2-adrenergic, α_2-adrenergic, and chicken heart muscarinic receptors about equally well, but rhodopsin to a much lesser degree [178,182–184]. A second βARK-like kinase, termed βARK-2, has very similar properties, although it recognizes rhodopsin even less [169]. Thus, it could be that the true physiological substrate for βARK is another, unknown receptor rather than the β_2-adrenoceptor. βARK may phosphorylate a variety of G-protein-coupled receptors, provided they are in their active conformations.

The main consequence of βARK-mediated phosphorylation seems to be the increase of the affinity of the receptors for β-arrestin by at least one order of magnitude [166]. This may be enough to enable β-arrestin to compete effectively with G_s to bind the receptor, thus interrupting receptor-G_s coupling. Although this scheme is in line with similar hypotheses for the visual system [174], direct evidence for such a competitive binding reaction is still missing. There is preliminary evidence for the existence of several β-arrestins, but with unknown specificities [184a].

Recent evidence suggests that βARK-mediated phosphorylation of receptors uses two separate triggers. The first, as detailed above, is the requirement for agonist occupancy of the receptors in order to become substrates for βARK. The second is the activation of the corresponding G-protein. This leads to the formation of free $\beta\gamma$-subunits (i.e. dissociated from their α-subunit). A stimulatory effect of free G-protein $\beta\gamma$-subunits was first described for the so-called muscarinic receptor kinase (MURK), which is similar or identical to βARK [184b,c]. Very recently, it was shown that the free $\beta\gamma$-subunits can serve as a membrane anchor for the kinase, allowing it to translocate from the cytosol to the plasma membrane in the immediate vicinity of the agonist-occupied receptor [184d].

Protein kinase A (PKA)-mediated phosphorylation of the β_2-adrenergic receptor might be conceptualized as negative feedback, whereby the effector enzyme (PKA) shuts off its own activation. The phosphorylation occurs in two regions [185–187], one is in the third intracellular loop in a region that is known to be essential for coupling to G_s, and the other is at the N-terminal part of the C-terminus [124,127,188,189]. Phosphorylation in this strategic region is apparently sufficient to impair the ability of the receptors to couple with G_s [167]. β-Arrestin can not distinguish between PKA-phosphorylated and unphosphorylated $β_2$-receptors and, thus, is apparently not involved in PKA-mediated desensitization [166]. However, the PKA-mediated impairment of $β_2$-receptor function is apparent only under certain assay conditions, in particular when the free Mg^{2+} concentration is low (< 1 mM) [167,177,190]. Protein kinase C (PKC) is also capable of phosphorylating the $β_2$-adrenergic receptor, and there is reason to believe that the sites of phosphorylation are the same for PKA and PKC [167,171]. The consequences of PKC-mediated receptor phosphorylation are therefore like those described for PKA-mediated phosphorylation [167].

The effects of the three different kinases and their mutual contributions to the desensitization of $β_2$-receptors differ. The βARK-mediated process is rapid ($t_{1/2} ≈ 15$ sec), quite large (> 50% desensitization), strictly receptor-specific (homologous), and requires high agonist concentrations [176,177,179,191]. PKA-mediated phosphorylation causes a generalized form of receptor desensitization [192]. This form of heterologous desensitization may also come into play when receptors activate phospholipase C, of which the reaction products activate PKC which in turn desensitizes receptors that are coupled to adenylyl cyclase. The observations described here are valid for the $β_2$-adrenergic receptor, but to what extent, if at all, they also apply to other receptors is not clear. For example, Zhou and Fishman [193] found no effects of βARK inhibitors on homologous desensitization of $β_1$-receptors, although the same kinase inhibitors completely prevented homologous desensitization of $β_2$-receptors. Nevertheless, one must take into consideration that βARK-mediated receptor phosphorylation might involve the translocation of this kinase from the cytosol to the plasma membrane and that the translocation might require simultaneous activation of several G-protein-coupled receptors [194,195].

βARK can phosphorylate several G-protein-coupled receptors in a reconstituted system. Furthermore, agonist-dependent homologous desensitization by receptor phosphorylation is a widespread phenomenon and includes the α-mating-factor receptor in yeast [196] and the cAMP receptor in dictyostelium discoideum [197,198]. And finally, the cDNAs of four different βARK-like kinases have recently been isolated from *Drosophila* [199]. All these findings suggest that homologous desensitization through phosphorylation by kinases specific for an agonist-activated receptor is a rather widespread phenomenon of general importance.

4.1.2. Sequestration
Uncoupling of β-receptors from G_s can also be effected by removal of the receptors from the cell surface, a process called receptor sequestration. Although this was the first mechanism originally proposed as an explanation for receptor desensitization [200,201], its importance is still questionable. Evidence that receptors in the course of

desensitization have left the cell surface is provided mainly by the following observations: (a) the receptors are no longer accessible to surface-reactive hydrophilic ligands such as the agonist isoproterenol or the antagonist CGP 12177, but remain accessible to hydrophobic ligands such as iodocyanopindolol which can penetrate the membrane [202,203] and (b) desensitized receptors are located in a membrane fraction that is lighter than the bulk of the plasma membrane and which can be separated from the latter by sucrose-gradient centrifugation [200,201]. Is has therefore been suggested that the receptors are translocated into a 'light vesicle' fraction where they are separated and spatially uncoupled from G_s [200,204]. The nature of this vesicle fraction has been controversial. β-Adrenoceptors have been located in preparations of clathrin-coated vesicles from bovine brain [205,360] and requirements for sequestration of β-receptors and internalization of EGF receptors via clathrin-coated pits in human A431 cells appear to be similar [206]. In contrast, immunological studies have suggested that sequestration does not occur via the coated-pit pathway [207]. Furthermore, experiments to replicate the radioligand-binding data on sequestration with receptor antibodies have yielded equivocal results, ranging from total disappearance from the plasma membrane of immunologically detectable receptors after agonist stimulation [208,209,360] to no apparent change of immunoreactive receptors [210]. From such studies it has even been suggested that disappearance of receptors from the cell surface merely reflects changes of the receptors' binding properties, rather than being a consequence of translocation, sequestration, or degradation of receptors (summarized in Wang et al. [211]). Thus, it remains an open question where the desensitized receptors are located and how they get there. The signals initiating sequestration are also not known. Translocation of receptors is strictly homologous, i.e. it requires agonist occupancy of the receptors. Thus, like in the case of βARK-mediated phosphorylation, the concentration-response curves for receptor occupancy by an agonist and for receptor sequestration are superimposable [177]. G_s is not required and receptors in cells deficient in G_s show normal sequestration patterns [212]. Moreover, a series of β_2-receptor mutants impaired in coupling to G_s have been shown to sequester normally [213], but delayed sequestration of similar mutants has also been observed and was thought to reflect structural changes of the receptor which are not directly related to G_s-coupling [129,214]. It has thus become evident that changes in the interaction between agonist-occupied receptor and G_s cannot be the signal for sequestration. Phosphorylation of the receptors by βARK and PKA can likewise be ruled out as signals for sequestration [176,177,188]. This suggests that the signals leading to sequestration belong to pathways different from the usual signalling pathway. It should also be remembered that sequestration is slow compared to receptor desensitization by phosphorylation ($t_{1/2}$ of >10 min [191]) and usually involves no more than 20–30% of the cell surface receptors. This makes it unlikely that sequestration plays a major role in receptor desensitization [177]. An alternative role for sequestration may be to allow the effective dephosphorylation of desensitized receptors [215]. In that case, sequestration would be an important step in the recycling pathway of receptors to the cell surface necessary for resensitization of the receptor signalling system.

4.2. β-Receptor expression

Modulation of receptor function is not the only way by which signal transfer can be regulated. Another way is the regulation of the total number of receptors expressed in a cell. While receptor function is regulated rapidly (over seconds to minutes), regulation of receptor number takes much longer, several hours, in most cases. Mechanisms that can change the number of β-receptors involve either receptor degradation or receptor synthesis. Since they operate superimposed on the basal turnover of receptors, they are not easy to quantitate. Whereas receptor activity is always decreased as a consequence of desensitization, receptor expression can either be enhanced (upregulation) or decreased (downregulation). Thus, the actual state of the β-adrenergic receptor signalling system in intact cells and tissues under physiological conditions may not necessarily be the most effective state since it can be further upregulated by increased receptor synthesis.

4.2.1. Turnover

Expression of receptors is, like expression of any cellular protein, the result of a dynamic state determined by the rates of degradation and synthesis. This topic has been extensively reviewed by Mahan et al. [216]. Most studies have indicated a turnover of β-receptors under basal conditions with half-lives of receptor degradation and synthesis of about 24 h or longer. However, in many instances, much longer half-lives have been estimated. For these studies agents, such as cycloheximide, were used that inhibit protein synthesis to measure receptor degradation. To study synthesis, receptors were inactivated covalently by affinity-labeling-receptor ligands. In some studies receptor reappearance and recovery following agonist- or cAMP-dependent downregulation was estimated (see also below). Based on such studies it appears that in most cases recovery from down-regulation occurs at a faster rate than basal receptor turnover [212,217], although the opposite has also been observed [218].

In cultured cells, an estimate of receptor turnover is complicated since the multiplying cells in the growing population synthesize receptors in order to equip their new cells with an adequate number of receptors. The situation is further complicated by differences in receptor numbers between pre- and postconfluent cultures. Taking that into consideration, it appears that in growing cells receptor turnover matches or is only slightly larger than the rate of growth. Thus, like in exponentially growing bacterial cultures, protein (receptor) turnover is minimal. For example, in S49 lymphoma cells, a receptor half-life of 30 h was reported, which is even longer than the doubling time of this cell [219]. Other studies indicated comparably long half-lives of many hours [220]. Accordingly, levels of receptor mRNA correspond to less than one copy per cell [221]. In most confluent cell lines, receptor turnover is even slower with half-lives of a week or more. It is therefore not surprising that receptor reappearance after irreversible blockade and inactivation is often incomplete due to the too slow turnover [222,223].

In whole animals, receptor turnover times are comparably slow. Basal turnover of β_1- and β_2-receptors in rat heart and lung was found to have half-lives of several days. Turnover was somewhat faster in young compared to old animals [224]. In guinea pig lung, recovery of receptors after injection of an irreversible receptor blocker took several days and recovery of receptor function was even slower [225]. In rat renal cortex, recovery of β_2-receptors from down-regulation has a half-life of 18 h and that of β_1- receptors of 45 h [226]. The important point to remember is that basal turnover of β-adrenergic receptors is too slow to play a major role in the rapid regulation of receptor expression after agonist stimulation.

4.2.2. Down-regulation

Down-regulation of receptor numbers is the consequence of either more rapid degradation or, in the long term, decreased synthesis. It appears that both mechanisms are operative. Consequently, recovery from down-regulation would be expected to require protein synthesis, what, however, has not always been observed (summarized in Mahan et al. [216], see also Allen et al. [227]). As is the case for the regulation of receptor function, multiple pathways also seem to be involved in the regulation of receptor expression, and most of these processes are not understood at least not in greater detail. This is in large part because down-regulation of β_2-adrenergic receptors was studied under very different conditions making comparisons difficult, if not impossible. Such studies have involved measurements of receptor levels and of the receptor mRNAs in cell lines that were defective in specific components of the signalling cascade from β-receptors to PKA, or cell lines expressing mutant β_2-receptors with altered capabilities to couple to G_s or to be phosphorylated by βARK or PKA. From the data available, three different pathways of β_2-adrenergic receptor down-regulation can be distinguished: (a) an agonist-dependent, but cAMP-independent pathway that may or may not be associated with reduced levels of the β_2-receptor mRNA; (b) a PKA-dependent pathway represented by enhanced degradation of PKA-phosphorylated receptor; and (c) a PKA-dependent pathway associated with unstable mRNA resulting in a reduction in the level of the receptor mRNA.

Agonist-dependent but cAMP-independent down-regulation is apparent from the observation that agonists can bring about down-regulation in cells incapable of responding to receptor stimulation with increased cAMP production [212,228,229]. Furthermore, in wild-type cells agonist-induced down-regulation sometimes exceeds down-regulation due to increased intracellular cAMP [230]. Agonists can also trigger β_2-receptor down-regulation in the H21a mutant of a S49 mouse lymphoma cell line, which is defective in G_s-cyclase coupling [212,231]. Finally agonist-dependent β_2-receptor down-regulation has been observed in cells lacking adenylyl cyclase [229]. However, others have not seen down-regulation, or not much of it, when receptor-G_s

coupling was impaired, either because of mutations in the receptor [213] or in G_s [212,228,229,231]. Downregulation has also been reported by some researchers [212,227,228] but not by others [231] in cells devoid of PKA activity. The mechanism of an agonist-induced cAMP-independent down-regulation is a matter of speculation. Based on their findings of a reduced receptor mRNA in the H21a mutant of S49 cells, Hadcock et al. [231] have made a G_s-mediated pathway responsible for mRNA reduction that is independent of the G_s-adenylyl cyclase pathway, because otherwise, it would have to be assumed that the defect in the H21a cell line is restricted to G_s-cyclase coupling. This is, however, not the case, since the defect actually seems to be an inability of G_s to bind GTP [232], which would mean that G_s is not able to propagate any signal to any effector. It is more likely therefore that such a signal is mediated by another G-protein. Alternatively, one could speculate that β_2-receptor-G_s complexes are degraded preferentially, which would explain why physical coupling to G_s without propagation of a signal can lead to down-regulation [213]. Such a mechanism could also explain why down-regulation occurs over a few hours, while basal receptor turnover is much slower.

The other pathways seem to require activation of PKA. The effects can therefore be mimicked by membrane-permeable analogs of cAMP or by forskolin. One pathway appears to involve degradation of the PKA-phosphorylated receptor [186,213]. When the PKA-phosphorylation sites in the β_2-receptor are removed by site-directed mutations, cAMP-induced or isoproterenol-induced down-regulation proceeds slower and to a lesser extent than in the case of wild-type receptors. This is in favor of the assumption that the PKA-phosphorylated G_s-coupled receptor might be a preferential target for degradation. On the other hand, βARK-dependent receptor phosphorylation does not seem to play a role in down-regulation [213].

The other PKA-dependent down-regulation seems to be due to a reduction of the receptor mRNA [186,231,233]. This reduction is apparently caused by a PKA-dependent destabilization of the mRNA, such that its half-life is shortened by one half [234]. The mechanism for this PKA-dependent destabilization of mRNA is unknown. The decrease of mRNA precedes the loss of receptors [230,233]. Agents such as forskolin, which elevate cAMP levels more directly by stimulation of adenylyl cyclase, also cause a decrease in mRNA levels, but receptor-specific agonists cause greater loss of receptors, a finding that supports the concept of an additional agonist-dependent rather than a cAMP-dependent pathway of down-regulation. The fact that in the case of β_2-receptors there appears to be a relatively satisfactory correlation between cAMP levels and loss of β_2-mRNA [231] does not exclude additional cAMP-independent pathways which may also contribute to a reduction of receptor mRNA levels.

Changes of receptor mRNA levels have also been observed in cases of other G protein-coupled receptors, even those which are not linked to adenylyl cyclase and do not elevate cAMP levels. For example, agonist-induced down-regulation of receptor levels and of receptor mRNA have been observed for the TSH receptor [235], the

α_{1B}-adrenergic receptor [236], and the M_1-, M_2-, and M_3-muscarinic receptors [237,238]. In the first case the reduction in receptor mRNA levels could also be achieved by agents that increase intracellular cAMP, while in the latter two cases the operation of cAMP-independent pathways is more likely.

However, the actual contribution of each of the various pathways to down-regulation are at present a matter of debate. The rather slow basal turnover of receptors in most cells (>24 h) compared with the much faster down-regulation (1–4 h) argues against a major role of inhibition of receptor synthesis in down-regulation. Furthermore, an argument against a major contribution of cAMP-dependent pathways is that down-regulation is a receptor-specific phenomenon. Thus, it is more likely that an agonist-dependent degradation of the receptors is the important step in this process. Physical coupling to G_s may be necessary for this mechanism.

4.2.3. Up-regulation

Expression of β-adrenergic receptors cannot only be decreased but also enhanced above basal levels. In all cases reported so far this appears to be due to increased transcription, which can be caused by several agents: glucocorticoids, androgens, thyroid hormones, and, paradoxically, increased intracellular cAMP. Enhanced receptor synthesis has also been observed during recovery from agonist-induced receptor down-regulation. Up-regulation is best documented for glucocorticoids and β_2-receptors (reviewed by Collins et al. [239], Hadcock and Malbon [240]). An improved response to β-receptor agonists after administration of glucocorticoids has long been known from both clinical experience and animal experiments. The 5'-flanking region of the β_2-adrenergic receptor gene contains several sequences that might represent glucocorticoid response elements [241]. These elements are indeed functional in cultured cell lines where glucocorticoids enhance synthesis of the receptor mRNA [242,243]. The increase in mRNA levels precedes the increase in receptor levels and receptor-mediated cAMP formation [242]. When β-receptor agonists and glucocorticoids are applied simultaneously [234], the glucocorticoid-induced increased transcription can offset the agonist-induced down-regulation.

Increased β-receptor synthesis induced by thyroid hormones may be the basis for the many sympathomimetic effects accompanying hyperthyroidism [244] (see also review by Bilezikian and Loeb [245]). The DNA-sequence elements responsible for up-regulation by thyroid hormones have not been identified with certainty. It should also be kept in mind that in intact animals, indirect effects may contribute to thyroid hormone-induced changes in adrenoceptor expression [244,246]. This is discussed in more detail in section 5.2.1. β-Receptor synthesis can also be increased by androgens in androgen-sensitive tissues such as the prostate gland [247].

A seemingly paradoxical mechanism leading to enhanced transcription of β_2-adrenergic receptor mRNA is mediated by enhanced cAMP levels and involves activation of PKA [230,248]. Exposure of cells to β-agonists (or other agents which elevate cAMP levels) causes a rapid (over minutes) 3–5-fold increase which is followed

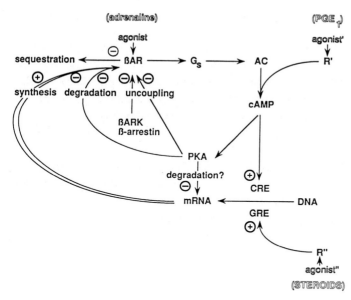

Fig. 4. Mechanisms of β-adrenoceptor regulation.

later by a prolonged decrease of the mRNA. These fast effects are mediated by a cAMP-responsive element which is located in the promoter region of the β_2-receptor gene. The corresponding transcription factor, the cAMP-response element binding protein (CREB), is a substrate for PKA and is sensitive to cAMP. It also responds to intracellular free calcium. Hence, this might be another example for a cross talk between cAMP- and calcium-dependent systems in this case at the level of the gene. The function of this mechanism may be to enhance stimulation by agonists and counteract the effects of desensitization and down-regulation. This mechanism might also account for the observation [212,217] that following receptor down-regulation the rate of receptor reappearance is faster than expected from basal receptor turnover. A similar enhancement of transcription via CREB may also play a role in the regulation of the expression of other G-protein-linked receptors, such as the α_{2A}-adrenergic receptor [249] and possibly the substance P receptor [250]. This points to the existence of common regulatory principles governing function and expression of G-protein-coupled receptors. The various mechanisms involved in the regulation of β-adrenergic receptors are summarized in Figure 4.

5. Clinical aspects

The β-adrenergic system plays an important role in many diseases. In the following sections its role in some cardiovascular and endocrinological diseases as well as one of its rare congenital alterations will be discussed.

5.1. Cardiovascular diseases

In the regulation of the cardiovascular system, the β-adrenergic system plays a pivotal role. Moreover, it is the cardiovascular system where the β-adrenergic function has been best characterized from a clinical point of view. Opposing modes of regulation of the β-adrenergic system will be exemplified in the two most important cardiovascular diseases, i.e. chronic heart failure and acute myocardial infarction.

5.1.1. Chronic heart failure and cardiomyopathy

Chronic heart failure is the inability of the heart to supply peripheral organs sufficiently with oxygen and nutrients as required by their metabolic needs. Although different causes, such as cardiomyopathy or ischemic heart disease, may be responsible for chronic heart failure, all these processes share common compensatory mechanisms primarily involving the β-adrenergic system. These mechanisms result in characteristic changes of β-adrenergic receptors which further promote the vicious cycle of the disease. The activation of the sympathetic system as a primary compensatory mechanism serves to maintain adequate peripheral perfusion [251,252]. This activation eventually results, however, in the depletion of endogenous cardiac catecholamines [253]. The early increase in circulating catecholamines [251,252,254] is due both to the increased release [255] and to the reduced neuronal uptake of catecholamines [256]. As a consequence this results in a persistent activation of cardiac β-adrenergic receptors, which in turn leads to desensitization of the β-adrenergic system in the failing heart.

For the failing heart the β-agonist-induced desensitization is not yet sufficiently characterized in molecular and mechanistic terms. Thus, in chronic heart failure the density of β-adrenergic receptors at the cell surface is reduced by about one half [257–261]. Not only redistribution, internalization, and sequestration of receptors, but mainly loss of total cellular receptors is involved in these changes. This loss of receptors has been made responsible for the impairment of the failing heart to respond even to extraordinarily high concentrations of catecholamines [257,259,262–264]. The reduced responsiveness of the β-adrenergic system is not restricted to the heart, but may also apply to peripheral smooth muscle cells [265]. However, the mechanisms involved in β-agonist-induced desensitization which have been defined for isolated cells are only beginning to be clarified in the failing heart. Recent studies suggest that in severe heart failure the mRNA levels as well as the enzymatic activity β-adrenergic receptor kinase are increased [363]. However, no data are available on phosphorylation of β-adrenergic receptors in failing heart.

In chronic heart failure due to dilated cardiomyopathy, primarily β_1-adrenergic receptors are down-regulated [258], resulting in about a 60% loss of this receptor subtype. In contrast, cardiac β_2-adrenergic receptors are not significantly reduced. The situation at the receptor level appears to be paralleled by the mRNA levels; in the failing heart, β_1-receptor mRNA levels are decreased whereas those for β_2-receptors are unchanged [362]. The mechanisms responsible for such selective regulation are not understood. The elevated level of circulating norepinephrine has been named responsible for the selective downregulation of β_1-adrenergic receptors [258]. The greater responsiveness of the β_1-adrenergic receptor to chronic downregulation might also reside in properties of the promoter regions of the receptor subtype gene. However, the promoter region of the β_1-adrenergic receptor has not been characterized completely.

Interestingly, mild β-blockade is able to partially revert the loss of β-adrenergic receptors in chronic heart failure [266], suggesting in fact that chronic β-agonist-induced activation is one of the major causes for the loss of β-adrenergic receptors. Therefore, chronic heart failure is perhaps the most important clinical correlate and example of the consequences of β-agonist-promoted desensitization. The positive inotropic effect of β-agonists is mediated predominantly by β_1-adrenergic receptors [10,267–269]. Actually, it is the inotropic force which is lost as a consequence of the loss of β_1-adrenergic receptors in the course of heart failure. In contrast, even in the late stage of the disease, the weak positive inotropic effect and the increase in heart rate which may be mediated by β_2-adrenergic receptors persists as a compensatory mechanism. Moreover, β_2-agonists have been shown to remain therapeutically effective longer than β_1-adrenergic agonists [259].

The unresponsiveness of the adrenergic system in chronic heart failure might also be due or might at least be aggravated by an impairment of some postreceptor step of the signal transduction chain. It has been claimed that the stimulatory G-protein is defective in some forms of experimental heart failure in dogs [270]. However, in other experimental models for chronic heart failure and in patients with dilated cardiomyopathy, the stimulatory G-protein was found functionally intact [271]. The reduced responsiveness of adenylyl cyclase to adrenergic stimuli in heart failure could also be due to an increase of receptor-G-protein-mediated inhibition. In experimental models and in human cardiomyopathy, several groups could show that inhibitory G-proteins are more effective and expressed to a greater extent under these circumstances [272,279,274]. Regulation at the level of the inhibitory G-proteins was considered to be a consequence of chronic catecholamine-induced loss of function of β-adrenergic receptors [275–278]. This regulatory response induces increased levels of mRNA specific for one of the α-subunits of the inhibitory G-proteins in heart [272,273,279].

Recent data provide evidence that the catalytic subunit of adenylyl cyclase may also be functionally impaired in chronic heart failure. Thus, the cardiac adenylyl cyclase is less responsive to forskolin and Mn^{2+} in chronic heart failure [271].

A chronic activation of β-adrenergic receptors by autoantibodies against the receptor has been named responsible for certain forms of cardiomyopathy [280]. However, convincing characterization of these antibodies is still missing. Therefore, at present it is unclear what role chronic activation of the β-adrenergic system actually plays in the course of chronic heart failure.

At present, for therapeutic purposes opposite approaches using either high concentrations of exogenous catecholamines or low doses of β-blockade [266,281–283] have been applied, but only with limited clinical success. It is hoped, that in the future, medical therapy for chronic heart failure and dilated cardiomyopathy can profit from a better and more discerning understanding of the different pathogenetic mechanisms operative in the β-adrenergic system.

5.1.2. Myocardial infarction
Acute myocardial infarction is another pathophysiological situation where the β-adrenergic system plays a central role. In contrast to chronic heart failure, in myocardial ischemia mechanisms for sensitization rather than for β-agonist-induced desensitization are operative. Actually an unresponsiveness would be beneficial, because it would protect the endangered myocardial tissue from the deleterious effects of adrenergic overstimulation [284]. However, both clinical [285,286] and experimental [287] observations of persistent malignant arrhythmias [288,289] in acute myocardial infarction, which are in part sensitive to β-blockade [290], point to a prolonged responsiveness and activity of the β-adrenergic system in the ischemic zone. This is supported by elevated intracellular cAMP levels in the ischemic heart [291,292]. This sensitization involves several components of the signal transduction pathway, i.e. the receptors, the G-proteins, and the effector enzyme, adenylyl cyclase. The density of β-adrenergic receptors in the plasma membrane rapidly increases in acute myocardial ischemia. This increase has been demonstrated both *in vivo* and *in vitro* [293–295]. The earliest time measured was after 15 minutes of ischemia [296]. The initial increase in β-adrenergic receptors by about 30–40% after a 15-minute period of ischemia was followed by a further rise after 30 and 50 minutes of ischemia. Even after 5 days the increase of β-adrenergic receptors persisted, although the functional responsiveness to β-adrenergic stimulation had already dramatically declined in the infarcted zone [296–298]. The molecular mechanism which leads to this increase in β-adrenergic receptors in the ischemic heart is not completely understood. Interestingly, in the ischemic zone not the total number of receptors is increased but the density of receptors at the cell surface. This increase occurs at the expense of intracellular receptors [299,300]. Accordingly a shift of receptors from an intracellular compartment to the plasma membrane has been cited responsible for the increased density of β-adrenergic receptors at the cell surface [299,300]. However, direct proof of a shift of receptors is missing. Moreover, redistribution of receptors, which is an energy-dependent process would seem to be unlikely in acute myocardial ischemia associated with the dramatic loss of high energy phosphates [301]. To test the hypothesis that not an increased "externalization" but perhaps an impairment of physiological agonist-induced internalization is responsible for a shift towards an increase in receptors at the cell surface, a model for ischemia with energy depletion despite continuous perfusion was developed using isolated perfused rat hearts. In analogy to acute ischemia, perfusion with cyanide but without glucose induced a rapid loss of high energy phosphates

and an increase in β-adrenergic receptors in the plasma membranes [297]. Additional superfusion with β-agonists, which under normoxic conditions leads to a complete desensitization of the β-adrenergic system, failed to promote internalization of β-adrenergic receptors [297]. Even at highest concentrations the density of β-adrenergic receptors remained persistently increased in the plasma membranes. These data indicate that in the energy-depleted ischemic hearts a defect in receptor internalization rather than "externalization" is responsible for the shift of receptors to the cell surface.

Even the first step of β-agonist-induced desensitization, i.e. the functional uncoupling of the receptors from the stimulatory G-protein, G_s, seems to be impaired in acute myocardial ischemia. In normoxic hearts all the receptors are in the so-called high affinity state [302,303] and, after perfusion with desensitizing concentrations of catecholamines, are converted to receptors only capable of forming the so-called low affinity state, suggesting that with desensitization these receptors become functionally uncoupled from G_s [297]. In acute myocardial ischemia, however, even at high concentrations of norepinephrine present at the site of the receptor, these β-adrenergic receptors remain in the β-agonist-promoted high affinity state. Consequently, this might result in a persistent sensitization and may add to the supersensitivity of the adenylyl cyclase in the ischemic heart [297].

Aside from activation of the adenylyl cyclase via the stimulatory G-protein, G_s, a blockade of an inhibitory pathway [60] was also postulated. This inhibitory pathway comprises inhibitory receptors, such as the muscarinic M_2 receptors, which couple to one of the inhibitory G-proteins, G_i [304]. In acute myocardial ischemia both types of G-proteins are altered, however, at different times. This leads not only to the sensitization at the receptor level but also to the sensitization of the adenylyl cyclase system at the G-protein level, also by a different mechanism. Within 15 minutes of the onset of ischemia, the binding of $[\gamma\text{-}^{35}S]GTP$ to G-proteins was found to be decreased by about 30% [305]. Binding of $[\gamma\text{-}^{35}S]GTP$ in the heart is mainly representative of binding to inhibitory G-proteins, since the amount of $G_{i\alpha}$ in the heart exceeds that of other α-subunits by more than tenfold (cf: [306,307]. Furthermore, the rapid decline of $[\gamma\text{-}^{35}S]GTP$ binding is accompanied by an equally rapid decrease in both basal and muscarinic-stimulated-GTPase activities. Thus, the α-subunit of G_i is becoming rapidly inactivated in the ischemic heart. The defect at the level of G_i is completely reversible upon reperfusion. Western blot analysis has indicated that the amount of $G_{i\alpha}$ is unchanged, suggesting that a post-translational modification may be responsible for the functional defect of the inhibitory G-proteins in acute myocardial ischemia. The loss of tonic inhibition of adenylyl cyclase would add to the transient sensitization of the adenylyl cyclase system in the infarcted heart.

In contrast to the G_i-proteins, the stimulatory G-protein, G_s, becomes more slowly inactivated after the ischemic insult. After 15 minutes of global ischemia, G_s is still capable of transduction of even more β-adrenergic signals [297]. Only after one hour of ischemia was the amount of the α-subunit of the G_s-protein decreased [308]. Using

Western blot analysis and functional reconstitution, it could be demonstrated that at this time both the total amount and function of G_s were reduced by about 25–30% [308].

These data show that only after prolonged periods of ischemia, of one hour or more, the response to β-adrenergic stimulation becomes reduced despite the persistent increase in β-adrenergic receptors at the cell surface [309]. However, in the early period of ischemia when the rapid loss of G_i precedes the slower loss of G_s, the increased responsiveness of the β-adrenergic system may contribute to the genesis of malignant arrhythmias. The electrical discharge may trigger additional responses with deleterious effects by means of direct activation of voltage-gated calcium channels [310,311], sodium channels [312], or potassium channels [65,313].

The third site of regulation of the adenylyl cyclase system in acute myocardial ischemia is adenylyl cyclase itself. In the very early phase of global ischemia (5–15 min), the response of adenylyl cyclase to direct stimulation is rapidly increased [297]. This sensitization persists even after solubilization and partial purification of the enzyme [297]. The molecular mechanism of this unique sensitization process of adenylyl cyclase remains to be elucidated. But whatever the mechanism may be, it is rapidly reversible upon reperfusion. These data suggest that this sensitization might be due to a covalent modification of the enzyme. With prolonged ischemia the supersensitivity of the adenylyl cyclase system is followed by decreased activity [297,309,314,315]. This decrease becomes already apparent after 20–30 minutes of global ischemia [297]. It is noteworthy that isoproterenol-stimulated adenylyl cyclase activity is greatly reduced even though the β-adrenergic receptor level in the plasma membrane remains elevated and the receptors remain sensitized. Consequently, the longer the duration of ischemia the less important the effect of sensitization at the receptor level becomes. However, it cannot be excluded that other effectors such as channels are activated by the persistently sensitized β-adrenergic receptors.

In summary, chronic heart failure represents a clinical model for β-agonist-induced desensitization of the β-adrenergic system involving all components of the adenylyl cyclase system. Acute myocardial ischemia is, on the other hand, a clinical model where supersensitivity of adenylyl cyclase is an important factor which further aggravates the pathophysiological situation.

5.2. Endocrinological diseases

Hyper- and hypothyroidism are treated as examples of endocrine modulation of the β-adrenergic system in the following section. These frequent clinical conditions result in opposite effects, sensitization or unresponsiveness, respectively, of the β-adrenergic system.

5.2.1. Hypo- and hyperthyroidism
Studies with intact animals and isolated cells have demonstrated that thyroid

hormones exert part of their chronic effects by modulating the expression of specific proteins including receptors at the level of the gene [316]. Hyperthyroidism is characterized by clinical features which indicate increased adrenergic activity [317]. Hypothyroidism, in contrast, is characterized by clinical signs of decreased adrenergic activity [318]. Thus, thyroid hormones modulate the adrenergic effects of catecholamines in many organs including the heart [317,319–322], liver [323], and adipose tissue [324]. These effects are not related to changes in the levels of circulating catecholamines indicating that thyroid hormones do not act at the release or neuronal uptake of catecholamines [245,325]. Several studies have shown that thyroid hormones act primarily at the level of β-adrenergic receptors both in vivo and in vitro [326,327]. Thyroid hormones induce an up-regulation of β_1-adrenergic receptors [328], which is not affected by blockade of β-adrenergic receptors [329]. Thus, the thyroid hormone-promoted regulation of β-adrenergic receptors occurs independently of receptor activation quite in contrast to the processes involved in β-agonist-promoted desensitization. The increased expression of β_1-adrenergic receptors in hyperthyroidism may result from activation of a thyroid hormone-responsive element at the β_1-adrenoceptor gene [316]. In fact, recent data obtained with cultured ventricular cardiomyocytes indicate that thyroid hormones up-regulate the β_1-adrenergic receptor gene at the level of transcription [328]. Thus, thyroid hormones induce a rapid increase in β_1-receptor mRNA levels with a maximal increase within 2 hours. These effects of thyroid hormone are β-receptor-, cell- and tissue-specific [330]. This primary effect of thyroid hormones on the expression of β_1-adrenergic receptors is further enhanced by a coordinated up-regulation of the other components of the adrenergic receptor system. In cultured thyroid cells the thyroid-stimulating hormone (TSH) is capable of inducing an increased expression of the stimulatory G-protein, G_s [331]. Inversely, receptor-linked inhibition of adenylyl cyclase was more effective in adipocytes from hypothyroid rats and patients [332,333]. The increased responsiveness of the inhibitory pathway was associated with an increased expression of each of the pertussis toxin-sensitive α-subunits of G_i (G_{i1}, G_{i2}, G_{i3}) in adipocytes [334,335]. Hyperthyroidism, in contrast, caused about a 30% decrease of $G_{i\alpha}$ [336] which is accompanied by an only slight increase of the stimulatory G-protein, G_s [337].

The up-regulation of α-subunits of inhibitory G-proteins of the adenylyl cyclase system is associated with the increased expression of β-subunits in hypothyroidism [336]. Inversely, thyroid hormones or hyperthyroidism inhibit the expression of G-protein β-subunits [338].

In summary, thyroid hormones have impressive and quite unique effects on the β-adrenergic signal transduction pathway, which involves β-adrenergic receptors and stimulatory and inhibitory G-proteins such that hypothyroidism results in a decreased responsiveness of the β-adrenergic system. In contrast, hyperthyroidism induces a sensitization of the adenylyl cyclase system with an increased responsiveness to stimulatory signals and a decreased sensitivity to inhibitory stimuli.

5.3. Congenital diseases

Among the rare congenital dysfunctions of the β-adrenergic system, the McCune Albright syndrome has been well-characterized and will therefore be discussed here as one example of a congenital disease caused by an inborn defect of the adenylyl cyclase system.

5.3.1. McCune Albright syndrome

The McCune Albright syndrome is an inborn error that leads to a persistent activation of adenylyl cyclase [339]. This disease is characterized by polyostotic fibrous dysplasia, café au lait pigmentation, and multiple endocrinopathies, including sexual precocity, hyperthyroidism, pituitary adenomas, and autonomous adrenal hyperplasias. In this genetic disease a mutation of the stimulatory G-protein has been found and characterized. Substitution and replacement of the amino acid residues Arg^{201} or Gln^{227} of G_{sa} by other amino acids results in a constitutive activation of G_{sa} [340,341]. Such mutations, referred to as 'gsp' mutations, have been identified in human growth hormone-secreting pituitary adenomas and in thyroid tumors [340,342–344]. The addition of the adenosine diphosphate ribose group to the amino acid Arg^{201} of G_{sa} by the exotoxin of *Vibrio cholerae* is a post-translational modification, known to inhibit the GTPase activity of G_s [27,345,346]. The targeted expression of the ADP-ribosylating subunit of cholera toxin in transgenic mice produces pituitary somatotrophic hyperplasia and gigantism, presumably as a consequence of the constitutive activation of G_{sa} [347]. This constitutive activation of the G_s-protein is not due to an altered activation by the β-adrenergic receptor since replacement of the Arg^{201} residue in G_{sa} with cysteine or histidine was solely responsible for a 30-fold decrease in GTPase activity and a corresponding activation of G_s. Cell membranes containing these mutated proteins produce cAMP levels at an elevated rate even in the absence of any stimulatory hormone [340].

Increased cAMP levels have been known to be associated with proliferation in certain cells (see: [348,349]). The McCune Albright syndrome is an impressive example of the role of G_s-signalling pathways in cell proliferation and human disease. This is a field that should receive increasing attention in the future.

6. Outlook

The wealth of information now available on β-receptor-G-protein effector systems has greatly influenced our thinking on hormone action. The structural and functional diversity of receptors, G-proteins, and effectors point to a high level of complexity. Although this complexity might have been expected considering the pleiotropy of hormone actions, it creates new problems and requires new approaches so that one can understand the complex regulatory networks operative in receptor activation and amplification of hormone action, including the mechanisms regulating receptor expression and desensitization. At the cellular level, use can be made of mutant cells where one or the other component of the hormonal signal chain is mutated or deleted or its expression is repressed by anti-genes (antisense oligonucleotides). On a molecular level, reconstituted proteoliposomes can be employed for the study of interactions between well-defined isolated G-proteins, β-adrenoceptors, and effectors made available by genetic engineering techniques. However, only when structural information including three-dimensional structures of receptors, G-proteins, target enzymes, and ion channels become available (see Perutz [350]), will it become possible to reduce the complexity which, at present, is overwhelming to a few basic patterns of structure-function relationships. This expectation is based on the assumption that the hormonal-G-protein-coupled signal transmission chains have evolutionary conserved structures and interaction domains because of their physiological importance and their wide distribution in eukaryotes. Once we understand these structures, it might become possible to reconstruct the molecular dynamics responsible for the complex versatile system of hormone action and its control.

Considering the breadth and vitality of this field of research and the wide attention it receives, it can be anticipated that research will continue to flourish in the future as it has in the past. Not only basic research but also pharmaceutical applications and clinical medicine should greatly benefit from further advances and new insights in the regulation of hormonal responses.

Acknowledgements

Research of the authors is supported by DFG-grants No. Lo386/3-1 and Lo386/5-1, the Bundesministerium für Forschung und Technologie and Fonds der Chemischen Industrie (M.J.L.), the SFB 320 and Hermann-Lilly-Schilling-Stiftung (R.H.S.), and SFB 355, DFG-grants No. He 22/36-4 and He22/44-1 and Fonds der Chemischen Industrie (E.J.M.H.). We gratefully acknowledge the help of Mrs. Sylvia Richter who typed the manuscript.

References

1. Cooke, B.A., King, R.J.B. and van der Molen, H.J. (Eds) (1988) Hormones and Their Actions, Parts I and II, New Comprehensive Biochemistry, Vols. 18A and 18B, Elsevier, Amsterdam.
2. Ahlquist, R.P. (1948) Am. J. Physiol. 153, 586–600.
3. Lands, A.M., Arnold, A., McAuliff, J.B., Luduena, A.P. and Brown, T.G. (1967) Nature (London) 214, 597–598.
4. Ablad, B., Carlsson, B., Dahlöf, C., Ek, L. and Hultberg, E. (1974) Adv. Cardiol. 12, 290–302.
5. Stiles, G.L., Taylor, S. and Lefkowitz, R.J. (1983) Life Sci. 33, 467–473.
6. Brodde, O.E., Karad, K., Zerkowski, H.R., Rohm, N. and Raidemeister, J.C. (1983) Circ. Res. 53, 752–758.

7. Heitz, A., Schwartz, J. and Velly, J. (1983) Br. J. Pharmacol. 80, 711–717.
8. Strader, C., Candelore, M.R., Rands E. and Dixon, R.A.F. (1987) Mol. Pharmacol. 32, 179–183.
9. Motomura, S., Zerkowski, H.R., Daul, A. and Brodde, O.E. (1990) Am. Heart. J. 119, 608–619.
10. Brodde, O.E., O'Hara, N., Zerkowski, H.R. and Rohm, N. (1984) J. Cardiovasc. Pharmacol. 6, 1184–1191.
11. Brodde, O.E. (1991) Pharmacol. Rev. 43., 203–242.
12. Zaagsma, J. and Nahorski, S.R. (1990) Trends Pharmacol. Sci. 11, 3–7.
13. Bojanic, D., Jansen, J.D., Nahorski, S.R. and Zaagsma, J. (1985) Br. J. Pharmacol. 84, 131–137.
14. Bojanic, D. and Nahorski, S.R. (1984) J. Receptor. Res. 4, 21–35.
15. Bahout, S.W. and Malbon, C.C. (1988) Mol. Pharmacol. 34, 318–326.
16. Arch, J.R.S., Ainsworth, A.T., Cawthorne, M.A., Piercy, V., Sennitt, M.V., Thody, V.E., Wilson, C. and Wilson, S. (1984) Nature (London) 309, 163–165.
17. Wilson, C., Wilson, S., Piercy, V., Sennitt, M.V. and Arch, J.R.S. (1984). Eur. J. Pharmacol. 100, 309–319.
18. Bahout, S.W., Hadcock, J.R. and Malbon, C.C. (1988) J. Biol. Chem. 263, 8822–8826.
19. Hollenga, C. and Zaagsma, J. (1989) Br. J. Pharmacol. 98, 1420–1424.
20. Hollenga, C., Haas, M., Deinum, J.T. and Zaagsma, J. (1990) Horm. Metab. Res. 22, 17–21.
21. Emorine, L.J., Marullo, S., Briend-Sutren, M.M., Patey, G., Tate, K., Delavier-Klutchko, C. and Strosberg, A.D. (1989) Science 245, 1118–1121.
22. Fève, B., Emorine, L.J., Lasnier, F., Blin, N., Baude, B., Nahmias, C., Strosberg, A.D. and Pairault, J. (1991) J. Biol. Chem. 266, 20329–20336.
23. Boege, F., Neumann, E. and Helmreich, E.J.M. (1991) Eur. J. Biochem. 199, 1–15.
24. Ashkenazi, A., Winslow, J.W., Peralta, E.G., Peterson, G.L., Schimerlik, M.I., Capon, D.J. and Ramachandran, J. (1987) Science 238, 672–675.
25. Rooney, T.A., Hager, R. and Thomas, A.P. (1991) J. Biol. Chem, 266., 15068–15074.
26. Birnbaumer, L., Mattera, R., Yatani, A., Codina, J., Van Dongen, A.M.J. and Brown, A.M. (1990) In: ADP-Ribosylating Toxins and G-Proteins (J. Moss and M. Vaughan, eds), Am. Soc. for Microbiology, Ch. 13, pp 225–266.
27. Cassel, D. and Selinger, Z. (1977) Proc. Natl Acad. Sci. USA 74, 3307–3311.
28. Cassel, D. and Selinger, Z. (1978) Proc. Natl Acad. Sci. USA 75, 4155–4159.
29. Gilman, A.G. (1987) Annu. Rev. Biochem. 56, 615–649.
30. Kent, R.S., De Lean, A. and Lefkowitz, R.J. (1980) Mol. Pharmacol. 17, 14–23.
31. Maguire, M.E., Van Arsdale, P.M. and Gilman, A.G. (1976) Mol. Pharmacol. 12, 335–339.
32. Shorr, R.G.L., Lefkowitz, R.J. and Caron, M.G. (1981) J. Biol. Chem. 256, 5820–5826.
33. Cerione, R.A., Codina, J., Benovic, J.L., Lefkowitz, R.J., Birnbaumer, L. and Caron, M.G. (1984) Biochemistry 23, 4519–4525.
34. Ferguson, K.M., Higashijima, T., Smigel, M. and Gilman, A.G. (1986) J. Biol. Chem. 261, 7393–7399.
35. Murayama, T. and Ui, M. (1984) J. Biol. Chem. 259, 761–769.
36. Pfister, C., Chabre, M., Plouet, J., Tuyen, V.V., De Kozak, Y., Faure, J.P. and Kuehn, H. (1985) Science 228, 891–893.
37. Howlett, A.C. and Gilman, A.G. (1980) J. Biol. Chem. 260, 2861–2866.
38. Northup, J.K., Sternweis, P.C. and Gilman, A.G. (1983) J. Biol. Chem. 258, 11361–11378.
39. Northup, J.K., Smigel, M.D., Sternweis, P.C. and Gilman, A.G. (1983) J. Biol. Chem. 258, 11369–11376.
40. Codina, J., Hildebrandt, J.D., Birnbaumer, L. and Sekura, R.D. (1984) J. Biol. Chem. 259, 11408–11418.
41. Heithier, H., Froehlich, M., Dees, C., Baumann, M., Haering, M., Gierschik, P., Schiltz, E., Vaz, W.L.C., Hekman, M. and Helmreich, E.J.M. (1992) Eur. J. Biochem. 204, 1169–1181.

42. Pfeuffer, E., Dreher, R.M., Metzger, H. and Pfeuffer, T. (1985) Proc. Natl Acad. Sci. USA 82, 3086–3090.
43. Pfeuffer, E., Mollner, S. and Pfeuffer, T. (1985) EMBO J. 4, 3675–3679.
44. Mollner, S., Simmoteit, R., Palm, D. and Pfeuffer, T. (1991) Eur. J. Biochem. 195, 281–286.
45. Yatani, A., Imoto, Y., Codina, J., Hamilton, S.L., Brown, A.M. and Birnbaumer, L. (1988) J. Biol. Chem. 263, 9887–9895.
46. Hamilton, S.L., Codina, J., Hawkes, M.J., Yatani, A., Sawada, T., Stickland, F.M., Froehner, S.C., Spiegel, A.M., Toro, L., Stefani, E., Birnbaumer, L. and Brown, A.M. (1991) J. Biol. Chem. 266, 19528–19535.
47. Asano, T., Pedersen, S.E., Scott, C.W. and Ross, E.M. (1984) Biochemistry 23, 5460–5467.
48. Brandt, D.R. and Ross, E.M. (1986) J. Biol. Chem. 261, 1656–1664.
49. Katada, T., Bokoch, G.M., Northup, J.K., Ui, M. and Gilman, A.G. (1984) J. Biol. Chem. 259, 3568–3577.
50. Katada, T., Bokoch, G.M., Smigel, M.D., Ui, M. and Gilman, A.G. (1984) J. Biol. Chem. 259, 3586–3595.
51. Cerione, R.A., Staniszewski, C., Gierschik, P., Codina, J., Somers, R.L., Birnbaumer, L., Spiegel, A.M., Caron, M.G. and Lefkowitz, R.J. (1986) J. Biol. Chem. 261, 9514–9520.
52. Hekman, M., Holzhoefer, A., Gierschik, P., Im, M.-J., Jakobs, K.-H., Pfeuffer, T. and Helmreich, E.J.M. (1987) Eur. J. Biochem. 169, 431–439.
53. Birnbaumer, L., Hildebrandt, J.D., Codina, J., Mattera, R., Cerione, R.A., Sunyer, T., Rojas, F.J., Caron, M.G., Lefkowitz, R.J. and Iyengar, R. (1985) In: Molecular Mechanisms of Signal Transduction, (Cohen, P. and Houslay, M.D., Eds.) pp 131–182, Elsevier/North Holland Biomedical Press, Amsterdam, Netherlands.
54. Tang, W.-J. and Gilman, A.G. (1991) Science 254, 1500–1503.
55. Kurstjens, N.P., Froehlich, M., Dees, C., Cantrill, R.C., Hekman, M. and Helmreich, E.J.M. (1991) Eur. J. Biochem. 197, 167–176.
56. Cerione, R.A., Staniszewski, C., Benovic, J.L., Lefkowitz, R.J. and Caron, M.G. (1985) J. Biol. Chem. 260, 1493–1500.
57. Tolkovsky, A.M. and Levitzki, A. (1978) Biochemistry 17, 3795–3810.
58. Tolkovsky, A.M., Braun, S. and Levitzki, A. (1982) Proc. Natl Acad. Sci. USA 79, 213–217.
59. Guy, P.M., Koland, J.G. and Cerione, R.A. (1990) Biochemistry 29, 6954–6964.
60. Katada, T., Oinuma, M. and Ui, M. (1986) J. Biol. Chem. 261, 5215–5221.
61. Marbach, I., Bar-Sinai, A., Minich, M. and Levitzki, A. (1990) J. Biol. Chem. 265, 9999–10004.
62. Logothetis, D.E., Kurachi, Y., Galper, J., Neer, E.J. and Clapham, D.E. (1987) Nature 325, 321–326.
63. Logothetis, D.E., Kim, D., Northup, J.K., Neer, E.J. and Clapham, D.E. (1988) Proc. Natl Acad. Sci. USA 85, 5814–5818.
64. Brown, A.M. and Birnbaumer, L. (1988) Am. J. Physiol. 254, H.401–H.410.
65. Birnbaumer, L., Abramowitz, J. and Brown, A.M. (1990) Biochem. Biophys. Acta 1031, 163–224.
66. Im, M.-J., Holzhöfer, A., Böttinger, H., Pfeuffer, T. and Helmreich, E.J.M. (1988) FEBS Lett. 227, 225–229.
67. Levitzki, A. and Helmreich, E.J.M. (1979) FEBS Lett. 101, 213–219.
68. Cuatrecasas, P., Jacobs, S. and Bennet, V. (1975) Proc. Natl Acad. Sci. USA 72, 1739–1743.
69. Citri, Y. and Schramm, M. (1980) Nature 287, 279–300.
70. Henis, Y.I., Hekman, M., Elson, E.L. and Helmreich, E.J.M. (1982) Proc. Natl Acad. Sci. USA 79, 2907–2911.
71. Helmreich, E.J.M. and Elson, E.L. (1984) In: Advances in Cyclic Nucleotide and Protein Phosphorylation Research (P. Greengard and G.A. Robison, Eds) Vol. 18, pp. 1–62, Raven Press, New York.
72. Hekman, M., Feder, D., Keenan, A.K., Gal, A., Klein, H.W., Pfeuffer, T., Levitzki, A. and Helmreich, E.J.M. (1984) EMBO J. 3, 3339–3345.

73. Feder, D., Im, M.-J., Klein, H.W., Hekman, M., Holzhoefer, A., Dees, C., Levitzki, A., Helmreich, E.J.M. and Pfeuffer, T. (1986) EMBO J. 5, 1509–1514.
74. Pfeuffer, T. and Helmreich, E.J.M. (1988) In: Current Topics in Cellular Regulation (B.L. Horecker and E.R. Stadtman, Eds) Vol. 29, 129–216, Academic Press Inc., New York.
75. Liebman, P.A., Sitaramayya, A., Parkes, J.H. and Buzdygon, B. (1984) Trends Pharmacol. Sci. 5, 293–296.
76. Stryer, L. (1991) J. Biol. Chem. 266, 10711–10714.
77. Sternweis, P.C., Northup, J.K., Smigel, M.D. and Gilman, A.G. (1981) J. Biol. Chem. 256, 11517–11526.
78. Vuong, T.M., Chabre, M. and Stryer, L. (1984) Nature 311, 659–661.
79. Baylor, D.A., Nunn, B.J. and Schnap, J.L. (1984) J. Physiol. (London) 357, 575–607.
80. Schleicher, A., Kuehn, H. and Hofmann, K.P. (1989) Biochemistry 28, 1770–1775.
81. Mc Naughton, P.A. (1990) Physiol. Rev. 70, 847–883.
82. Koch, K.-W. and Stryer, L. (1988) Nature (London) 334, 64–66.
83. Dizhoor, A.M., Ray, S., Kumar, S., Niemi, G., Spencer, M., Brolley, D., Walsh, K.A., Philipov, P.P., Hurley, J.B. and Stryer, L. (1991) Science 251, 915–918.
84. Fung, B.K.-K. (1983) J. Biol. Chem. 258, 10495–10502.
85. Yamanaka, G., Eckstein, F. and Stryer, L. (1985) Biochemistry 24, 8094–8101.
86. Arshavsky, V.Yu., Antoch, M.P. and Philipov, P.P. (1987) FEBS Lett. 224, 19–22.
87. Sitaramayya, A., Casadevall, C., Bennett, N. and Hakki, S. (1988) Biochemistry 27, 4880–4887.
88. Phillips, W.J. and Cerione, R.A. (1988) J. Biol. Chem. 253, 15498–15505.
89. Vuong, T.M. and Chabre, M. (1990) Nature (London) 346, 71–74.
90. Vuong, T.M. and Chabre, M. (1991) Proc. Natl Acad. Sci. USA 88, 9813–9817.
91. Arshavsky, V.Yu., Gray-Keller, M.P. and Bownds, M.D. (1991) Biophys. J. 59, 407a (abstr.).
92. Trahey, M. and Mc Cormick, F. (1987) Science 238, 542–545.
93. Bauer, P.H., Müller, S., Puzicha, M., Pippig, S., Obermaier, B., Helmreich, E.J.M. and Lohse, M.J. (1992) Nature (London) 358, 73–76.
94. Shorr, R.G.L., Lefkowitz, R.J. and Caron, M.G. (1981) J. Biol. Chem. 256, 5820–5826.
95. Shorr, R.G.L., Strohsacker, M.W., Lavin, T.N., Lefkowitz, R.J. and Caron, M.G. (1982) J. Biol. Chem. 257, 12341–12350.
96. Feder, D., Arad, H., Gal, A., Hekman, M., Helmreich, E.J.M. and Levitzki, A. (1984) In: Advances in Cyclic Nucleotide and Protein Phosphorylation Research (P. Greengard and G.A. Robison, Eds) Vol. 17, 61–71.
97. Burgermeister, W., Hekman, M. and Helmreich, E.J.M. (1982) J. Biol. Chem. 257, 5306–5311.
98. Frielle, T., Collins, S., Daniel, K.W., Caron, M.G., Lefkowitz, R.J. and Kobilka, B.K. (1987) Proc. Natl Acad. Sci. USA 84, 7920–7924.
99. Machida, C.H., Bunzow, J.R., Searles, R.P., Van Tol, H., Tester, B., Neve, K.A., Teal, P., Nipper, V. and Civelli, O. (1990) J. Biol. Chem. 265, 12960–12965.
100. Dixon, R.A.F., Kobilka, B.K., Strader, D.J., Benovic, J.L., Dohlmann, H.G., Frielle, T., Bolanowski, M.A., Bennett, C.D., Rands, E., Diehl, R.E., Mumford, R.A., Slater, E.E., Sigal, I.S., Caron, M.G., Lefkowitz, R.J. and Strader, C.D. (1986) Nature 321, 75–79.
101. Kobilka, B.K., Dixon, R.A.F., Frielle, T., Dohlman, H.G., Bolanowski, M.A., Sigal, I.S., Yana-Feng, T.L., Francke, U., Caron, M.G. and Lefkowitz, R.J. (1987) Proc. Natl Acad. Sci. USA 84, 46–50.
102. Allen, J.M., Baetge, E.E., Abrass, I.B. and Palmiter, R.D. (1988) EMBO J. 7, 133–138.
103. Gocayne, J., Robinson, D.A., Fitz-Gerald, M.G., Chung, F.-Z., Kerlavage, A.R., Lentes, K.-U., Lai, J., Wang, C.-D., Fraser, C.M. and Venter, J.C. (1987) Proc. Natl Acad. Sci. USA 84, 8296–8300.
104. Buckland, P.R., Hill, R.M., Tidmarks, S.F. and Mc Guffin, P. (1990) Nucleic Acid Res. 18, 1053.
105. Henderson, R. and Unwin, P.N.T. (1975) Nature 257, 23–32.

106. Henderson, R., Baldwin, J.M., Ceska, T.A., Zemlin, F., Beckmann, E. and Downing, K.H. (1990) J. Mol. Biol. 213, 899–929.
107. Kobilka, B.K., Kobilka, T.S., Daniel, K., Regan, J.W., Caron, M.G. and Lefkowitz, R.J. (1988) Science 240, 1310–1316.
108. Dohlman, H.G., Thorner, J., Caron, M.G. and Lefkowitz, R.J. (1991) Annu. Rev. Biochem. 60, 653–688.
109. Minneman, K.P., Hedberg, A. and Molinoff, P.B. (1979) J. Pharmacol. Exp. Ther. 211, 502–508.
110. Harden, T.K. (1983) Pharmacol. Rev. 35, 5–32.
111. Wong, S.K., Parker, E.M. and Ross, E.M. (1990) J. Biol. Chem. 265, 6219–6224.
112. Lavin, T.N., Nambi, P., Heald, S.L., Jeffs, P.W., Lefkowitz, R.J. and Caron, M.G. (1982) J. Biol. Chem. 257, 12332–12340.
113. Rashidbaigi, A. and Ruoho, A.E. (1982) Biochem. Biophys. Res. Commun. 106, 139–148.
114. Hekman, M., Schiltz, E., Henis, Y.I., Elson, E.L. and Helmreich, E.J.M. (1984) Adv. Cyclic Nucleotide Protein Phosphorylation Res. 17, 47–60.
115. Jürss, R., Hekman, M. and Helmreich, E.J.M. (1985) Biochemistry 24, 3349–3354.
116. Boege, F., Jürss, R., Cooney, D., Hekman, M., Keenan, A. and Helmreich, E.J.M. (1987) Biochemistry 26, 2418–2425.
117. Beutler, E., Carson, D., Dannawi, H., Forman, L., Kuhl, W., West, C. and Westwood, B. (1983) J. Clin. Invest. 72, 648–655.
118. Dunkel, F.-G., Muench, G., Boege, F., Cantrill, R. and Kurstjens, N.P. (1989) Biochem. Biophys. Res. Commun. 165, 264–270.
119. Wong, S., K.-F., Slaughter, C., Ruoho, A.E. and Ross, E.M. (1988) J. Biol. Chem. 263, 7925–7928.
120. Hertel, C., Nunnally, H.H., Wong, S.K.-F., Murphy, E.A., Ross, E.M. and Perkins, J.P. (1990) J. Biol. Chem. 265, 17988–17994.
121. Parker, E.M. and Ross, E.M. (1991) J. Biol. Chem. 266, 9987–9996.
122. Luxembourg, A., Hekman, M., and Ross, E.M. (1991) FEBS Lett. 283, 155–158.
123. Strader, C.D., Dixon, R.A.F., Cheung, A.H., Candelore, M.R., Blade, A.D. and Sigal, I.S. (1987) J. Biol. Chem. 262, 16439–16443.
124. O'Dowd, B.F., Hnatowich, M., Regan, J.W., Leader, W.M., Caron, M.G. and Lefkowitz, R.J. (1988) J. Biol. Chem. 263, 15985–15992.
125. Palm, D., Muench, G., Dees, D. and Hekman, M. (1989) FEBS Lett. 254, 89–93.
126. Palm, D., Muench, G., Malek, D., Dees, C. and Hekman, M. (1990) FEBS Lett. 261, 294–298.
127. Muench, G., Dees, C., Hekman, M. and Palm, D. (1991) Eur. J. Biochem. 198, 357–364.
128. Koenig, B., Arendt, A., Mc Dowell, J.H., Kahlert, M., Hargrave, P.A. and Hofmann, K.P. (1989) Proc. Natl Acad. Sci. USA 86, 6878–6882.
129. Cheung, A.H., Sigal, I.S., Dixon, R.A.F. and Strader, C.D. (1989) Mol. Pharmacol. 35, 132–138.
130. Strader, C.D., Sigal, I.S. and Dixon, R.A.F. (1989) FASEB J. 3, 1825–1832.
131. Hausdorff, W.P., Hnatowich, M., O'Dowd, B.F., Caron, M.G. and Lefkowitz, R.J. (1990) J. Biol. Chem. 265, 1388–1393.
132. Frielle, T., Daniel, K.W., Caron, M.G. and Lefkowitz, R.J. (1988) Proc. Natl. Acad. Sci. USA 85, 9494–9498.
133. Marullo, S., Emorine, L.J., Strosberg, A.D. and Delavier-Klutchko, C. (1990) EMBO J. 9, 1471–1476.
134. Strader, C.D., Candelore, M.R., Hill, W.S., Sigal, I.S. and Dixon, R.A.F. (1989) J. Biol. Chem. 264, 13572–13578.
135. Strader, C.D., Sigal, J.S., Candelore, M.R., Rands, E., Hill, W.S. and Dixon, R.A.F. (1988) J. Biol. Chem. 263, 10267–10271.
136. Eimerl, S., Schramm, M., Lok, S., Goodman, M., Khan, M. and Melmon, K. (1987) Biochem. Pharmacol. 36, 3523–3527.
137. Portenier, M., Hertel, C., Mueller, P. and Staehelin, M. (1984) J. Receptor Res. 4, 103–111.

138. Mohell, N. and Dicker, A. (1989) Biochem. J. 261, 401–405.
139. Strader, C.D., Candelore, M.R., Hill, W.S., Dixon, R.A.F. and Sigal, I.S. (1989) J. Biol. Chem. 264, 16470–16477.
140. Heithier, H., Jaeggi, K.A., Ward, L.D., Cantrill, R.C. and Helmreich, E.J.M. (1988) Biochimie 70, 687–694.
141. Tota, M.R. and Strader, C.D. (1990) J. Biol. Chem. 265, 16891–16897.
142. O'Dowd, B.F., Hnatowich, M., Caron, M.G., Lefkowitz, R.J. and Bouvier, M. (1989) J. Biol. Chem. 264, 7564–7569.
143. Ovchinnikov, Yu.A., Abdulaev, N.G. and Bogachuk, A.S. (1988) FEBS Lett. 230, 1–5.
144. Dohlman, H.G., Caron, M.G., De-Blasi, A., Frielle, A. and Lefkowitz, R.J. (1990) Biochemistry 29, 2335–2342.
145. Fraser, C.M. (1989) J. Biol. Chem. 264, 9266–9270.
146. Dixon, R.A.F., Sigal, I.S., Candelore, M.R., Register, R.B., Scattergood, W., Rands, E. and Strader, C.D. (1987) EMBO J. 6, 3269–3275.
147. Benovic, J.L., Shorr, R.G.L., Caron, M.G. and Lefkowitz, R.J. (1984) Biochemistry 23, 4510–4518.
148. Lomasney, J.W., Leeb-Lundberg, L.M., Cotecchia, S., Regan, J.W., De Bernardis, J.F., Caron, M.G. and Lefkowitz, R.J. (1986) J. Biol. Chem. 261, 7710–7716.
149. Regan, J.W., Nakata, H., De Maninis, R.M., Caron, M.G. and Lefkowitz, R.J. (1986) J. Biol. Chem. 261, 3894–3900.
150. Repaske, M.G., Nunnari, J.M. and Limbird, L.E. (1987) J. Biol. Chem. 262, 12381–12386.
151. Dohlman, H.G., Bouvier, M., Benovic, J.L., Caron, M.G. and Lefkowitz, R.J. (1987) J. Biol. Chem. 262, 14282–14288.
152. Rands, E., Candelore, M.R., Cheung, A.H., Hill, W.S., Strader, C.D. and Dixon, R.A.F. (1990) J. Biol. Chem. 265, 10759–10764.
153. Stiles, G.L., Benovic, J.L., Caron, M.G. and Lefkowitz, R.J. (1984) J. Biol. Chem. 259, 8655–8663.
154. Benovic, J.L., Staniszewski, C., Cerione, R.A., Codina, J., Lefkowitz, R.J. and Caron, M.G. (1987) J. Receptor. Res. 7, 257–281.
155. Boege, F., Ward, M., Jürss, R., Hekman, M. and Helmreich, E.J.M. (1988) J. Biol. Chem. 263, 9040–9049.
156. George, S.T., Ruoho, A.E. and Malbon, C.C. (1986) J. Biol. Chem. 261, 16559–16564.
157. Clark, A.J. (1933) The Mode of Action of Drugs on Cells, E. Arnold & Co., London.
158. Stephenson, R.P. (1956) Br. J. Pharmacol. 11, 379–394.
159. Furchgott, R.F. (1966) Adv. Drug Res. 3, 21–56.
160. Lohse, M.J. (1990) J. Biol. Chem. 265, 3210–3211.
161. Green D.A. and Clark R.B. (1981) J. Biol. Chem. 256, 2105–2108.
162. Kassis, S. and Fishman, P.H. (1984) Proc. Natl Acad. Sci. USA 81, 6686–6690.
163. Kassis, S., Olasmaa, M., Sullivan, M. and Fishman, P.H. (1986) J. Biol. Chem. 261, 12233–12237.
164. Strulovici, B., Cerione, R.A., Kilpatrick, B.F., Caron, M.G. and Lefkowitz, R.J. (1984) Science 225, 837–840.
165. Keenan, A.K., Cooney, D., Holzhöfer, A., Dees, C. and Hekman, M. (1987) FEBS Lett. 217, 287–291.
166. Lohse, M.J., Andexinger, S., Pitcher, J., Trukawinski, S., Codina, J., Faure, J-P., Caron, M.G. and Lefkowitz, R.J. (1992) J. Biol. Chem. 267, 8558–8664.
167. Pitcher, J., Lohse, M.J., Codina, J., Caron, M.G. and Lefkowitz, R.J. (1992) Biochemistry 31, 3193–3197.
168. Benovic, J.L., DeBlasi, A., Stone, W.C., Caron M.G. and Lefkowitz, R.J. (1989) Science 246, 235–240.
169. Benovic, J.L., Onorato, J.J., Arrizza, J.L., Stone, C.W., Lohse, M.J., Jenkins, N., Gilbert, N.G., Caron, M.G. and Lefkowitz, R.J. (1991) J. Biol. Chem. 266, 14939–14946.

170. Benovic, J.L., Pike, L.J., Cerione, R.A., Staniszewski, C., Yoshimasa, T., Codina, J., Caron, M.G. and Lefkowitz, R.J. (1985) J. Biol. Chem. 260, 7094–7101.
171. Bouvier, M., Leeb-Lundberg, L.M.F., Benovic, J.L., Caron, M.G. and Lefkowitz, R.J. (1987) J. Biol. Chem. 262, 3106–3113.
172. Lohse, M.J., Benovic, J.L., Codina, J., Caron, M.G. and Lefkowitz, R.J. (1990) Science 248, 1547–1550.
173. Lorenz, W., Inglese, J., Palczewski, K., Onorato, J.J., Caron, M.G. and Lefkowitz, R.J. (1991) Proc. Natl Acad. Sci. USA 88, 8715–8719.
174. Wilden, U., Hall, S.W. and Kühn, H. (1986) Proc. Natl Acad. Sci. USA 83, 1174–1178.
175. Strasser, R.H., Sibley, D.R. and Lefkowitz, R.J. (1986) Biochemistry 25, 1371–1377.
176. Hausdorff, W.P., Bouvier, M., O'Dowd, B.F., Irons, G.P., Caron, M.G. and Lefkowitz, R.J. (1989) J. Biol. Chem. 264, 12657–12665.
177. Lohse, M.J., Benovic, J.L., Caron, M.G. and Lefkowitz, R.J. (1990) J. Biol. Chem. 265, 3202–3209.
178. Benovic, J.L., Strasser, R.H., Caron, M.G. and Lefkowitz, R.J. (1986) Proc. Natl Acad. Sci. USA 83, 2797–2801.
179. Lohse, M.J., Lefkowitz, R.J., Caron, M.G. and Benovic, J.L. (1989) Proc. Natl. Acad. Sci. USA 86, 3011–3015.
180. Benovic, J.L., Onorato, J.J., Lohse, M.J., Dohlman H.G., Caron, M.G. and Lefkowitz, R.J. (1990) Br. J. Clin. Pharmacol. 30, 3S–12S.
181. Onorato, J.J., Regan, J.W., Caron, M.G., Lefkowitz, R.J. and Benovic, J.L. (1991) Biochemistry 30, 5118–5125.
182. Benovic, J.L., Mayor, F., Somers, R.L., Caron, M.G. and Lefkowitz, R.J. (1986) Nature 322, 869–872.
183. Benovic, J.L., Regan, J.W., Matsui, H., Mayor, F., Cotecchia, S., Leeb-Lundberg, L.M.F., Caron, M.G. and Lefkowitz, R.J. (1987) J. Biol. Chem. 262, 17251–17253.
184. Kwatra, M.M., Benovic, J.L., Caron, M.G., Lefkowitz, R.J. and Hosey, M.M. (1989) Biochemistry 28, 4543–4547.
184a. Attramadal, H., Lohse, M.J., Caron, M.G. and Lefkowitz, R.J. (1992) Clin. Res. 40, A190.
184b. Haga, K. and Haga, T. (1990) FEBS Lett. 268, 43–47.
184c. Haga, K. and Haga, T. (1990) J. Biol. Chem. 267, 2222–2227.
184d. Pitcher, J.A., Inglese, J., Higgins, J.B., Arrizza, J.L., Casey, P.J., Kim, C., Benovic, J.L., Kwatra, M.M., Caron, M.G. and Lefkowitz, R.J. (1992) Science 257, 1264–1267.
185. Blake, A.D., Mumford, R.A., Strout, H.V., Slater, E.G. and Strader, C.D. (1987) Biochem. Biophys. Res. Commun. 147, 168–173.
186. Bouvier, M., Collins, S., O'Dowd, B.F., Campbell, P.T., DeBlasi, A., Kobilka, B.K., MacGregor, C., Irons, G.P., Caron, M.G. and Lefkowitz, R.J. (1989) J. Biol. Chem. 264, 16786–16792.
187. Clark, R.B., Kunkel, M.W., Friedman, J., Goka, T.J. and Johnson, J.A. (1988) Proc. Natl Acad. Sci. USA 85, 1442–1446.
188. Strader, C.D., Sigal, I.S., Blake, A.D., Cheung, A.H., Register, R.B., Rands, E., Zemcik, B.A., Candelore, M.R. and Dixon, R.A.F. (1987) Cell 49, 855–863.
189. Okamoto, T., Murayama, Y., Hayashi, Y., Inagaki, M., Ogata, E. and Nishimoto, I. (1991) Cell 67, 723–730.
190. Clark, R.B., Friedman, J., Johnson, J.A. and Kunkel, M.W. (1987) FASEB J. 1, 289–297.
191. Roth, N., Campbell, P.T., Caron, M.G., Lefkowitz, R.J. and Lohse, M.J. (1991) Proc. Natl Acad. Sci. USA 88, 6201–6204.
192. Toews, M.L., Liang, M. and Perkins, J.P. (1987) Mol. Pharmacol. 32, 737–742.
193. Zhou, X.-M. and Fishman, P.H. (1991) J. Biol. Chem. 266, 7462–7468.
194. Strasser, R.H., Benovic, J.L., Caron, M.G. and Lefkowitz, R.J. (1986) Proc. Natl. Acad. Sci. USA 83, 6362–6366.
195. Mayor, F. Jr., Benovic, J.L., Caron, M.G. and Lefkowitz, R.J. (1987) J. Biol. Chem. 262, 6468–6471.

196. Reneke, J.E., Blumer, K.J., Courchesne, W.E. and Thorner, J. (1988) Cell 55, 221–234.
197. Vaughan, R.A. and Devreotes, P.N. (1988) J. Biol. Chem. 263, 14538–14543.
198. Meier, K. and Klein, C. (1988) Proc. Natl. Acad. Sci. USA 85, 2181–2185.
199. Cassill, J.A., Whitney, M., Joazeiro, C.A.P., Becker, A. and Zuker, C.S. (1991) Proc. Natl. Acad. Sci. USA 88, 11067–11070.
200. Harden, T.K., Cotton, C.V., Waldo, G.L., Lutton, J.K. and Perkins, J.P. (1980) Science 210, 441–443.
201. Waldo, G.L., Northup, J.K., Perkins, J.P. and Harden, T.K. (1983) J. Biol. Chem. 258, 13900–13908.
202. Hertel, C., Staehelin, M. and Perkins, J.P. (1983) J. Cyclic Nucleotide Protein Phosphorylation Res. 9, 119–128.
203. Toews, M.L. and Perkins, J.P. (1984) J. Biol. Chem. 259, 2227–2235.
204. Stadel, J.M., Nambi, P., Shorr, R.G.L., Sawyer, D.F., Caron, M.G. and Lefkowitz, R.J. (1983) Proc. Natl. Acad. Sci. USA 80, 3173–3177.
205. Chuang, D.-M., Dillon-Carter, O., Spain, J.N., Laskowski, M.W., Roth, B.L. and Coscia, C.J. (1986) J. Neurosci. 6, 2578–2584.
206. Hertel, C., Coulter, S.J. and Perkins, J.P. (1986) J. Biol. Chem. 261, 5974–5980.
207. Raposo, G., Dunia, I., Delavier-Klutchko, C., Kaveri, S., Strosberg, A.D. and Benedetti, E.L. (1989) Eur. J. Cell. Biol. 50, 340–352.
208. Guillet, J.G., Kaveri, S.V., Durieu, O., Delavier-Klutchko, C., Hoebeke, J. and Strosberg, A.D. (1985) Proc. Natl. Acad. Sci. USA 82, 1781–1784.
209. Zemcik, B.A. and Strader, C.D. (1988) Biochem. J. 251, 333–339.
210. Wang, H., Berrios, M. and Malbon, C.C. (1989) Biochem. J. 263, 533–538.
211. Wang, H.-Y., Berrios, M., Hadcock, J.R. and Malbon, C.C. (1991) Int. J. Biochem 23, 7–20.
212. Mahan, L.C., Koachman, A.M. and Insel, P.A. (1985) Proc. Natl. Acad. Sci. USA 82, 129–133.
213. Campbell P.T., Hnatowich, M., O'Dowd, B.F., Caron, M.G., Lefkowitz, R.J. and Hausdorff, W.P. (1991) Mol. Pharmacol. 39, 192–198.
214. Cheung, A.H., Dixon, R.A.F., Hill, W.S., Sigal, I.S. and Strader, C.D. (1990) Mol. Pharmacol. 37, 775–779.
215. Sibley, D.R., Strasser, R.H., Benovic, J.L., Daniel, K. and Lefkowitz, R.J. (1986) Proc. Natl. Acad. Sci. USA 83, 9408–9412.
216. Mahan, L.C., McKernan, R.M. and Insel, P.A. (1987) Annu. Rev. Pharmacol. Toxicol. 27, 215–235.
217. Hughes, R.J. and Insel, P.A. (1986) Mol. Pharmacol. 29, 521–530.
218. Neve, K.A. and Molinoff, P.B. (1986) Mol. Pharmacol. 30, 104–111.
219. Mahan, L.C. and Insel, P.A. (1986) Mol. Pharmacol. 29, 7–15.
220. Jasper, J.R., Motulsky, H.J., Mahan, L.C. and Insel, P.A. (1990) Am. J. Physiol. 259, C41–C46.
221. Hughes, R.J., Howard, M.J., Allen, J.M. and Insel, P.A. (1991) Mol. Pharmacol. 40, 974–979.
222. Doss, R.C., Perkins, J.P. and Harden, T.K. (1981) J. Biol. Chem. 256, 12281–12286.
223. Homburger, V., Pantaloni, C., Lucas, M., Gozlan, H. and Bockaert, J. (1984) J. Cell. Physiol. 121, 589–597.
224. Pitha, J., Hughes, B.A., Kusiak, J.W., Dax, E.M. and Baker, S.P. (1982) Proc. Natl. Acad. Sci. USA 79, 4424–4427.
225. Nelson, C.A., Muther, T.F., Pitha, J. and Baker, S.P. (1986) J. Pharmacol. Exp. Ther. 237, 830–836.
226. Snavely, M.D., Ziegler, M.G. and Insel, P.A. (1985) Mol. Pharmacol. 27, 19–26.
227. Allen, J.M., Abrass, I.B. and Palmiter, R.D. (1989) Mol. Pharmacol. 36, 248–255.
228. Shear M., Insel P.A., Melmon K.L. and Coffino P. (1976) J. Biol. Chem. 251, 7572–7576.
229. Su, Y.F., Harden, T.K. and Perkins, J.P. (1980) J. Biol. Chem. 255, 7410–7419.
230. Collins, S., Bouvier, M., Bolanowski, M.A., Caron, M.G. and Lefkowitz, R.J. (1989) Proc. Natl. Acad. Sci. USA 86, 4853–4857.
231. Hadcock, J.R., Ros, M. and Malbon, C.C. (1989) J. Biol. Chem. 264, 13956–13961.

232. Miller, R.T., Masters, S.B., Sullivan, K.L., Beiderman, B. and Bourne, H.R. (1988) Nature 334, 712–714.
233. Hadcock, J.R. and Malbon, C.C. (1988) Proc. Natl. Acad. Sci. USA 85, 5021–5025.
234. Hadcock, J.R., Wang, H. and Malbon, C.C. (1989) J. Biol. Chem. 264, 19928–19933.
235. Akamizu, T., Ikuyama, S., Saji, M., Kosugi, S., Kozak, C., McBridge, O.W. and Kohn, L. (1990) Proc. Natl. Acad. Sci. USA 87, 5677–5681.
236. Izzo, N.J., Seidman, C.E., Collins, S. and Colucci, W.S. (1990) Proc. Natl. Acad. Sci. USA 87, 6268–6271.
237. Wang, S.-Z., Hu, J., Long, R.M., Pou, W., Forray, C. and El-Fakahany, E.E. (1990) FEBS Lett. 276, 185–188.
238. Fukamauchi, F., Hough, C. and Chuang, D.-M. (1991) J. Neurochem. 56, 716–719.
239. Collins, S., Lohse, M.J., O'Dowd, B., Caron, M.G. and Lefkowitz, R.J. (1991) Vitam. Horm. (NY) 46, 1–39.
240. Hadcock, J.R. and Malbon, C.C. (1991) Trends Neurosci 14, 242–247.
241. Evans, R.M. (1988) Science 240, 889–895.
242. Collins, S., Caron, M.G. and Lefkowitz, R.J. (1988) J. Biol. Chem. 263, 9067–9070.
243. Hadcock, J.R. and Malbon, C.C. (1988) Proc. Natl. Acad. Sci. USA 85, 8415–8419.
244. Lazarwesley, E., Hadcock, J.R., Malbon, C.C., Kunos, G. and Ishac, E.J.N. (1991) Endocrinology 129, 1116–1118.
245. Bilezikian, J.P. and Loeb, J.N. (1983) Endocr. Rev. 4, 378–388.
246. Sundaresan, P.R. and Banerjee, S.P. (1987) Horm. Res. 27, 109–118.
247. Collins, S., Quarmby, V., French, F., Lefkowitz, R.J. and Caron, M.G. (1988) FEBS Lett. 233, 173–176.
248. Collins, S., Altschmied, J., Herbsman, O., Caron, M.G. and Lefkowitz, R.J. (1990) J. Biol. Chem. 265, 19330–19335.
249. Sakaue, M. and Hoffman, B.B. (1991) J. Biol. Chem. 266, 5743–5749.
250. Hershey, A.D., Dykeman, P.D. and Krause, J.E. (1991) J. Biol. Chem. 266, 4366–4374.
251. Thomas, J.A. and Marks, B.H. (1978) Am. J. Cardiol. 41, 223–233.
252. Francis, G.S., Goldsmith, S.R. and Cohn, J.N. (1982) Am. Heart J. 104, 725–731.
253. Chidsey, C.A., Braunwald, E. and Morrow, A.G. (1965) Am. J. Med. 39, 442–451.
254. Cohn, J.N., Levine, T.B., Olivari, M.T., Garberg, V., Lura, D., Francis, G.S., Simon, A.B. and Rector, T. (1984) N. Engl. J. Med. 311, 819–823.
255. Hasking, G.J., Esler, M.D., Jennings, G.L., Burton, D., Johns, J.A. and Korner, P.I. (1986) Circulation 73, 615–621.
256. Petch, M.C. and Nayler, W.G. (1979) Br. Heart J. 41, 336–339.
257. Bristow, M.R., Ginsburg, R., Minobe, W., Cubicciotti, R.S., Sageman, W.S., Lurie, K., Billigham, M.E., Harrison, D.C. and Stinson, E.B. (1982) N. Engl. J. Med. 307, 205–211.
258. Bristow, M.R., Ginsburg, R., Umans, V., Fowler, M., Minobe, W., Rasmussen, R., Zera, P., Menlove, R., Shah, P., Jamieson, S. and Stinson, E.B. (1986) Circ. Res. 59, 297–309.
259. Fowler, M.B., Laser, J.A., Hopkins, G.L., Minobe, W. and Bristow, M.R. (1986) Circulation 74, 1290–1302.
260. Brodde, O.-E., Zerkowski, H.R., Doetsch, N., Motomura, S., Khamssi, M. and Michel, M.C. (1989) J. Am. Coll. Cardiol. 14, 323–331.
261. Frey, M.J. and Molinoff, P.B. (1989) J. Cardiovasc. Pharmacol. 14(5), S13–S18.
262. Ginsburg, R., Bristow, M.R., Billingham, M.E., Stinson, E.B., Schroeder, J.S. and Harrison, D.C. (1983) Am. Heart J. 106, 535–540.
263. Ginsburg, R., Esserman, L.J. and Bristow, M.R. (1983) Ann. Intern. Med. 98, 603–606.
264. Baumann, G., Mercader, D., Busch, U., Felix, S.B., Loher, U., Ludwig, L., Sebening, H., Heidecke, C.D., Hagl, S., Sebening, F. and Blömer, H. (1983) J. Cardiovasc. Pharmacol. 5, 618–625.

265. Fan, T.-H.M., Liang, C.-S., Kawashima, S. and Banerjee, S.P. (1987) Eur. J. Pharmacol. 140, 123–132.
266. Heilbrunn, S.M., Shah, P., Bristow, M.R., Valantine, H.A., Ginsburg, R. and Fowler, M.B. (1989) Circulation 79, 483–490.
267. Vanhees, L., Aubert, A., Fagard, R., Hespel, P. and Amery, A. (1986) J. Cardiovasc. Pharmacol. 8, 1086–1091.
268. Strauss, M.H., Reeves, R.A., Smith, D.L. and Leenen, F.H.H. (1986) Clin. Pharmacol. Ther. 40, 108–115.
269. Brodde, O.-E., Daul, A., Wellstein, A., Palm, D., Michel, M.C. and Beckeringh, J.J. (1988) Am. J. Physiol. 254, H199–H206.
270. Longabaugh, J.P., Vatner, D.E., Vatner, S.F. and Homcy, C.J. (1988) J. Clin. Invest. 81, 420–424.
271. Marzo, K.P., Frey, M.J., Wilson, J.R., Liang, B.T., Manning, D.R., Lanoce, V. and Molinoff, P.B. (1991) Circ. Res. 69, 1546–1556.
272. Feldman, A.M., Cates, A.E., Veazey, W.B., Hershberger, R.E., Bristow, M.R., Baughman, K.L., Baumgartner, W. and Van Dop, C. (1988) J. Clin. Invest. 82, 189–197.
273. Neumann, J., Scholz, H., Döring, V., Schmitz, W., v. Meyeninck, L. and Kalmar, P. (1989) Lancet ii, 105–106.
274. Böhm, M., Gierschik, P., Jakobs, K.-H. Pieske, B., Schnabel, P., Ungerer, M. and Erdman, E. (1990) Circulation 82, 1249–1265.
275. Reithmann, C., Gierschik, P., Werdan, K. and Jakobs, K.H. (1989) Eur. J. Pharmacol. 172, 211–221.
276. Marquetant, R., Schwencke, C., Gierschik, P. and Strasser, R.H. (1990) Z. Kardiol. 79, 378.
277. Marquetant, R., Brehm, B. and Strasser, R. (1992) J. Mol. Cell. Cardiol. 24, 535–548.
278. Reithmann, C., Gierschik, P., Werdan, K. and Jakobs, K.H. (1990) Br. J. Clin. Pharmacol. 30 Suppl. 1, 118S–120S.
279. Feldman, A.M., Ray, P.E., Silan, C.M., Mercer, J.A., Minobe, W. and Bristow, M.R. (1991) Circulation 83, 1866–1872.
280. Limas, C.J., Goldenberg, I.F. and Limas, C. (1989) Circ. Res. 64, 97–103.
281. Waagstein, F., Hjalmarson, A., Varnauskas, E. and Wallentin, I. (1975) Br. Heart J. 37, 1022–1036.
282. Engelmeier, R.S., O'Connell, J.B., Walsh, R., Rad, N., Scanlon, P.J. and Gunnar, R.M. (1985) Circulation 72, 536–546.
283. Gilbert, E.M., Anderson, J.L., Deitchman, D., Bartholomew, M., Mealey, P., Yanowitz, F.G., O'Connell, J.B., Renlund, D.G. and Bristow, M.R. (1987) Circulation 76, 1423.
284. Rona, G. (1985) J. Mol. Cell. Cardiol. 17, 291–306.
285. Hjalmerson, A., Elmford, D., Herlitz, J., Holmberg, S., Malek, I., Ryden, L., Svedberg, K., Vedin, A., Waagstein, F., Waldenstroem, A., Waldenstroem, J., Wedel, H., Wilhelmson, L. and Wilhelmson, C. (1981) Lancet ii, 823–827.
286. Beta-Blocker Heart Attack Trail Research Group (1982) J. Am. Med. Assoc. 247, 1707–1714.
287. Menken, U., Wiegand, V., Bucher, P. and Meesman, W. (1979) Cardiovasc. Res. 13, 588–594.
288. Corr, P.B., Witkowski, F.X. and Sobel, B.E. (1978) J. Clin. Invest. 61, 109–119.
289. Penny, W.J. (1984) Eur. Heart J. 5, 960–973.
290. E'Puddu, P., Jouve, R., Langlet, F., Guillen, J.C., Fornaris, M., Torresani, J. and Reale, A. (1986) Cardiovasc. Res. 20, 721–726.
291. Nokin, P., Clinet, M. and Schoenfeld, P. (1983) Arch. Int. Physiol. Biochem. 91, B110–B111.
292. Wollenberger, A., Krause, E.G. and Heier, G. (1969) Biochem. Biophys. Res. Commun. 36, 664–670.
293. Freissmuth, M., Schütz, W., Weindlmayer-Göttel, M., Zimpfer, M. and Spiss, C.K. (1987) J. Cardiovasc. Pharmacol. 10, 568–574.
294. Devos, C., Robberecht, P., Waelbroeck, M., Clienet, M., Beaufort, P., Schoenfield, P. and Christophe, J. (1985) Naunyn-Schmiedeberg's Arch. Pharmacol. 331, 71–75.
295. Karliner, J.S., Stevens, M.B., Honbo, N. and Hoffman, J.I.E. (1989) J. Clin. Invest. 83, 474–481.
296. Strasser, R.H., Krimmer, J. and Marquetant, R. (1988) J. Cardiovasc. Pharmacol. 12, S15–S24.

297. Strasser, R.H., Krimmer, J., Dullaens-B., R., Marquetant, R. and Kübler, W. (1990) J. Mol. Cell. Cardiol. 22, 1405–1423.
298. Vatner, D.E., Young, M.A., Knight, D.R. and Vatner, S.F. (1990) Am. J. Physiol. 258, H140–H144.
299. Maisel, A.S., Motulsky, H.J. and Insel, P.A. (1985) Science 230, 183–186.
300. Maisel, A.S., Motulsky, H.J., Ziegler, M.G. and Insel, P.A. (1987) Am. J. Physiol. 253, H1159–H1166.
301. Jennings, R.B. and Steenbergen, C. (1985). Annu. Rev. Physiol. 47, 727–749.
302. De Lean, A., Stadel, J.M. and Lefkowitz, R.J. (1980) J. Biol. Chem. 255, 7108–7117.
303. Stadel, J.M., De Lean, A. and Lefkowitz, R.J. (1980) J. Biol. Chem. 255, 1436–1441.
304. Goyal, R.K. (1989) N. Engl. J. Med. 321, 1022–1029.
305. Rauch, B., Weinbrenner, C., Marquetant, R., Schwencke, C., Beyer, T., Kübler, W., Hasselbach, W. and Strasser, R.H. (1990) J. Mol. Cell. Cardiol. 22, 17.
306. Gilman, A.G. (1989). G proteins and regulation of adenylyl cyclase. J. Am. Med. Assoc. 262, 1819–1825.
307. Freissmuth, M., Casey, P.J. and Gilman, A.G. (1989) FASEB J. 3, 2125–2131.
308. Susanni, E.E., Manders, W.T., Knight, D.R., Vatner, D.E., Vatner, S.F. and Homcy, C.J. (1989) Circ. Res. 65, 1145–1150.
309. Strasser, R.H., Marquetant, R. and Kübler, W. (1990) Br. J. Pharmacol. 30, 27S–35S.
310. Yatani, A., Codina, J., Imoto, Y., Reeves, J.P., Birnbaumer, L. and Brown, A.M. (1987) Science 238, 1288–1292.
311. Yatani, A. and Brown, A.M. (1989) Science 245, 71–74.
312. Schubert, B., VanDongen, A.M.J., Kirsch, G.E. and Brown, A.M. (1989) Science 245, 516–519.
313. Yatani, A., Mattera, R., Codina, J., Graf, R., Okabe, K., Padrell, E., Iyengar, R., Brown, A.M. and Birnbaumer, L. (1988) Nature 336, 680–682.
314. Vatner, D.E., Vatner, S.F., Fujii, A.M. and Homcy, C. (1985) J. Clin. Invest. 76, 2259–2264.
315. Vatner, D.E., Knight, D.R., Shen, Y.T., Thomas, J.X.J., Homcy, C.J. and Vatner, S.F. (1988) J. Mol. Cell. Cardiol. 20, 75–82.
316. Samuels, H.H., Forman, B.M., Horowitz, Z.D. and Ye, Z.-S. (1988) J. Clin. Invest. 81, 957–967.
317. Wildenthal, K. (1972) J. Clin. Invest. 51, 2702–2709.
318. Kunos, G., Vermes-Kunos, I. and Nickerson, M. (1974) Nature 250, 779–781.
319. Tse, J., Wrenn, R.W. and Kuo, J.F. (1980) Endocrinology 107, 6–16.
320. Levey, G.S., Skelton, C.L. and Epstein, S.E. (1969) Endocrinology 85, 1004–1009.
321. Guarnieri, T., Filburn, C.R., Beard, E.S. and Lakatta, E.G. (1980) J. Clin. Invest. 65, 861–868.
322. Krawietz, W., Werdan, K. and Erdman, E. (1982) Biochem. Pharmacol. 31, 2463–2469.
323. Malbon, C.C. and Greenberg, M.L. (1982) J. Clin. Invest. 69, 414–426.
324. Malbon, C.C., Rapiejko, P.J. and Mangano, T.J. (1985) J. Biol. Chem. 260, 2558–2564.
325. Polikar, R., Kennedy, B., Maisel, A., Ziegler, M., Smith, J., Dittrich, H. and Nicod, P. (1990) J. Am. Coll. Cardiol. 15, 94–98.
326. Dillmann, W.H. (1989) Annu. Rev. Med. 40, 373–394.
327. Tsai, J.S. and Chen, A. (1978) Nature 275, 138–140.
328. Bahouth, S.W. (1991) J. Biol. Chem. 266, 15863–15869.
329. Bedotto, J.B., Gay, R.G., Graham, S.D., Morkin, E. and Goldman, S. (1989) J. Pharmacol. Exp. Ther. 248, 632–636.
330. Elks, M.L. and Manganiello, V.C. (1985) Endocrinology 117, 947–953.
331. Delemer, B., Dib, K., Saunier, B., Haye, B., Jacquemin, C. and Corrèze, C. (1991) Mol. Cell. Endocrinol. 75, 123–131.
332. Mills, I., Garcia Sainz, J.A. and Fain, J.N. (1986) Biochim. Biophys. Acta 876, 619–630.
333. Saggerson, E.D. (1986) Biochem. J. 238, 387–394.
334. Milligan, G., Spiegel, A.M., Unson, C.G. and Saggerson, E.D. (1987) Biochem. J. 247, 223–227.
335. Milligan, G. and Saggerson, E.D. (1990) Biochem. J. 270, 765–769.

336. Ros, M., Northup, J.K. and Malbon, C.C. (1988) J. Biol. Chem 263, 4362–4368.
337. Rapiejko, P.J., Northup, J.K., Evans, T., Brown, J.E. and Malbon, C.C. (1986) Biochem. J. 240, 35–40.
338. Rapiejko, P.J., Watkins, D.C., Ros, M. and Malbon, C.C. (1989) J. Biol. Chem. 264, 16183–16189.
339. Weinstein, L.S., Shenker, A., Gejman, P.V., Merino, M.J., Friedman, E. and Spiegel, A.M. (1991) N. Engl. J. Med. 325, 1688–1695.
340. Landis, C.A., Masters, S.B., Spada, A., Pace, A.M., Bourne, H.R. and Vallar, L. (1989) Nature 340, 692–696.
341. Masters, S.B., Miller, R.T., Chi, M.-H., Chang, F.-H., Beiderman, B., Lopez, N.G. and Bourne, H.R. (1989) J. Biol. Chem. 264, 15467–15474.
342. Clementi, E., Malgaretti, N., Meldolesi, J. and Taramelli, R. (1990) Oncogene 5, 1059–1061.
343. Lyons, J., Landis, C.A., Harsh, G., Vallar, L., Grenewald, K., Feichtinger, H., Duh, Q.Y., Clark, O.H., Kawasaki, E., Bourne, H.R. et al. (1990) Science 249, 655–659.
344. Suarez, H.G., Du Villard, J.A., Caillou, B., Schlumberger, M., Parmentier, C. and Monier, R. (1991) Oncogene 6, 677–679.
345. Cassel, D. and Pfeuffer, T. (1978) Proc. Natl. Acad. Sci. USA 75, 2669–2673.
346. Fishman, P.H. (1990) In: ADP-ribosylating Toxins and G-Proteins (J. Moss and M. Vaughan, Eds) Ch. 8, pp. 127–140, Am. Soc. Microbiology, Washington, D.C.
347. Burton, F.H., Hasel, K.W., Bloom, F.E. and Sutcliffe, J.G. (1991) Nature 350, 74–77.
348. Dumont, J.E., Jauniaux, J.-C. and Roger, P.P. (1989) TIBS 14, 67–71.
349. Maenhaut, C., Roger, P.P., Reuse, S. and Dumont, J.E. (1991) Biochimie 73, 29–36.
350. Perutz, M.F. (1992) Protein Structure. New Approaches to Disease and Therapy. New York, W.H. Freeman and Company.
351. Yarden, Y., Rodriguez, H., Wong, S.K.-F., Brandt, D.R., May, D.C., Burnier, J., Harkins, R.N., Chen, E.J., Ramachandran, J., Ullrich, A. and Ross, E.M. (1986) Proc. Natl Acad. Sci. USA 83, 6795–6799.
352. Birnbaumer, L. (1990) Annu. Rev. Pharmacol. Toxicol. 30, 675–705.
353. Sternweis, P.C. (1986) J. Biol. Chem. 261, 631–637.
354. Lohse, M.J., Klotz, K.-N. and Schwabe, U. (1991) Mol. Pharmacol. 39, 517–523.
355. Gross, W. and Lohse, M.J. (1991) Mol. Pharmacol. 39, 524–530.
356. Ji, I. and Ji, T.H. (1991) J. Biol. Chem. 266, 14953–14957.
357. Horstman, D., Brandon, S., Wilson, A.L., Guyer, C.A., Cragoe, E.J. and Limbird, L.E. (1990) J. Biol. Chem. 265, 21590–21595.
358. Zastrow, M.V. and Kobilka, B.K. (1992) J. Biol. Chem. 267, 3530–3538.
359. Phillips, W.J. and Cerione, R.A. (1992) J. Biol. Chem. 267, 17032–17039.
360. Phillips, W.J., Wong, S.C. and Cerione, R.A. (1992) J. Biol. Chem. 267, 17040–17046.
361. Kleuss, C., Scherübl, H., Hescheler., J., Schultz, G. and Wittig, B. (1992) Nature 358, 424–426
362. Kleuss, C., Scherübl, H., Hescheler, J., Schultz, G. and Wittig, B. (1993) Science 259, 832–834.
363. Ungerer, M., Böhm, M., Elce, J.S., Erdmann, E. and Lohse, M.J. (1993) Circulation 87, 454–463.
364. Schertler, G.F.X., Villa, C., Henderson, R. (1993) Nature 362, 770–772.

III. Receptors for 'Classic' Neurotransmitters

F. Hucho (Ed.) *Neurotransmitter Receptors*
© 1993 Elsevier Science Publishers B.V. All rights reserved.

CHAPTER 6

GABA$_A$ and glycine receptors

F. ANNE STEPHENSON

Department of Pharmaceutical Chemistry, School of Pharmacy, 29/39 Brunswick Square, London WC1N 1AX, U.K.

1. Introduction

The amino acids, γ-aminobutyric acid (GABA) and glycine, are both inhibitory neurotransmitters that mediate fast postsynaptic inhibition in the nervous system. Their action is to bind specifically to GABA$_A$ and glycine receptors, respectively. This is followed within milliseconds by the gating or opening of an integral chloride ion channel which results, in general, in the hyperpolarisation of the recipient neuronal cell. The GABA$_A$ and glycine receptors are structurally related and their amino acid sequences identify them as members of the ligand-gated ion-channel superfamily of which the excitatory nicotinic acetylcholine receptor is the prototype. Their pharmacological, functional, and biochemical properties are described in this Chapter.

2. GABA$_A$ and glycine receptor molecular pharmacology

2.1. GABA$_A$ receptors

GABA receptors predominate in the brain where they have a widespread distribution. GABA$_A$, as distinct from GABA$_B$, receptors are defined by their competitive antagonism by the plant alkaloid, bicuculline, and the noncompetitive antagonism by the plant-derived convulsant, picrotoxin. In contrast, the pharmacological subclass of GABA receptors, the GABA$_B$ receptor, is bicuculline- and picrotoxin-insensitive, but sensitive to the GABA$_B$ receptor agonist, (−)baclofen. The GABA$_B$ receptor belongs to the G-protein-coupled receptor superfamily; it is distinct at all levels from the GABA$_A$ receptor and will not be discussed further here. However, for a review, see [1].

Abbreviations: BZ, Benzodiazepine (Figures only); DMCM, Methyl-4-ethyl-6,7-dimethoxy-β-carboline-3-carboxylate; GABA, γ-Aminobutyric acid; TBPS, t-Butyl bicyclophosphorothionate

GABA$_A$ receptor function is regulated by several distinct classes of compounds, most notably the anxiolytic benzodiazepines, such as Valium and Librium, the barbiturates, neurosteroids, anthelminthic agents like the avermectins, and Zn^{2+}. The allosteric interaction of these regulatory sites was demonstrated using radioligand-binding studies. These findings have led to the proposal that the actions of benzodiazepines and barbiturates are mediated by the specific interaction with endogenous GABA$_A$ receptors. They are summarised below: more detail is found in [2].
1. GABA agonists enhance benzodiazepine binding to brain synaptic membranes in a dose-dependent manner to saturation. Scatchard analysis of benzodiazepine binding in the absence and presence of GABA showed that this enhancement is due to a decrease in the dissociation constant (K_d), i.e. a shift to higher affinity. It is commonly referred to as the 'GABA shift'. The enhancement is blocked by the competitive GABA$_A$ receptor antagonist, bicuculline. Physiologically, benzodiazepines have no intrinsic action. However, in the presence of GABA, they potentiate the resultant chloride ion conductance. The mechanism by which this occurs is by an increase in the frequency of channel opening in the presence of benzodiazepines [3].
2. Barbiturates such as (−)pentobarbital, at physiologically relevant concentrations, enhance both GABA and benzodiazepine binding to brain synaptic membranes in a dose-dependent, chloride ion-dependent manner to saturation. Scatchard analysis of benzodiazepine binding in the presence and absence of barbiturate showed that the enhancement is due to a decrease in K_d as for the effect of GABA on benzodiazepine binding. The analogous experiments for the effect of barbiturates on GABA binding show decreases in both the K_d values that are found in the control membranes as well as changes in the relative B_{max} values. Barbiturate enhancement is blocked by picrotoxin competitively, by bicuculline noncompetitively, and for benzodiazepine binding, the enhancement by GABA and by barbiturates is additive indicating two distinct allosteric sites. Barbiturates facilitate GABA neurotransmission by increasing the mean open time of the chloride ion channel [3].
3. Neurosteroids, e.g. 5β-pregnan-3α-ol-20-one, potentiate both GABA and benzodiazepine binding in a stereospecific manner which is similar to that found for the barbiturates. Barbiturate and steroid enhancements, however, are additive showing a distinction between the two sites. The facilitation by steroids is associated with a prolongation of the individual channel openings [4].
4. The anthelminthic avermectins, glycosidic macrocyclic lactones isolated from *Streptomyces avermitilis* bacteria, also potently enhance GABA- and benzodiazepine-binding sites with an EC_{50} = 40 nM. These enhancements again are similar but not identical to those found for the barbiturates, thus suggesting a unique site of interaction [5].

Benzodiazepine pharmacology is complex and subclasses of benzodiazepine-binding sites were delineated initially by radioligand-binding assay. It is now known that these have a genetic basis (see Section 5.1). Thus agonist benzodiazepines are anxiolytic and have a positive GABA shift, whereas antagonist benzodiazepines, such as Ro-15-1788, have no GABA shift but they block the physiological effects of agonists. Inverse agonists act at the benzodiazepine recognition site even though they are anxiogenic and in radioligand-binding studies, they have a negative GABA shift, i.e. in the presence of GABA they bind to their site with reduced affinity. An example of an inverse agonist is the β-carboline, methyl-4-ethyl-6,7-dimethoxy-β-carboline-3-carboxylate (DMCM). Benzodiazepine partial agonists have also been described. Some ligands can discriminate between benzodiazepine-binding sites. Oxoquazepam and nonbenzodiazepines, such as CL-218872 and ethyl β carboline in contrast to classical benzodiazepines which show standard sigmoidal one-site displacement curves, inhibit conventional agonist benzodiazepine binding with a shallow displacement curve. This was interpreted as the existence of two affinity states of the receptor. The subpopulations were called the Type I and the Type II benzodiazepine recognition sites which had high and low affinity respectively for the discriminating compounds.

2.2. Glycine receptors

Glycine receptors are found predominantly in the spinal cord and in the brainstem. The pharmacology of the glycine receptors is less extensive than that of the $GABA_A$ receptors. They are defined by the antagonism by the convulsive alkaloid, strychnine, in contrast to the strychnine-insensitive glycine-binding site that is associated with the excitatory N-methyl-D-aspartate subclass of glutamate receptor. There are no known allosteric modulators of the glycine receptor which are analogous to the benzodiazepine and barbiturate regulatory sites of the $GABA_A$ receptor. However, avermectin B1a and the steroid-related compound RU-5135, both of which are potently active at $GABA_A$ receptors, also inhibit [^{3}H] strychnine binding with the same order of potency [6]. Some of these structures are shown in Figure 1.

3. $GABA_A$ and glycine receptor biochemistry and molecular biology

3.1. $GABA_A$ receptors

The biochemical analysis of $GABA_A$ receptors was facilitated by their isolation from mammalian brain by agonist benzodiazepine affinity chromatography [7,8]. These studies showed that:
1. $GABA_A$ receptor sites and t-butyl-bicyclophosphorothionate (TBPS), a ligand for the picrotoxin chloride channel-related site, copurified with the agonist, antagonist, and inverse agonist benzodiazepine-binding activity from bovine cerebral cortex.
2. The isolated receptor was a hetero-oligomeric glycoprotein with a molecular size

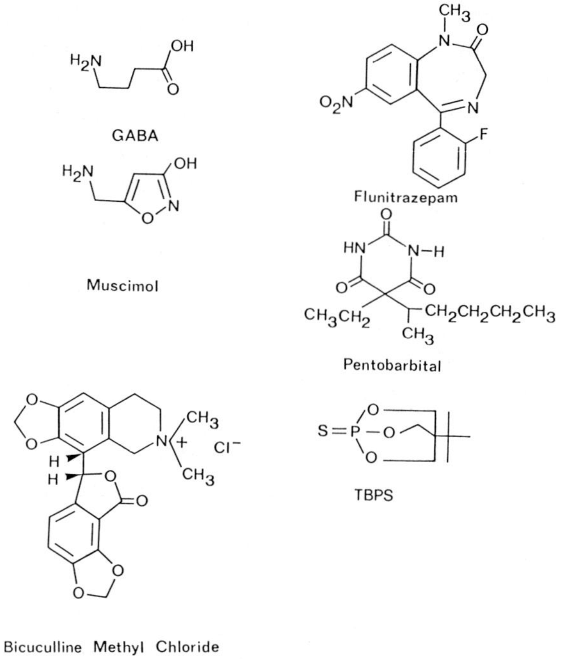

Fig. 1. A representative selection of the structures of some of the drugs that interact with the GABA$_A$ receptor. TBPS, t-butyl bicyclophosphorothionate.

of 250,000 daltons. Two polypeptides were identified on the basis of molecular weight differences, the α- and β-subunits with M$_r$ 53,000 and 58,000 respectively (however, see Section 3.1).
3. Both GABA- and benzodiazepine-binding activities were coimmunoprecipitated by monoclonal antibodies directed at either the α- or the β-subunits showing that both were integral components of the same structure [9].
4. The α-subunit was specifically labelled by the agonist benzodiazepine photoaffinity probe, [^{3}H]-flunitrazepam [7]. The β-subunit was predominantly photoaffinity labelled by the specific high affinity GABA$_A$ receptor agonist, [^{3}H]-muscimol (see [10] for a summary).
5. Importantly, however, the availability of the purified receptor permitted the determination of partial amino acid sequence information which led to the molecular cloning of GABA$_A$ receptor cDNAs [11].

Now, in common with the other neurotransmitter receptor families and with the availability of cDNA probes, multiple GABA$_A$ receptor genes have been discovered by screening cDNA libraries under conditions of reduced stringency (e.g. [12]). GABA$_A$ receptor genes currently number sixteen. The protein products of the GABA$_A$ receptor genes have been divided into five subunit classes, namely α, β, γ, δ, and ρ on the basis of their amino acid sequence identity. Within each subunit class isoforms exist, i.e. $\alpha 1-\alpha 6, \beta 1-\beta 4, \gamma 1-\gamma 3$, and $\rho 1-\rho 2$ where the enumeration is simply the historical order of discovery. Alternative splice variants of the $\gamma 2$ and $\beta 4$ subunits exist. For a more extensive review, see [13]. All the $GABA_A$ receptor polypeptides share structural features not only between themselves, but also with the other members of the ligand-gated ion-channel superfamily which includes the glycine receptor subunits, the nicotinic acetylcholine receptor subunits, and more recently the serotonin$_3$ (5-HT$_3$) receptor. The same domain organisation is found between the $GABA_A$ and glutamate receptor subunits, but here, there is no conservation of amino acid sequence similarity. Figure 2 shows the transmembrane model for the $GABA_A$ receptor polypeptides which was predicted from the deduced amino acid sequences.

$GABA_A$ receptor subunits consist of 424–521 amino acids with corresponding M_rs 48,500–64,000 for the core, i.e. nonglycosylated, polypeptides. They are predicted to have a large, hydrophilic N-terminal extracellular domain (all $\alpha 1$ numbering, 1–225), four transmembrane-spanning regions M1–M4 (M1–M3 226–307; M4 395–416), and an intracellular loop region between M3 and M4 (308–394). At the level of the polypeptide sequences, isoforms of one subunit class share at least 75% amino acid sequence identity, whereas the identity found between any two distinct subunit classes is of the order of 35%. The most divergent regions between the polypeptides of one subclass are found generally at the extreme N- and C-terminal domains and within the respective intracellular loops. Amino acid sequence similarity with the nicotinic acetylcholine receptor is found within the four transmembrane regions and within a region termed the 'cys-cys β loop' in the N-terminal extracellular domain. GABA is membrane impermeable and studies in whole cells with charged benzodiazepines have shown potentiation of GABA inhibition. Therefore, both these binding sites are found in the extracellular region (see also Section 5.1). Different combinations of five subunits coassemble to form functional receptors. However, these complements have not yet been determined (see Section 4.2).

3.2. Glycine receptors

Glycine receptors were purified by amino-strychnine affinity chromatography from the vertebrate spinal cord by the group of Dr Heinrich Betz in Germany [14]. Analysis of the purified preparations showed that:
1. The receptor is a pentameric glycoprotein with a molecular size of 250,000 daltons. Two polypeptides with M_r 48,000 (α-subunit) and M_r 58,000 (β-subunit) were identified. A subunit ratio of $\alpha_3\beta_2$ has been proposed.
2. The α-subunit is specifically photoaffinity labelled by [^{3}H]-strychnine and together with information from the characterisation of cloned glycine receptors, it is now known that this subunit contains the high affinity glycine recognition site.
3. A M_r 93,000 polypeptide that copurified with the glycine receptor was identified as a cytoplasmic peripheral membrane protein associated with the glycine

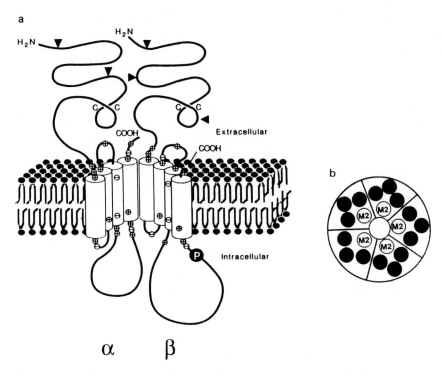

Fig. 2. A diagrammatic representation of the transmembrane organisation of the mammalian $GABA_A$ and glycine receptors. (a) The predicted transmembrane topology of the $GABA_A$ receptor α- and β-subunits seen from within the plane of the membrane. The same organisation is predicted for all $GABA_A$ and glycine receptor subunits and their isoforms. All the polypeptides are predicted to have four membrane-spanning helices which are shown as cylinders; ▼ represents the potential sites for N-glycosylation and P is the cyclic AMP-dependent protein kinase site of the β subunit. The charged residues shown are only those that are located adjacent to the ends of the membrane domains and that are thought to be involved in chloride ion binding. (b) A view of the predicted model perpendicular to the plane of the membrane. It represents each of the five subunits of the receptors as a segment within the annular structure where the hole in the centre represents the chloride ion channel. The four membrane-spanning regions within each polypeptide are shown as filled circles with the helix M2 lining the wall of the channel. The subunit complements and their ordering around the rossette for either the $GABA_A$ and glycine receptors have not yet been determined. (Reproduced in part (a) from Schofield et al. [11] by courtesy of the Editors of *Nature* (Copyright © 1987 Macmillan Magazines Ltd.))

postsynaptic membranes. The protein, named Gephyrin from the Greek meaning bridge, is a glycine receptor-tubulin linker protein.

The first glycine receptor cDNA to be identified was reported simultaneously with the molecular cloning of the $GABA_A$ receptor α- and β-subunits [15]. Like the $GABA_A$ receptors, multiple genes have now been identified, but so far they have been restricted to the two subunit classes which were identified by protein chemistry [16]. The glycine receptor subunit genes known are $\alpha 1, \alpha 2, \alpha 3, \alpha 4$, and β with alternative splice variants of the α1- and α2-subunits being found. These polypeptides share the same structural features as the $GABA_A$ receptor subunits (Fig. 2). Indeed, amino acid sequence comparison between a glycine receptor and a $GABA_A$ receptor subunit shows approximately the same percentage of conservation as found between pairs of $GABA_A$ receptor subunits of different subclasses. Some of the properties of $GABA_A$ and glycine receptors are compared in Table I.

4. Oligomeric receptor structures and their distribution

As described above, cross-linking studies in combination with monoclonal antibody detection has shown that the glycine receptor has a pentameric structure [17]. The quaternary structures of the $GABA_A$ receptors are not known, but by analogy it is presumed that they too are pentamers. In both cases, the actual polypeptide complements are unknown and their elucidation is complex with the existence of so many different subunits. Some points can, however, be made.

4.1. Glycine receptors

For the glycine receptor subunits, in-situ hybridisation studies have shown that each identified subunit mRNA has its own characteristic pattern of expression. α1 Transcripts are found at high levels in the spinal cord and brainstem characteristic of the classical distribution of glycine receptors determined by either [^{3}H]-strychnine receptor autoradiography or immunocytochemistry using anti-glycine receptor α subunit monoclonal antibodies. α2 and α3 Transcripts are found in higher brain regions, whereas β transcripts are found in higher brain regions and spinal cord. More re-

TABLE I
A comparison of $GABA_A$ and glycine receptors

Property	$GABA_A$	Glycine
Hill coefficient[1]	2	2.7
Conductance states (pS)	12,19,30*,44	12,20,30,46*
Effective pore (nm)[2]	0.56	0.52
Molecular size (kDa)	240	250
Subunit types	α,β,γ,δ,ρ	α,β
Quaternary structure	Not proven	Pentamer

*Denotes the most frequently observed conductance state in outside-out patches from mouse spinal neurons in culture [33].
[1]Hill coefficient for action of agonist to open the channel. Estimates may be complicated by desensitisation.
[2]The effective pore diameter of the respective open channels determined by the permeability properties for a series of large polyatomic cations.

cently, [^{3}H]-strychnine binding and glycine receptor subunit immunoreactivity have both been reported in higher brain regions. These findings of colocalisation are consistent with models for the receptor of $\alpha_3\beta_2$ combinations. It is not known if the different α subunit isoforms coexist within individual receptors. Interestingly, β-mRNA is expressed at high levels throughout the higher brain regions. This pattern is not compatible with the distribution of any of the α subunits and since the presence of an α subunit is required for the formation of the glycine-binding site, it has been suggested that the β subunit may coassemble with a GABA$_A$ receptor polypeptide. Levels of mRNA, however, are not always quantitatively correlated with translated protein, which may explain this apparent mismatch.

4.2. GABA$_A$ receptors

The determination of the oligomeric structures of the GABA$_A$ receptors is more difficult simply because the gene family is more extensive than the one for the glycine receptors. Extensive in-situ hybridisation studies using subunit-specific oligonucleotide probes have shown that each isoform of each subunit class has its own specific distribution (e.g. [18] and Fig. 3). For example, each α subunit isoform mRNA has a unique but partially overlapping pattern, with the α1-subunit being the most abundant with its occurrence in many different cell types throughout the brain. The other α subunits are more limited in their distribution. α2 and α3 Transcripts are found in relative abundance in rat brain in the hippocampal formation and the olfactory bulb, respectively, as well as in cerebral cortex for both; α4 is rare; α5 is in the hippocampus and α6 is restricted to the granule cells of the cerebellum. The δ-subunit mRNA is highly expressed in the granule cells of the cerebellum and ρ-subunit mRNA is found almost exclusively in the retina. Overall, the most abundant mRNAs are the α1, β2, and γ2.

These findings correlate well with the distribution of the respective translated products. That is, there is a unique brain-region distribution for each of the α1, α2, α3, α5, and α6 polypeptides [19,20]. Immunoaffinity purification of the different GABA$_A$ receptor subpopulations has demonstrated the existence of α subunit iso-oligomers as well as minor populations which contain more than one type of α sub-

---

Fig. 3. Macroscopic images of the radioautographic distribution of GABA$_A$ receptor α-subunit mRNAs in rat brain revealed by in-situ hybridisation. On the left are sagittal sections, on the right frontal sections of the distribution of the α1, α2, α3, α5, and α6 GABA$_A$ receptor subunit mRNAs in adult brain localised by in-situ hybridisation using ^{35}S-labelled oligonucleotide probes complementary to the least-conserved sequences in the cytoplasmic loops between the membrane-spanning domains M3 and M4. White areas represent brain regions with high levels of expression of receptor mRNA. Scale bars = 2.5 mm. Note the differential but partially overlapping pattern of distribution as discussed in the text. Reproduced in part from Persohn et al. [18] by courtesy of the Editors of J. Comp. Neurol. and with special gratitude to Dr Grayson Richards of F. Hoffmann-La Roche Ltd., Basle, Switzerland.

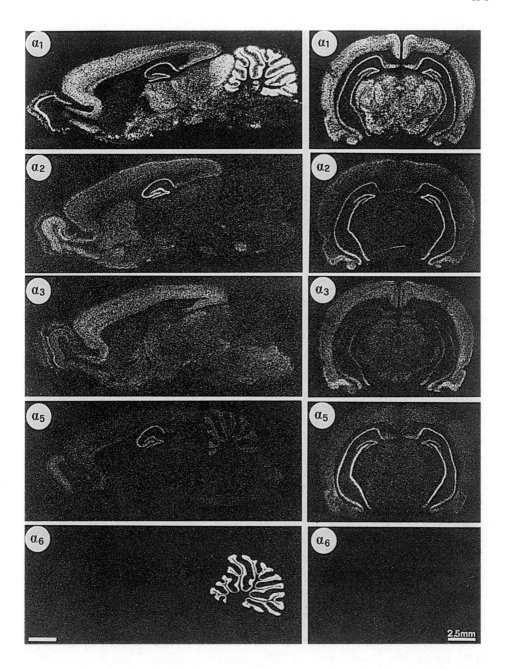

unit [21]. Furthermore, the most abundant receptor subunits are α1, β2/β3, and γ2 [22]. Thus, the latter is a prime candidate for a $GABA_A$ receptor, but the subunit stoichiometry and indeed the number of ligand-binding sites per receptor are not known. Other types of receptors in the main would contain only one type of α subunit. There are, however, examples where two different α subunits coassemble but their remaining subunit complements are again still to be determined.

5. Functional properties of cloned $GABA_A$ and glycine receptors

The functional significance of the diversity of the $GABA_A$ and glycine receptors is not understood. To understand why a particular cell type expresses a particular receptor subtype at a particular subcellular location at a specific point in development, we need to know what the differences are between various receptor subtypes. In general terms, these properties may be differences in channel activation and inactivation, pharmacological differences, differential regulation by either or both, intracellular messengers, and endogenous ligands such as the neurosteroids. Much information has been recently accrued with respect to these alternatives by the characterisation of cloned receptors expressed in either the *Xenopus* oocyte or mammalian cells. This section will summarise current knowledge.

5.1. $GABA_A$ receptors

Each $GABA_A$ receptor polypeptide when expressed alone can form a GABA-gated ion channel with selectivity for chloride anions. Where these properties have been analysed at the single channel level i.e. for α1 and β1 homo-oligomers, four conductance states have been found [23]. Native receptors also show four conductance states (Table I and [24]) thus this must be a property of each channel rather than one conductance state reflecting one subtype of receptor. With the exception of the ρ subunit which forms a robust homo-oligomer [25], the coexpression of two different subunits produces enhanced channel activity. A comparison of various cloned receptors shows differences in desensitisation rates, differential sensitivity to agonist activation, and differential sensitivity to inactivation by Zn^{2+}. The responses recorded from cloned receptors so far have not been identical to those from native receptors. In particular, native receptors have a Hill coefficient for GABA activation of ~ 2 which is interpreted as a cooperative binding of two molecules of GABA to open the channel. Most cloned receptors have Hill coefficients < 2; the most correspondence with native receptors is found with the α5β2γ2-subunit combination which shows higher Hill numbers [26].

The pharmacology of recombinant $GABA_A$ receptors has focused primarily on the benzodiazepine recognition site. In summary, the coexpression of a γ2 or γ3 polypeptide with an $\alpha\beta$ combination is requisite for the production of a reproducible benzodiazepine response with the full spectrum of agonist/antagonist/inverse agonist responsiveness [12]. α Variant receptors, i.e. $\alpha 1\beta 2\gamma 2$, $\alpha 2\beta 2\gamma 2$ etc., show differences in benzodiazepine pharmacological specificity [27]. Thus $\alpha 1$-subunit-containing receptors display Type I benzodiazepine receptor pharmacology (see Section 2.1), while $\alpha 2$-,$\alpha 3$- and $\alpha 5$-subunit-containing receptors display Type II specificity. However, $\alpha 5$-subunit-containing receptors have a lower affinity for the imidazolopyridazine, Zolpidem, which is known to act at benzodiazepine sites. These differences in affinity are due to a point mutation since substitution of a glycine in the $\alpha 3$-subunit for a glutamic acid found in the corresponding aligned position in the $\alpha 1$-subunit sequence (201 $\alpha 1$ numbering) converts the $\alpha 3$-subunit to a Type I receptor [28]. The recombinant $\alpha 4\beta 2\gamma 2$ and $\alpha 6\beta 2\gamma 2$ receptors have high affinity for the benzodiazepine partial inverse agonist Ro-15-4513 only. The loss of high affinity benzodiazepine binding in $\alpha 6$-subunit-containing receptors is also due to a point mutation since replacement of arginine by histidine 101. (rat $\alpha 1$-subunit numbering; 102 in bovine $\alpha 1$-subunit) in the $\alpha 6$-subunit restores high affinity agonist binding [30]. Expression of $\alpha 1\beta 2\gamma 1$ gives a receptor with high affinity for benzodiazepine agonists only.

There is a good correlation between the brain-region distribution of the subsets of benzodiazepine-binding sites determined by quantitative receptor autoradiography, the differential pattern of expression of α subunit mRNAs by in-situ hybridisation, and the protein products themselves by quantitative immunoprecipitation using isoform-specific antibodies. Note that whilst these three parameters concur in the main and thus point to the identification of $GABA_A$ receptor subpopulations, there is no endogenous ligand which is universally accepted as the one acting at these benzodiazepine sites, thus their physiological significance must be questioned.

Each $GABA_A$ receptor homo-oligomer is potentiated by barbiturates and neurosteroids. Whilst no differences in pharmacology have been reported at these sites, there are subtle differences in efficacy. Note that $\rho 1$ homo-oligomers are bicuculline-insensitive. Coexpression of the splice variant $\gamma 2L$ with an $\alpha\beta$ combination endows the $GABA_A$ receptor with ethanol sensitivity [31]. These points are summarised in Table II and Figure 4.

With regard to the significance of the differential regulation by intracellular messengers, several of the $GABA_A$ receptor subunits contain consensus sequences for various protein kinases in their respective intracellular loop domains (Table III and [32]). The $GABA_A$ receptor is phosphorylated in vitro by both protein kinase A and protein kinase C. The rundown of the GABA-activated conductance is prevented by magnesium ATP, implying that phosphorylation of the receptor is required to sustain function. As discussed above, the $\alpha\beta\gamma 2L$-cloned receptor is ethanol-sensitive and since this differs from the $\alpha\beta\gamma 2S$ receptor by eight amino acids which encode a protein kinase C consensus sequence, it is suggested that it is the cross talk between signalling pathways that results in phosphorylation [31].

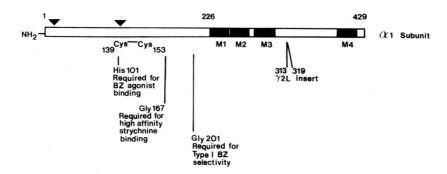

Fig. 4. Diagrammatic representation and inter-relationship of important domains of the $GABA_A$ and glycine receptor polypeptides. The figure shows the linear amino acid sequence of the bovine $GABA_A$ receptor α1-subunit highlighting significant residues and regions within this and other $GABA_A$ and glycine receptor polypeptides. ▼ Show the consensus sequences for N-glycosylation; M1–M4, are the four transmembrane domains and cys-cys, is the conserved extracellular motif common to all members of the ligand-gated ion-channel superfamily. BZ, benzodiazepine.

5.2. Glycine receptors

Native glycine receptors also have four conductance states [25]. The α subunit homo-oligomers expressed in *Xenopus* oocytes form robust chloride ion channels, but their single channel analysis has not yet been reported. The glycine-gated responses are

TABLE II
Characteristics of $GABA_A$ receptor polypeptides

Subunit type	Comments
α1	Type I benzodiazepine pharmacology in α1β1γ2 combinations
α2	Type II benzodiazepine pharmacology in α2β1γ2 combinations
α3	Type II benzodiazepine pharmacology in α3β1γ2 combinations
α4	α4β2γ2 receptors have high affinity for Ro-15-4513 only
α5	Type II benzodiazepine pharmacology and reduced Zolpidem affinity
α6	α6β2γ2 receptors have high affinity for Ro-15-4513 only
β1	
β2	
β3	
β4	Splice variant β4' found in chick
γ1	α1β2γ1 combinations bind benzodiazepine agonists only
γ2	This subunit confers a complete spectrum of benzodiazepine activities with αβ combinations; confers Zn^{2+} insensitivity; splice variants γ2S and γ2L found
γ3	This subunit confers a complete spectrum of benzodiazepine activities with αβ combinations
δ	Enriched in cerebellar granule cells
ρ1	Identified in the retina; forms robust homo-oligomers that are bicuculline-insensitive

TABLE III
A summary of potential phosphorylation sites of $GABA_A$ and glycine receptor subunits

Subunit type	Potential phosphorylation site
$GABA_A$ α4	2 Protein kinase A
	1 Protein kinase C
$GABA_A$ α5	3 Protein kinase C
$GABA_A$ α6	1 Protein kinase A
	2 Protein kinase C
$GABA_A$ β1, β2, β3, β4 (chick)	1 protein kinase A
$GABA_A$ γ1	1 Tyrosine kinase
$GABA_A$ γ2S	1 Tyrosine kinase
γ2L	1 Tyrosine kinase
	1 Protein kinase C
$GABA_A$ ρ1	2 Protein kinase C
Glycine α1L	1 Protein kinase C

blocked by strychnine. The α2* variant of the α2-subunit forms strychnine-insensitive responses. A comparison of this sequence with α2 together with site-directed mutagenesis studies showed that the substitution of glycine 167 with glutamate resulted in the loss of strychnine sensitivity [33]. Interestingly, this strychnine-insensitive glycine receptor subunit is developmentally regulated since it is highly expressed in rodent neonatal spinal cord. However, it is replaced two to three weeks after birth by the adult form α2 which correlates with the appearance of [^{3}H]-strychnine binding [34].

The glycine receptor β subunit will only show glycine gating of chloride ion channels after overexpression in oocytes and even then the responses are of low affinity.

The splice variant of the α1 glycine-receptor subunit contains an insertion of eight amino acids in the intracellular loop which has a putative protein kinase consensus sequence, similar to that found for the $GABA_A$ receptor γ2L form. Functional studies relating phosphorylation phenomena have still to be investigated.

6. *$GABA_A$ and glycine receptors in disease*

6.1. *Glycine receptors*

The mutant mouse *spastic* develops a characteristic motor disorder within two weeks of birth which manifests clinically with similar symptoms to sublethal poisoning with strychnine. Reduced levels of high affinity strychnine binding and glycine-mediated chloride ion conductances are found in these mice. It has been shown that the spastic mutation, localised to chromosome 3, selectively interferes with the accumulation of the adult glycine receptor α1-subunit [34]. Similarly, an inherited myoclonus of the Poll Hereford calves has been described which is also characterised by a specific loss of high affinity strychnine-binding sites. Both of these animal disorders of motor function resemble human diseases such as hyperkinesia and spastic paraplegia, implicating therefore deficiencies in glycine receptors in neurological disorders.

6.2. $GABA_A$ receptors

$GABA_A$ receptors have been implicated in mechanisms of anxiety by virtue of the fact that the benzodiazepines mediate their therapeutic effects via interaction with these proteins. Indeed, anxiety may be elicited in animal models by the pharmacological blockade of GABA neurotransmission at either the GABA, benzodiazepine, or chloride ion channel recognition sites. This stress is associated with a reduction in cortical receptors. More recently, the chromosomal localisations of the human $GABA_A$ receptor genes have been determined. So far, the α1-subunit maps to chromosome 5; the α2-subunit and the β1-subunit to chromosome 4; the α3-subunit to the X chromosome, and the β3-subunit to chromosome 15. The α3-subunit gene maps close to a region associated with a heritable form of manic-depressive illness [35]. Interestingly, deletions in the β3-subunit gene are found in the rare neurological disorder, Angelman's syndrome [36] which is characterised by seizures and mental retardation.

7. Concluding remarks

Nearly ten years ago, *the* $GABA_A$ and *the* glycine receptors were both isolated to apparent homogeneity. Advances in the last few years, however, have shown that this early work was premature and we now know that there exists a large gene family for both the related receptors. The protein products of these genes together with their spliced variants could give rise to possibly hundreds of $GABA_A$ and glycine receptor subpopulations. It is now a major challenge to researchers to decipher the functional significance of all the different subsets of receptors that exist in both normal and neurological disorders where defects in inhibitory neurotransmission are implicated.

References

1. Bowery, N.G. (1989) Trends Pharmacol. Sci. 10, 401–407.
2. Olsen, R.W. (1981) J. Neurochem. 37, 1–13.
3. Study, R.E. and Barker, J.L. (1981) Proc. Natl. Acad. Sci. USA 78, 7180–7184.

4. Simmonds, M.A. (1991) Semin. Neurosci. 3, 231–239.
5. Olsen, R.W. and Snowman, A.M. (1985) J. Neurochem. 44, 1074–1082.
6. Betz, H. (1987) Trends Neurosci. 10, 113–117.
7. Sigel, E., Stephenson, F.A., Mamalaki, C. and Barnard, E.A. (1983) J. Biol. Chem. 258, 6965–6971.
8. Sigel, E. and Barnard, E.A. (1984) J. Biol. Chem. 259, 7219–7223.
9. Schoch, P., Richards, J.G., Haring, P., Takacs, B., Stahl, C., Staehelin, T., Haefely, W. and Mohler, H. (1985) Nature 314, 168–170.
10. Stephenson, F.A. (1988) Biochem. J. 249, 21–32.
11. Schofield, P.R., Darlison, M.G., Fujita, N., Burt, D.R., Stephenson, F.A., Rodriguez, H., Rhee, L.M., Ramachandran, J., Reale, V., Glencorse, T.A., Seeburg, P.H. and Barnard, E.A. (1987) Nature 328, 221–227.
12. Pritchett, D.B., Sontheimer, H., Shivers, B.D., Ymer, S., Kettenmann, H., Schofield, P.R. and Seeburg, P.H. (1989) Nature 338, 582–585.
13. Olsen, R.W. and Tobin, A.J. (1990) FASEB J. 4, 1469–1480.
14. Pfeiffer, F., Graham, D. and Betz, H. (1982) J. Biol. Chem. 257, 9389–9393.
15. Grenningloh, G., Rienitz, A., Schmitt, B., Methfessel, C., Zensen, M., Beyreuther, K., Gundelfinger, E.D. and Betz, H. (1987) Nature 328, 215–220.
16. Betz, H. (1991) Trends Neurosci. 14, 458–461.
17. Langosch, D., Thomas, L. and Betz, H. (1988) Proc. Natl. Acad. Sci. USA 85, 7394–7398.
18. Persohn, E., Malherbe, P. and Richards, J.G. (1992) (in press). J. Comp. Neurol.
19. Duggan, M.J. and Stephenson, F.A. (1990) J. Biol. Chem. 265, 3831–3835.
20. McKernan, R.M., Quirk, K., Prince, R., Cox, P.A., Gillard, N.P., Ragan, C.I. and Whiting, P. (1991) Neuron, 7, 667–676.
21. Duggan, M.J., Pollard, S. and Stephenson, F.A. (1991) J. Biol. Chem. 266, 24778–24784.
22. Benke, D.S., Mertens, S., Treciak, A., Gillesen, D. and Mohler, H. (1991) J. Biol. Chem. 266, 4478–4483.
23. Blair, L.A.C., Levitan, E.S., Marshall, J., Dionne, V.E. and Barnard, E.A. (1988) Science 242, 577–579.
24. Hammill, O.P., Bormann, J. and Sakmann, B. (1983) Nature 305, 805–808.
25. Cutting, G.R., Lu, L., O'Hara, B.F., Kasch, L.M., Montrose-Rafizader, C., Donovan, D.M., Shimada, S., Antonarakis, S.E., Guggino, W.B., Uhl, G.R. and Kazazian, H.H. (1991) Proc. Natl Acad. Sci. USA 88, 2673–2677.
26. Sigel, E., Baur, R., Trube, G., Mohler, H. and Malherbe, P. (1990) Neuron 5, 703–711.
27. Pritchett, D.B., Luddens, H. and Seeburg, P.H. (1989) Science 245, 1389–1392.
28. Pritchett, D.B. and Seeburg, P.H. (1991) Proc. Natl Acad. Sci. USA 88, 1421–1425.
29. Wieland, H.A., Luddens, H. and Seeburg, P.H. (1992) J. Biol. Chem. 267, 1426–1429.
30. Wafford, K.A., Burnett, D.M., Leidenheimer, N.J., Burt, D.R., Wang, J.B., Kofuji, P., Dunwiddie, T.V., Harris, R.A. and Sikela, J.M. (1991) Neuron 7, 27–33.
31. Leidenheimer, N.J., Browning, M.D. and Harris, R.A. (1991) Trends Pharmacol. Sci. 12, 84–87.
32. Kuhse, J., Schmieden, V. and Betz, H. (1990) Neuron 5, 867–873.
33. Bormann, J. (1988) Trends Neurosci. 11, 112–116.
34. Becker, C.-M., Schiemeden, V., Tarroni, P., Strasser, U. and Betz, H. (1992) Neuron 8, 283–289.
35. Buckle, V.J., Fujita, N., Ryder-Cook, A.S., Derry, J.M.J., Barnard, P.J., Lebo, R.V., Schofield, P.R. et al. (1989) Neuron 3, 647–654.
36. Wagstaff, J., Knoll, J.H.M., Fleming, J., Kirkness, E.F., Martin-Gallardo, A., Greenberg, F., Graham, J.M., Menninger, J., Ward, D., Venter, J.C. and Lalande, M. (1991) Am. J. Hum. Gen. 49, 330–337.

F. Hucho (Ed.) *Neurotransmitter Receptors*
© 1993 Elsevier Science Publishers B.V. All rights reserved.

CHAPTER 7

Muscarinic acetylcholine receptors

JAAK JÄRV and AGO RINKEN

Laboratory of Bioorganic Chemistry, University of Tartu, 2 Jakobi Str, Tartu, Estonia

1. Introduction

Muscarinic acetylcholine receptors were defined by Sir Henry Dale [1] as receptors which are stimulated by acetylcholine and muscarine and blocked by atropine. Until now these effects have provided the best definition of the muscarinic receptor action in physiological, pharmacological, and biochemical meaning. Muscarinic receptors generate slow biological responses upon binding of agonists with a latency period of 100 ms [2] and are abundant not only in the central nervous system, heart, gastrointestinal tract, smooth muscle, and secretory organs, but also in circulating cells, such as erythrocytes and T lymphocytes [3]. The contamination of the receptor protein in tissues remains, in general, between 0.1 and 1 pmol/mg protein. However, even lower concentrations, like 0.02 pmol/mg protein in human erythrocytes and 0.01 pmol/mg protein in N1E-115 neuroblasoma cells, were investigated [3]. Besides the postsynaptic localization of muscarinic receptors, strong evidence exists about the regulation of acetylcholine release via presynaptic muscarinic receptors [4].

The major cholinergic pathways are the septohippocampal pathway and the habenulointerpedunclar tract (for more details see [5]). In the central nervous system most of the cholinergic innervation is of the muscarinic type and thus these receptors are thought to play a key role not only in neural mechanisms underlying memory, learning, and control of movements, but also in cognitive, emotional, and associative processes. The defects in muscarinic synaptic transmission can lead to variety of pathological states and neurological and mental disorders. Therefore, the muscarinic acetylcholine receptors have been the objects of extensive study by a large number of investigators and many crucial aspects of their function and structure have been clarified during the last decade.

2. Phenomena used for receptor assay

Different physiological and biochemical effects, triggered off by agonist binding with muscarinic receptor, or inhibition of these effects by antagonists have been used to characterize this receptor and the events which follow the receptor-ligand interaction, as well as to determine affinity of drugs by using the binding isotherm

$$B = \frac{B_{max}[L]}{K_d + [L]} \qquad (eq.\ 1)$$

where B is a quantitative measure of the effect observed at ligand concentration [L], B_{max} is the maximal effect achieved at saturation of the receptor sites by ligand L, and K_d is the drug-receptor complex dissociation constant. In principle, other affinity-related parameters like IC_{50}, EC_{50}, etc. can be measured depending on the experimental procedure and conditions used (see Chapter 2).

The most well-known organ-bath assay procedures are the muscarinic receptor-mediated contraction of the guinea pig, rat, or rabbit ileum [6], effects of muscarinic agonists on the isolated rat or rabbit heart [7], and muscarinic agonist-stimulated secretion of salivary and gastric juice quantified by measurement of some enzyme activity [8]. All these responses are atropine-sensitive, pointing to their connection with the muscarinic receptor system. On the other hand, however, it is important to emphasize that the possibility of negative feedback via presynaptic muscarinic receptors may make the meaning of the bioassay observations less clear. This shortcoming can be avoided in biochemical or biophysical assays by using isolated cells like oocytes or cell cultures and cell-free systems like synaptosomes or cell-membrane fragments.

The classical biochemical responses related to muscarinic receptors are activation of phospholipase C (EC 3.1.4.11), which cleaves phosphatidylinositol 4,5-bis-phosphate into inositol 1,4,5-triphosphate and diacylglycerol [9], activation of guanosine 3',5'-cyclic monophosphate synthesis by guanylate cyclase (EC 4.6.1.2) [10], and inhibition of synthesis of adenosine 3',5'-cyclic monophosphate by adenylate cyclase (EC 4.6.1.1) [11]. The activity of these enzymes can be measured by monitoring concentrations of the reaction products and regulated by muscarinic drugs in a dose-dependent manner.

The activation of muscarinic receptors is also related to electrophysiological responses like the suppression of the M-current in neurons [12] and an increase in the permeability of the cardiac muscle membranes to potassium ions through regulation of potassium channels [13]. In all these examples the regulatory effect is achieved by coupling the enzyme or channel system to muscarinic receptors via different signal-transducing G-proteins (see Section 7). As a result of that the activity of the whole system can be inhibited either by muscarinic antagonists, by some inhibitors of the coupled enzymes, by blockers of the coupled channels, or by effectors whose action is directed to the linking G-protein. In the latter case the inhibitory effect of cholera toxin and pertussis toxin through ADP-ribosylation of G-protein is widely used in studies of the receptor-mediated biochemical responses [14].

The first attempt of direct determination of muscarinic-binding sites in membrane fragments was made using radioactive atropine [15]. Although the results of this study were not clear-cut, a list of radioactive muscarinic ligands was begun and presently it includes [^{3}H]dexetimide [16], [^{3}H]N-methyl-4-piperidinyl benzilate [17], [^{3}H]N-methylatropine [18], [^{3}H]N-methylscopolamine [18], [^{3}H]pirenzepine [19], [^{3}H]N-propylbenzilylcholine mustard [20], [^{3}H]3-quinuclidinyl benzilate [21], [^{3}H]scopolamine [22], [^{3}H]N-methylquinuclidinyl benzilate [23], and [^{125}I]3-quinuclidinyl benzilate [24]. The determination of muscarinic receptors with these radioligands, which all appear to be muscarinic antagonists (see in Fig. 1), is based on their stoichiometric complex formation with the receptor and on the possibility of effective separation of the complex from the excess of radioligand. Most commonly the filtration or rapid centrifugation methods are used for this purpose [25]. Since both procedures dramatically shift the equilibrium state of the assay system, the radioligands should dissociate rather slowly from the receptor-ligand complex [26]. This requirement is fulfilled in the case of potent muscarinic antagonists, as shown by kinetic studies described in Section 6.2. If used as radioligands, most of the muscarinic antagonists reveal the same number of receptor sites [20]; only pirenzepine seems to give unclear results with receptor titration [19].

Attempts have been made to use radiolabelled agonists [^{3}H]acetylcholine [27], [^{3}H]oxotremorine-M [28], [^{3}H]pilocarpine [29], and cis-[^{3}H]methyldioxolane [30] in direct binding studies. The formulas of these ligands are shown in Figure 1. In these experiments it has been revealed that only part of the receptor sites can be detected using all of these ligands. Commonly, this fact is explained by the presence of at least two types of receptor sites, and the radioactive agonists usually label just the high affinity sites because of the practical limits on the ligand concentration used in the binding assay. On the other hand, the same situation can be explained by different kinetic mechanisms of agonist and antagonist binding with muscarinic receptors [31], since rapid dissociation of the agonist-receptor complex interferes with the applicability of the filtration or centrifugation methods in the case of these ligands. Thus, both affinity and kinetic parameters are critical for selection of radioligands for direct binding studies as shown below.

In summary, it should be emphasized that all these approaches assume that a true equilibrium between the ligand-receptor complex and the free ligand has been achieved. In many cases this is, however, rather problematic due to the slowness of dissociation of potent antagonists, used as radioactive 'reporter' ligands. This means that in these experiments the order of mixing of the ligands as well as the incubation time and temperature may play crucial role in obtaining meaningful results.

AGONISTS

ACETYLCHOLINE $\quad CH_3C(O)OCH_2CH_2N^+(CH_3)_3$
CARBACHOL $\quad NH_2C(O)OCH_2CH_2N^+(CH_3)_3$
ETHOXYETHYLTRIMETHYLAMMONIUM $\quad CH_3CH_2OCH_2CH_2N^+(CH_3)_3$

cis-METHYLDIOXOLANE

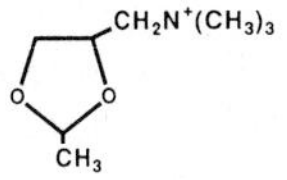

MUSCARINE

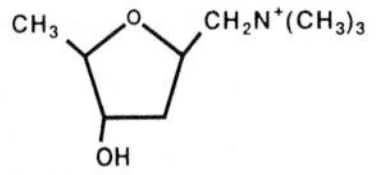

McN-A 343

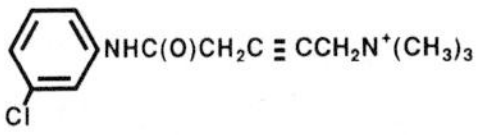

OXOTREMORINE

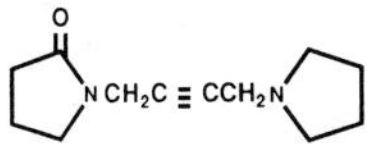

PENTYLTRIMETHYLAMMONIUM $\quad CH_3(CH_2)_4N^+(CH_3)_3$

PILOCARPINE

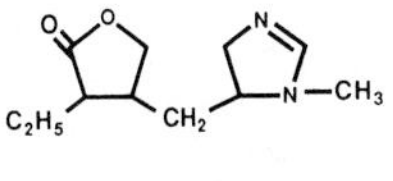

TETRAMETHYLAMMONIUM $\quad CH_3-N^+(CH_3)_3$

PARTIAL AGONISTS

ETHOXYETHYLDIMETHYLETHYLAMMONIUM $\quad CH_3CH_2OCH_2CH_2N^+(CH_3)_2C_2H_5$
ETHOXYETHYLTRIETHYLAMMONIUM $\quad CH_3CH_2OCH_2CH_2N^+(C_2H_5)_3$

ANTAGONISTS

AF-DX 116

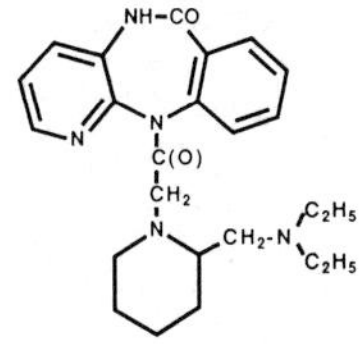

ATROPINE

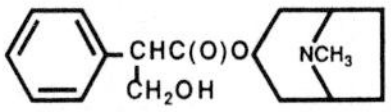

BENZOYL CHOLINE

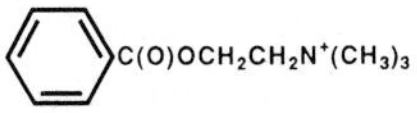

DEXETIMIDE

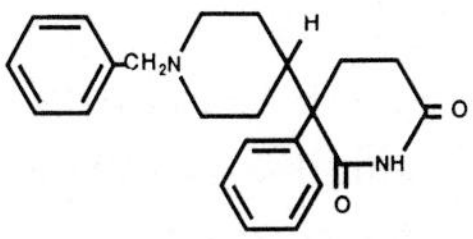

ETHOXYETHYLTRIETHYLAMMONIUM $\quad CH_3CH_2OCH_2CH_2N^+(C_2H_5)_3$

N-METHYLPIPERIDINYL BENZILATE

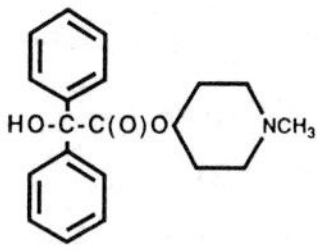

PIRENZEPINE

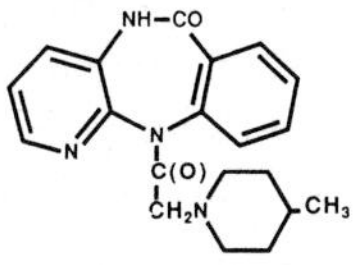

PENTYLTRIETHYLAMMONIUM $\quad CH_3(CH_2)_4N^+(C_2H_5)_3$

N-PROPYLBENZILYLCHOLINE MUSTARD

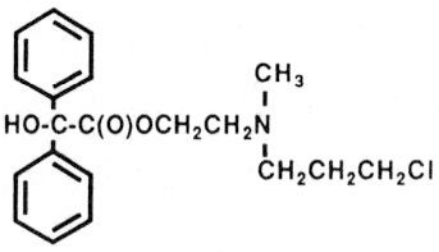

3-QUINUCLIDINYL BENZILATE

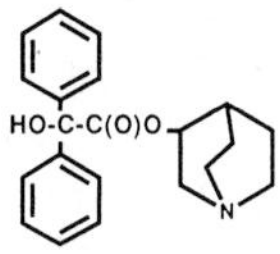

SCOPOLAMINE

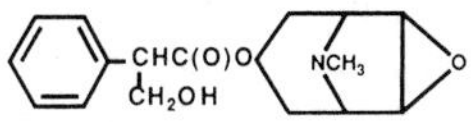

Fig. 1. Some compounds that interact with muscarinic receptors as agonists, partial agonists, and antagonists.

3. Compounds interacting with the receptor

3.1. Agonists and antagonists

Muscarinic agonists mimic the effect of acetylcholine on muscarinic nerve junctions and their peripherial effects are increased salivation, hypotension, and diarrhea. The central effects of agonists lead to restlessness, mania, and hallucination, all of which can be treated by muscarinic antagonists, for instance atropine [32]. The exogenic agonist muscarine, from which this type of acetylcholine receptor received its name, was isolated from the fungus *Amanita muscaria* by Oswald Schmiedeberg in Tartu University and described in 1869 [33]. In this classical study the nerve endings were identified as the action site of muscarine and the antagonizing effect of atropine against acute poisoning with muscarine was discovered. It should be emphasized that these conclusions were made considerably earlier than the synaptic theory of nerve transmission appeared.

The muscarinic antagonists cause intoxication known as the 'cholinergic syndrome', which presents with symptoms such as confusion, delirium, and coma [34]. The muscarinic antagonist quinuclidinyl benzilate was originally developed as a hallucinogenic chemical warfare agent with the code name BZ (Merck Index 8005). However, the extra high activity of this compound seems to be rather an exception, since many substances used as pharmaceutical agents possess moderate affinities for muscarinic receptor (see [3]). The agonists discussed in this Chapter are listed in Figure 1.

3.2. Partial agonists

A group of compounds has been identified, which are able to initiate weak muscarinic effects when compared with the 'true' agonists, although all receptor sites can be occupied by these drugs [35]. This phenomenon can be explained by the low 'intrinsic activity' or low efficacy of these compounds when compared with true agonists. These concepts remove the absolute distinction between agonists and antagonists; however, they both do not explain the molecular basis of the phenomenon. Moreover, in some cases bell-shaped dose-response curves were obtained for the partial agonists [36], formally showing the dependence of the properties of the ligand upon its concentration. The latter fact cannot be supported by the explanation cited above and calls for further modification of the receptor theory [37]. Secondly, the type of physiological activity of several compounds depends upon the tissue used in experiments. Therefore, the mechanism of differentiation of agonists, partial agonists, and antagonists by the receptor remains one of the key questions for understanding of the mechanisms of drug action on muscarinic receptor, as different types of ligands may reveal rather similar molecular structure (see in Fig. 1).

3.3. Structural requirements for muscarinic drugs

Based on the structure of acetylcholine, the structural fragment COCCN has been postulated as the basic element for muscarinic drugs, while structural variations around this 'backbone' lead to ligands of different affinity as well as different types of activity (see examples listed in Fig. 1). The variation of substituents at the C-end of the COCCN fragment can lead to two different results [38]. First, a moderate increase in the substituent size decreases the affinity of agonists. Second, large bulky groups at this C-atom transform the ligands into antagonists and the binding affinity of the latter compounds is, as a rule, rather high [35,38].

Similar effects can be observed at the N-end of the COCCN fragment. The replacement of methyl groups with ethyl groups in the acetylcholine molecule leads to a sharp decrease in the affinity of these agonists. A similar effect can be observed if the choline group of acetylcholine is substituted by quinuclidinole or piperidinole [35,39]. However, the variation of the ammonium group structure in the 1,3-dioxolane series transforms these ligands into antagonists [40]. Similar changes occur in the case of ethoxyethyltrimethylammonium ion and a change of the methyl groups by larger alkyl substituents yields partial agonists and then antagonists as illustrated in Figure 1 [35]. Thus, the agonistic and antagonistic properties of muscarinic ligands reveal opposite dependence on the 'bulkiness' of these molecules.

The oxygen atom of the COCCN group can be replaced by other basic atoms like tertiary nitrogen, indicating the possibility of donor-acceptor interaction between the ligand and receptor at this point [38]. On the other hand, this interaction is not absolutely necessary for ligand binding, since simple alkylammonium ions can also reveal quite a high activity on muscarinic receptor [35].

Replacement of the oxygen atom of the COCCN fragment by a triple bond yields 2-butynylamine derivatives, among which oxotremorine is one of the most effective muscarinic agonists known [41]. The stereoelectronic properties of the acetylenic group seem to be important for the high activity of these compounds, since the replacement of the triple bond by a double bond or -CH_2- group leads to remarkably less active derivatives. A variation of oxotremorine structure at both ends of the =$NCH_2C \equiv CCH_2N$= fragment gives either weaker agonists or transforms these compounds into antagonists, while again the size of the substituents seems to be critical [41].

Muscarinic receptor also recognizes chirality of ligands, most sensitively at two positions of the backbone, C*COCCN or CCOC*CN, respectively. However, the stereospecificity pattern for muscarinic agonists and antagonists is different, as these effects are rather large in the former case but weak for antagonists [42].

In summary, rather similar trends in structure-activity relationships can be observed for different types of compounds while interacting with muscarinic receptor. In virtually all cases the size of the drug molecule seems to be critical for the distinction between agonists and antagonists [37].

3.4. Snake venom toxins

Recently, two toxins have been isolated from green mamba *Dendroaspis angusticeps* venom, which interfere with the binding of [^{3}H]quinuclidinyl benzilate to rat cerebral cortex synaptososomal membranes, pointing to the possibility of their action on muscarinic receptors [43]. These toxins are compounds of peptide nature and their amino acid sequence reveals clear homologies with other short-chain neurotoxins acting on nicotinic acetylcholine receptors, including several positively charged groups and eight cysteine residues in analogous positions. Despite this similarity the muscarinic and nicotinic receptors differentiate these snake venom toxins.

4. *Pharmacological and molecular subtypes of muscarinic receptor*

The concept of muscarinic receptor subtypes emerged because of the findings of selective influence of some drugs on muscarinic responses in different peripheral tissues and brain, like the cardioselective antimuscarinic effect of gallamine, described in 1951 [44]. Later, more tissue-selective muscarinic antagonists were discovered, among which pirenzepine provided the major impetus for the definition of the receptor subclasses [45]. This compound binds weakly to muscarinic receptors in the heart (K_d = 683 nM) and submandibular glands (K_d = 323 nM) of rat and guinea pig ileum (K_d = 500 nM) and strongly to these receptors in rat cerebral cortex (K_d = 14 nM) [46]. Another muscarinic antagonist with code name AF-DX 116 differentiates receptors in rat submandibular glands (K_d = 2923 nM) from rat heart (K_d = 113 nM) and guinea pig ileum (K_d = 110 nM) receptors and has intermediate affinity for rat brain receptors (K_d = 746 nM) [46]. On the basis of these data, muscarinic receptors are classified as neuronal (M_1), cardiac (M_2), and glandural (M_3). This classification is supported by the binding data for several other compounds (see [46]), which all appear to be muscarinic antagonists.

Much less is known about subtype-specific agonists. One of these compounds with code name McN-A-343 shows a higher activity in sympathetic ganglia (M_1) when compared with atrial (M_2) tissue [47]. However, this phenomenon can be explained by different 'efficacy' of the compound in brain and heart, while the affinity of McN-A-343 for these receptors was rather similar [48]. There are also some other muscarinic agonists, like pilocarpine and *cis*-3-acetoxy-S-methylthiane, which show some selectivity in functional tests.

The pharmacologically defined receptor subtypes are distinct gene products, as shown by cloning of the cDNA for the M_1 [49] and M_2 [50] receptors from pig brain and heart. Subsequently, a family of five mammalian receptor genes, denoted as m1, m2, m3, m4, and m5 were isolated from rat and human cDNA, while as many as four or five more molecular forms have been proposed on the basis of genomic blot hybridization analysis [51]. The identity of each of the five receptor proteins has been established by expression of these clones in *Xenopus* oocytes or in mammalian cell lines. As these cell lines do not express endogenous muscarinic receptors, this system provided the possibility of testing the ligand-binding properties of the individual subtypes [51].

Since the first two receptors sequenced corresponded to the M_1 and M_2 pharmacological subtypes, there is a similarity between the numberings used for the structural and pharmacological subtypes. However, the pharmacological definition of receptors encoded by m3, m4, and m5 genes is not clear, even though some tissue specificity in the distribution of these receptors can be observed. The m2 receptor is abundant in heart, but is also present in small amounts in some regions of brain and smooth muscle. Broad and abundant distribution of m1, m3, and m4 receptor types can be found in brain. Exocrine glands are rich in m1 and m3 subtypes. The distribution of the m5 receptor is not well known. In summary, it is conceivable that the pharmacologically defined receptor subtypes, except the M_2 receptor in heart, comprise more than one receptor species, not distinguishable by pharmacological methods.

5. Receptor molecule

5.1. Purification and reconstitution

Muscarinic receptor is an integral membrane protein and it can be isolated by combining the methods of protein solubilization and purification. The list of detergents which could be used to extract the receptor protein from biomembrane with retention of its ability to bind specific ligands, is rather short [52] and involves mainly digitonin [53], 3-[(3-cholamidopropyl)-dimethylammonio]-1-propane sulfonate (CHAPS) [54], deoxycholate [55], and sucrose monolaurate [56]. The yields of solubilization generally range from 20% to 60% depending not only on the receptor source, but also on the ratio of detergent/protein concentration [57]. In some cases combinations of these detergents were found to be more effective than the individual compounds. The solubilized receptor particles have Stokes radius between 43 Å and 58 Å and contain a single peptide combined with carbohydrates. The solubilized receptor particles bind 1–3 g of detergent per gram of protein.

The receptor has been purified by affinity chromatography on 3-(2′-aminobenzhydroxy)tropane agarose [58] and by combination of this method with chromatography on ion exchange and hydroxylapatite columns and/or wheatgerm-agglutinin Sepharose [59]. The purified receptors can be reconstituted with purified G-proteins in phospholipid vesicles that restores the mutual regulatory interactions of these proteins, known from experiments with native biomembranes [60].

5.2. Antibodies

Purified muscarinic receptor was used to rise monoclonal antibodies. Two antibodies, which immunoprecipitated a peptide of 70 kDa and recognized denatured and native receptors, respectively, exhibited agonist-like activity in intact tissues. This activity could be blocked by atropine [61]. In another study [62] it has been shown that the monoclonal antibodies against purified porcine atrial receptor precipitated to the muscarinic receptors from pig and rat heart, but few if any to rat cerebral cortex, demonstrating specificity for the receptor subtype [62].

Antibodies for every subtype of muscarinic receptor (m1–m5) were obtained by using special soluble muscarinic receptor fusion proteins, incorporating large regions of the nonconserved third cytoplasmic (i3) loop of the corresponding subtype sequences [92]. Immunoprecipitation assay with detergent solubilized cloned receptors (m1–m5) indicated a high degree of subtype selectivity of these antibodies (cross-reactivity less than 3%) and precipitation efficiencies have been used to determine the localization of muscarinic receptor subtypes in different tissues [93,94].

No attempts are known of using antibodies for receptor purification.

5.3. Protein structure

Muscarinic receptors belong to the G-protein-linked receptor superfamily (see Chapter 6) and some examples of the primary structures of these proteins, obtained from porcine (m1p, [49]), human (m2h, m3h, m4h [63]), and rat (m5r [64]) tissues, are listed in Figure 2. As generally accepted [51], gaps are used to maximize the sequence identities of these primary structures, marked in Figure 2 with bold letters. A rather remarkable homology (ca. 20–30%) also exists between these sequences and the primary structure of α_2-adrenoceptors, conventionally considered as the 'prototype' for this superfamily. Much fewer amino acids are conserved across this entire superfamily of receptors and these positions are denoted by closed circles below the listed sequences in Figure 2.

The hydropathic analysis of the amino acid sequence data for muscarinic receptors has suggested a structural motif, common for the whole receptor superfamily. These proteins have seven hydrophobic regions, each long enough to span the plasma membrane in the form of an alfa-helix, and connecting extracellular and cytoplasmic loops [49,50,64]. The predicted transmembrane segments are denoted in Figure 2 by asterisks above the sequences, while i and e stand for intracellular and extracellular loops, respectively. The amino terminus of the peptide remains outside and the carboxyl terminus inside the cell.

It is obvious from these data that the conservation of the amino acid sequence is the greatest within the membrane-spanning regions, explaining the principal similarities in the secondary structure of the receptor proteins. On the other hand, the third cytoplasmic loop varies appreciably in length and also contains the most divergent amino acid sequences. Therefore, this region together with the C-terminal part of the protein is believed to play a crucial role in the coupling of muscarinic receptor subtypes to different effector systems that, in turn, determines the type of biochemical response to agonist binding. This was shown by constructing chimeric muscarinic receptors by exchanging the third cytoplasmic loops between the structural subtypes m1 and m2 [65]. The chimeric receptors were expressed in *Xenopus* oocytes and revealed changed coupling specificity, but the same ligand-binding specificity.

The three-dimensional structure of muscarinic receptors can be discussed based on their possible structural similarities with bacteriorhodopsin, another member of the same superfamily of membrane proteins, determined by a combination of electron microscopy and low-dose electron diffraction analysis. These data together with several structure-predictive techniques have lead to the 'helical wheel model' (Chapter 6). If this model is applied to the muscarinic receptor, the seven transmembrane peptide helixes can be arranged in a clockwise fashion when viewed from outside the cell so that most of the conserved residues are on the inside of the structure, while non-conserved and mostly hydrophobic residues remain facing the lipid bilayer without causing any packing constraints. Approximately 30 of the 47 hydrogen-bonding side chains face one another around the pore-like hydrophilic cavity, enclosed in the receptor structure. Some groups inside the cavity can be chemically modified by affinity labels, pointing to the possibility of their participation in ligand binding.

5.4. Biological modification

Muscarinic receptor is a glycoprotein and the carbohydrate part, containing mostly sialic acid combined with some amino sugars and common hexoses, makes up about $\frac{1}{4}$ by weight of the purified cardiac receptor [66]. Oligosaccharides are attached to the receptor via asparagine residues in the extracellular part of the N-terminal sequence and there are two to five such potential N-glycosylation sites in different receptor subtypes. The glycoprotein nature of muscarinic receptor explains the applicability of lectin chromatography for the protein purification [59]. On the other hand, enzymatic cleavage of the glycosyl part, as well as receptor complexation with wheatgerm agglutinin, has almost no effect on antagonist binding to the atrial muscarinic receptor [67]. N-glycosylation is thought to play an important role in receptor maintenance in the plasma membrane.

The intracellular serine and threonine residues of muscarinic receptors can be phosphorylated by protein kinase A, protein kinase C, and probably by some other more specific protein-phosphorylating enzymes, depending on the particular receptor subtype and tissue. These phosphorylation sites are most probably located in the C-terminal tail of the protein [68]. Phosphorylation of the receptor molecule decreased the number of the binding sites, as detected by [^{3}H]quinuclidinyl benzilate, and these binding sites can be restored through the enzymatic removal of the phos-

```
m1p    -------------MNTSAPP                                                7
m2h    -------------MNNST                                                  5
m3h    MTLHNNSTTSPLFPNISSSWIHSPSDAGLPPGTVTHFGSYNVSRAAGNFS                  50
m4h    -------------MANFTPVNGSSGNQ                                        14
m5r    -------------MEGESYNEST                                            10

        eeeeeeeeeeeeee***********************iiiiii
m1p    AVSPNITVLAPGKGP-WQVAFIGITTGLLSLATVTGNLLVLISFKVNTEL                  56
m2h    NSSNNSLALTSPYKTF-EVVFIVLVAGSLSLVTIIGNILVMVSIKVNRHL                  54
m3h    SPDGTTDDPLGGHTV-WQVVFIAFLTGILALVTIIGNILVVSFKVNKQL                   99
m4h    SVRLVTSSSHNRYETV-EMVFIATVTGSLSLVTVVGNILVMLSFKVNTQL                  63
m5r    VNGTPVNHQALERHGLWEVITIAVVTAVVSLMTIVGNVLVMISFKVNSQL                  60

        iiiiii***********************eeeeeeeeeeeeee*******
m1p    KTVNNYFLLSLACADLIIGTFSMNLYTTYLLMGHWALGTLACDLMLALDY                 106
m2h    QTVNNYFLFSLACADLIIGVFSMNLYTLYTVIGYWPLGPVVCDLWLALDY                 104
m3h    KTVNNYFLLSLACADLIIGAFSMNLFTTYIIMNRWALGNLACDLWLAIDY                 149
m4h    QTVNNYFLFSLACADLIIGAFSMNLTVYIIIKGYWPLGAVVCDLWLALDY                 113
m5r    KTVNNYYLLSLACADLIIGIFSMNLYTTYILMGRMVLGSLACDLMLALDY                 110

        ***********iiiii**********************eeeeee******
m1p    VASNASVMNLLLISFDRYFSVTRPLSYRAKRTPRRAAIMIGLAWLVSFVL                 156
m2h    VVSNASVMNLLIISFDRYFCVTKPLTYPVKRTTKMAGMMIAAAWVLSFIL                 154
m3h    VASNASVMNLLVISFDRYFSITRPLTYRAKRTTKRAGVMIGLAWVISFVL                 199
m4h    VVSNASVMNLLIISFDRYFCVTKPLTYPARRTTKMAGLMIAAAWVLSFVL                 163
m5r    VASNASVMNLLVISFDRYFSTRPLTYRAKRTPKRAGIMIGLAWLVSFIL                  160

        ***********eeeeeeeeeeeeeeeeeeee********************
m1p    WAPAILFWQYLVGERTVLAGQCYIQFLSQPIITFGTAMAAFYLPVTVMCT                 206
m2h    WAPAILFWQFIVGVRTVEDGECYIQFFSNAAVTFGTAIAAFYLPVIIMTV                 204
m3h    WAPAILFWQYFVGKRTVPPGECFIQFLSEPTITFGTAIAAFYMPVTIMTI                 249
m4h    WAPAILFWQFVVGKRTVPDNQCFIQFLSNPAVTFGTAIAAFYLPVVIMTV                 213
m5r    WAPAILCWQYLVGKRTVPPDECQIQFLSEPTITFGTAIAAFYIPVSVMTI                 210

        ****iiiiiiiiiiiiiiiiiiiiiiiiiiiiiiiiiiiiiii
m1p    LYWRIYRETENRARELAALQGSETPGKGGG------SSSSSERSQPG                   247
m2h    LYWHISRASKSRIKKDKKEPVANQDPVSPSLVQGRIVKPNNNM-PSSDD                  253
m3h    LYWRIYKETEKRTKELAGLQASGTEAETENFVHPTGSSRSCSSYELQQQS                 299
m4h    LYIHISLASRSRVHKHRPEGPKEKKAKTLAFLKSPLMKQSVKKPPPGEAA                 263
m5r    LYCRIYRETEKRTKDLADLQGSDSVAEAKKREPAQRTLLRSFFSCPRPSL                 260
```

m1p   AEGSPETPGRCRCCRAPRLLQAY-------SWKEEEEDEGSMESL 288
m2h   GLEHNKIQNGKAPRDPVTENCVQGEEKESSNDSTSVSAVASNMRDDEITQ 303
m3h   MKRSNRRKYGRCHFWFTTKSWKPSSEQMDQDHSSDSWNNNDAAASLENS 349
m4h   REELRNGKLEEAPPALPPPRPVADKDTNESSSGSA-TQNTKERPATE 312
m5r   AQRERNQASWSSRRSTSTTGKTTQATDL------SADWEKAEQVTTCSSY 305

```
         iiiiiiiiiiiiiiiiiiiiiiiiiiiiiiii
m1p    TSSEGEEPGSEVVIKMP------------------MV              307
m2h    DENTVSTSLGHSKDENSKQTCIRIGTKTPKSDSCTPTN---TTVEVVGS  349
m3h    ASSDEDIGSETRAIYSIVLKLPGHSTILNSTKLPSSDNLQVPEEELGMV  399
m4h    LSTTEATTPAMPAPLQPRALNPARSWSKIQIVTKQTGNECVTAIEIVP---  361
m5r    PSSEDEAKPTTDPVFQMVYKSEAKESPGKESNTQETKETVNTRTENSDY  355

         iiiiiiiiiiiiiiiiiiiiiiiiiiiiiiiiiiii******
m1p    DPEAQAPAKQPPRSSP-------------------                323
m2h    ------------------                                 349
m3h    DLERKADKLQAQKSVDDGGSFPKSFSKLPIQLESAVDTAKTSDVNSSVGK 449
m4h    -------------                                      361
m5r    DTPKYFLSPAAAHRLKSQKCVAYKFRLVVKADGTQETNNGCRKV------ 399

         **********iiiiiiiiiiiiiiiiiiiiiiii******
m1p    NTVKRPTRKGRERAGKGQKPRGKEQLAKRKTFSLVKEKKAARTLSAILLA 373
m2h    --SGQNGDEKQNIVARKIVKMTK-QPAKKKPPPS-REKKVTRTILAILLA 395
m3h    STATLPLSFKEATLAKRFALKTRSQITKRKRMSLVKEKKAAQTLSAILLA 499
m4h    --ATPAGMRPAANVARKFASIARNQVRKKRQMAA-RERKVTRTIFAILLA 408
m5r    KIMPCSFPVSKDPSTKGPDPNLSHQMTKRKRMVLVKERKAAQTLSAILLA 449

         *******************eeeeeeee***************i
m1p    FIVTWTPYNIMVLVSTFCKDCVPETLWELGYWLCYVNSTINPMCYALCNK 423
m2h    FIITWAPYNVMVLINTFCAPCIPNTVWTIGYWLCYINSTINPACYALCNA 445
m3h    FIITWTPYNIMVLVNTFCDSCIPKTFWNLGYWLCYINSTVNPVCYALCNK 549
m4h    FILTWTPYNVMVLVNTFCQSCIPDTVWSIGYWLCYVNSTINPACYALCNA 458
m5r    FIITWTPYNIMVLVSTFCDKCVPVTLWHLGYWLCYVNSTINPICYALCNR 499

         iiiiiiiiiiiiiiiiiiiiiiiiiiii
m1p    AFRDTFRLLLLCRWDKRRWRKIPKRPGSVHRTPSRQC              460
m2h    TFKKTFKHLLMCHYKNIGATR                              466
m3h    TFRTTFKMLLLCQCDKKKRRKQQYQQRQSVIFHKRAPEQAL          590
m4h    TFKKTFRHLLLCQYKNIGTAR                              479
m5r    TFRKTFKLLLLCRWKKKKVEEKLYWQGNSKLP                   531
```

Fig. 2. Primary sequence comparison of five structural subtypes of muscarinic receptors from porcine (p), man (h), and rat (r). Data are compiled from Bonner [51]. Gaps are introduced to maximize sequence identities and to compensate for the largely different sizes of the cytoplasmic loops. Asterisks, predicted transmembrane regions; i and e, intracellular and extracellular regions; closed circles below the sequences, amino acids that are conserved in the receptor superfamily.

phate groups [69]. Therefore, it is conceivable that phosphorylation may be involved in the mechanism of short-term desensitization of muscarinic receptor.

5.5. Functional groups

Chemical reagents, which selectively react with disulfide groups in proteins, inhibit specific binding of muscarinic antagonists and also decompose the receptor-ligand complex [70], emphasizing that the SS-bridges are important for stabilization of the protein fold. These disulfide bonds may form between sulfhydryl groups located within transmembrane regions as well as in the extracellular and intracellular loops of the receptor molecule. Indeed, it can be seen in Figure 2 that several cysteines are conserved not only in these regions of the receptor subtypes, but also on the level of the whole receptor superfamily. Evidence for disulfide-bond formation between conserved cysteines in the first and the second extracellular loops was obtained by protein chemistry methods [71]. At the same time differences in the number and location of other cysteines in receptor subtypes may explain somewhat the differential effects of chemical modification observed in different tissues [71].

Some cysteine residues of muscarinic receptors are not involved in formation of disulfide bridges, since reagents selective for the free sulfhydryl groups also inhibit antagonist binding with muscarinic receptor. However, in this case the crucial sulfhydryl group should be located in the vicinity of the antagonist-binding site as the receptor-bound ligands effectively protect the receptor against chemical modification [72].

Muscarinic antagonist [^{3}H]propylbenzilylcholine mustard, which has been used as an affinity label of the antagonist-binding site, alkylated two aspartic acid residues in or close to the third transmembrane helix of the receptor [71]. The main alkylation point was the carboxyl residue, buried in approximately one-third of the way into the transmembrane region and corresponding to position 105 in the sequence of m1 receptor. Less extensive alkylation was found at the aspartic acid residue at the N-terminus of the same transmembrane helix, corresponding to position 99 in m1 sequence. It can be seen in Figure 2 that both residues are conserved within the muscarinic receptor family.

The roles of the conserved aspartic acids in positions 71, 99, 105, and 122, located in or near the second and third transmembrane domains of the m1 muscarinic receptor, were examined by site-directed mutagenesis [73]. These aspartic acids were replaced with asparagine to eliminate the negative charge in each position of the receptor molecule. In position 71 this replacement produced a receptor with normal antagonist-binding properties, but 5.5 times higher affinity for the agonist carbachol. In position 99 the substitution produced a receptor with a 3–5-fold reduced affinity for both agonists and antagonists, while in position 105 this mutation lead to almost complete loss of antagonist binding. The latter fact interfered with binding studies with agonists. All these results are in good agreement with the affinity-labelling data discussed above. The mutation in position 122, located on the cytoplasmic side of the third transmembrane domain, had no influence on antagonist binding, but increased affinity for carbachol 1.9-fold, probably through some conformational change in the protein structure.

These data clearly point to the importance of the carboxylic groups around the second extracellular loop and the proximal portion of the third transmembrane domain for ligand-receptor interactions. On the other hand, the effects of mutations are not strictly parallel in the case of agonists and antagonists, pointing to the possibility of a different location of their binding sites on the receptor.

Chemical modification of muscarinic receptor with tetranitromethane implicates some role of tyrosine residues in agonist binding, since this treatment increases affinity of the receptor for these ligands [74]. Simultaneously, there is almost no influence of this reagent on antagonist binding, although nitration of the receptor can be prevented by the presence of the latter ligands. The location of modification in the muscarinic receptor has not yet been determined, but a tyrosine residue in the transmembrane segment 7 has been pinpointed as a possible candidate [75]. This suggestion is based on the comparison of the primary structures of different G-protein-dependent receptors that revealed a correlation between the presence of this tyrosine residue and the functionally important aspartic acid in transmembrane segment 3 (position 105 in subtype m1). This combination of tyrosine and aspartic acid appears in all receptors of biogenic amines, whose specific ligands are cationic, and therefore their participation in the ligand-binding mechanism seems possible [75].

6. *Ligand-receptor interactions*

6.1. *Equilibrium-binding studies*

Usually radioactively labelled potent muscarinic antagonists are used as tools to investigate the mechanism of drug interaction with the receptor. The conventional methods of radioligand binding and displacement studies are discussed in more detail elsewhere in this book. In all these assay methods it is assumed that the true equilibrium between the free and bound states of the reactants has been achieved during the incubation time used. In this case the simplest binding mechanism can be applied:

$$A + R \underset{k_{-1}}{\overset{k_1}{\rightleftharpoons}} RA, \qquad \text{(eq. 2)}$$

where A stands for ligand, R for receptor, and the dissociation constant $K_d = k_{-1}/k_1$ is the only parameter which characterizes the ligand-receptor interaction. In many cases this approach gives satisfactory results and the number of different ligands, the interaction of which with muscarinic receptor was investigated by means of radioligand analysis exceeds 500 (Bank of Binding Data for Muscarinic Receptor, compiled by the authors of this Chapter).

It is generally accepted that antagonist binding with a muscarinic receptor follows reasonably well the regular binding isotherm (eq. 1) for a single type of receptor sites [20]. However, steep radioligand displacement curves are obtained with agonists, pointing to their interaction with at least two binding sites of different affinity [28]. These different affinity states are most probably related to interaction between muscarinic receptor and G-protein, as the complex has higher affinity for agonists than the free receptor. This was directly demonstrated by reconstitution of the purified receptor with G-protein [76]. Antagonists do not differentiate between these two states of the receptor.

In spite of the apparent simplicity of the radioligand-binding studies, there is a critical point concerning the slowness of dissociation of the complex formed between the receptor and potent antagonist used as radioligand. For this reason [^{3}H]quinuclidinyl benzilate, which is incontestably the most widely used muscarinic radioligand, cannot be used in the displacement experiments which require the achievement of the true equilibrium state of the system [26]. Similar complications may also arise with some other radioligands. Thus, the knowledge of the kinetic aspects of radioligand interaction with the receptor is a prerequisite for any correct binding study, especially if this approach is used for the quantitative analysis of receptor-ligand interactions.

6.2. Kinetics of radioligand binding

Earlier studies with radioactively labelled potent muscarinic antagonists revealed several contradictions between the results of binding experiments and the simple reaction mechanism (eq. 2). First, different K_d values can be obtained from binding curves if calculated as the ratio of rate constants k_{-1} and k_1 according with the reaction mechanism (eq. 2) [17,21,22]. Second, the dissociation of radioligands from their complex with muscarinic receptor follows a bi-exponential kinetic curve, concealing a more complex reaction mechanism than the monomolecular dissociation of the receptor-ligand complex [17,82]. Third, a remarkable scattering of radioligand dissociation constants can be observed in the literature with a clear trend of the K_d values to decrease in parallel with the increase in the specific radioactivity of the radioligand used. This phenomenon has been analyzed in more detail with [^{3}H]quinuclidinyl benzilate [26]. All these circumstances lead to the necessity to apply more strict kinetic methods to analyze the mechanism of ligand-receptor interactions.

In this kinetic approach the observed rate constants k_{obs} of radioligand binding are determined from the time course of radioligand-receptor complex formation, measured under the pseudo-first-order conditions at the excess of ligand over the receptor concentration:

$$B_t = B_{nonsp} + B_{sp}(1 - \exp(-k_{obs} t)) \qquad (eq.\ 3)$$

where B_t stands for the bound radioligand at time t, B_{nonsp} and B_{sp} for nonspecifically and specifically bound radioligand, respectively. Further, the plots of k_{obs} versus ligand concentration can be analyzed as it is generally accepted in physical organic chemistry and enzymology [77]. It is important to stress that nonspecific radioligand binding equilibrates fast and can be considered a constant in the rate equation (eq. 3) [78].

For the radioactive antagonists [^{3}H]quinuclidinyl benzilate [78,79], [^{3}H]N-methyl-quinuclidinyl benzilate [26], [^{3}H]N-methylpiperidinyl benzilate [78], [^{3}H]N-methyl-scopolamine [80], and [^{3}H]pirenzepine [81], the plots of k_{obs} versus ligand concentration were not linear, as is predicted by the reaction scheme (eq. 2), but showed a more complex form, which regularly includes a hyperbolic part. The latter fact gives evidence for at least a two-step consecutive binding mechanism, which should involve a fast reversible step of complex formation (K_A) followed by slower step of the complex 'isomerization':

$$R + A \underset{}{\overset{K_A}{\rightleftharpoons}} RA \underset{k_{-i}}{\overset{k_i}{\rightleftharpoons}} (RA). \qquad (eq.\ 4)$$

The rate equation for this reaction scheme is:

$$B_{obs} = \frac{k_i [A]}{K_A + [A]} + k_{-i} \qquad (eq.\ 5)$$

The kinetic parameters calculated from this plot for some radioactive antagonists are listed in Table I. An important conclusion, based on the kinetic regularities observed, is that only the 'isomerized' complex (RA) can be experimentally measured, while the rapidly dissociating complex RA escapes determination with the common proce-

TABLE I
Kinetic data for radioactive antagonist binding with muscarinic receptor from rat brain (25°C, pH 7.40)

Antagonist	K_A nM	$10^2\ k_i$ s^{-1}	$10^4\ k_{-i}$ s^{-1}	K_d nM	K_i
[^{3}H]N-Methylpiperidinyl benzilate	5.5 + 3.1	1.4 + 0.3	6.7 + 0.8	0.40 + 0.03	0.05
[^{3}H]N-Methylquinuclidinyl benzilate	5.0 + 2.3	4.6 + 0.5	14 + 4	0.13 + 0.03	0.03
L-[^{3}H]Quinuclidinyl benzilate	1.3 + 0.5	1.2 + 0.3	1.3 + 0.4	–	–
[^{3}H]N-Methylscopolamine	9.7 + 2.1	11.9 + 0.8	9 + 4	0.082 + 0.008	0.008

Data compiled from Järv [31].

dures like filtration and centrifugation assay [78]. Thus, the presence of the 'isomerization' step seems to be the main criterion for selection of compounds to be used as radioactive 'tools' in receptor studies.

Secondly, it should be taken into consideration that the 'isomerization' step is a monomolecular reaction and therefore the equilibrium cannot be shifted by addition of an excess of antagonist [78]. As a result of this, a portion of the receptor sites, presented by the complex RA, cannot be detected even at very high antagonist concentrations. The latter circumstance is important to consider when measuring the number of receptor-binding sites by radioactive antagonists. In the case of radioligands listed in Table I, the equilibrium of 'isomerization' is shifted to the complex (RA), which can be deduced from the small K_i values.

Since both steps of the binding scheme (eq. 4) are reversible, the overall process can also be characterized by a dissociation constant K_d, which is, however, a combination of the constants K_A and $K_i = k_{-i}/k_i$, $K_d = K_A \cdot K_i$. The latter fact explains the particularly high potency of several muscarinic antagonists, reaching even the picomolar concentration range [26,80].

Under certain experimental conditions biphasic kinetic curves have been obtained for the dissociation of [^{3}H]quinuclidinyl benzilate [26,82], [^{3}H]N-methylpiperidinyl benzilate [22], and [^{3}H]N-methylscopolamine [80] from their complexes with muscarinic receptor. A more thorough kinetic analysis of these data has revealed a multistep 'isomerization' mechanism for these antagonists [26,80]:

$$R + A \underset{}{\overset{K_A}{\rightleftharpoons}} RA \underset{k_{-i}}{\overset{k_i}{\rightleftharpoons}} (RA) \underset{k'_{-i}}{\overset{k'_i}{\rightleftharpoons}} ((RA)) \quad \text{(eq. 6)}$$

As the isomerized complexes (RA) and ((RA)) dissociate slowly and thus can be detected by the common assay procedures, the transformation of (RA) into ((RA)) cannot be traced in the binding studies. On the other hand, this extra step of complex 'isomerization' may additionally decrease the apparent K_d value if K_{i2} (= k'_{-i}/k'_i) << 1, as $K_d = K_A K_{i1} K_{i2}$ [26,80].

In the case of [^{3}H]quinuclidinyl benzilate [26] and [^{3}H]N-methylpiperidinyl benzilate [83], cooperative regulation of the 'isomerization' rate by the excess of antagonist was discovered, pointing to an even more complex reaction mechanism.

6.3. Kinetic studies with nonradioactive ligands

The still-limited usage of kinetic methods in receptorology is obviously due to the traditional thinking based on the concept of ligand-receptor equilibria. On the other hand, this is certainly connected with the small number of the radioactively labelled ligands available for receptor studies. Therefore, efforts have been made to develop experimental procedures for kinetic measurements with nonradioactive ligands. The approach proposed [84–86] is based on analysis of the influence of a nonradioactive ligand on the binding kinetics of a radioactive 'reporter ligand' with a receptor. The list of compounds, that bind to a muscarinic receptor and can now be studied kinetically, is practically unlimited.

The kinetic studies with muscarinic antagonists have revealed that the process of 'isomerization' is initiated only by some compounds and, in general, the both binding mechanisms, (eq. 2) and (eq. 4), can be observed [31,85]. This means that the classical muscarinic antagonists can be divided into two subclasses depending on their receptor-binding mechanism. These different kinetic mechanisms seem to be directly correlated with antagonist potency in its pharmacological meaning, because with the simple binding mecahnism (eq. 2) the apparent K_d value is equal to the dissociation constant K_A, while with the 'isomerization' mechanism (eq. 4) this parameter is a combination of K_A and K_{i1} that increases the drug potency.

The presence or absence of the 'isomerization' step obviously determines the dynamics of the physiological effect of muscarinic antagonists, as both the onset rate and recovery of the receptor in pharmacological experiments depend upon the rate of ligand association and dissociation. Different behavior of muscarinic antagonists was described in experiments using ileum preparations, where some potent drugs were slow in onset and very slow to wear off after washing [35]. The same compounds were later found to initiate 'isomerization' according to the kinetic theory of ligand-receptor interaction.

The kinetic studies with muscarinic agonists revealed that these ligands do not initiate slow 'isomerization' of the agonist-receptor complex [84,87]. On the other hand, carbachol and oxotremorine have no effect on the affinity of the receptor for [^{3}H]N-methylpiperidinyl benzilate, which was used as radioactive 'reporter' ligand in these studies [87]. This means that the antagonist binds equally well with the free receptor and the agonist-receptor complex, supplying evidence for formation of the ternary agonist-antagonist-receptor complex. At the same time agonists inhibit in a dose-dependent manner the process of 'isomerization' of the antagonist-receptor complex. For this reason the displacement of radioactive antagonists by agonists can be observed in the conventional binding experiments, giving the impression of a competitive mechanism in the agonist-antagonist interaction on muscarinic receptor.

7. Mechanisms of signal transduction

Muscarinic receptors are coupled to different G-proteins which probably recognize the molecular identity of the receptor subtypes [76,88]. These G-proteins activate second messenger systems and probably regulate some ion channels. Inhibition of adenylate cyclase and stimulation of phospholipase C seem to be the two responses which are directly mediated via G-proteins [89]. In the first case the receptor subtypes m2 and m4 are involved and coupled with G-proteins that are sensitive to ADP-ribosylation by pertussis toxin and denoted as G_i. Presently, three different types of

G_i have been described and it is not clear which of them is coupled with the muscarinic receptor [90]. As a result of this signal transduction pathway, the intracellular level of cAMP decreases leading to inactivation of protein kinase A and inhibition of the appropriate phosphorylation processes. In heart tissue, the potassium channel is activated by m2 receptors via specific G-proteins [91].

The receptor subtypes m1, m3, and m5 generally mediate responses connected with stimulation of phospholipase C, catalyzing phosphoinositide breakdown [89]. The G-proteins coupled to this system are not, or are only partially, sensitive to blockade by pertussis toxin. Activation of phospholipase C leads to formation of the second messenger inositol 1,4,5-triphosphate which mediates the release of Ca^{2+} ions from intracellular calcium stores [75]. In turn, the increased calcium concentration may activate adenylate cyclase and thus even increase the level of adenosine 3′,5′-cyclic monophosphate, contrary to the muscarinic response mediated via adenylate cyclase-inhibiting G_i-proteins. The second product of phosphoinositide hydrolysis is diacylglycerol, which is an effective protein kinase C activator. Through this enzyme, hydrolysis may govern protein phosphorylation. Hydrolysis of diacylglycerol by phospholipase A_2 produces the important regulatory metabolite arachidonic acid. The same regulatory pathway comprising phosphoinositide hydrolysis may be involved in the regulation of DNA synthesis and induction of proto-oncogene transcription [89].

In summary, the neurotransmitter acetylcholine exerts intracellular effects conditioned by specificity of the consecutive interactions between receptors, signal transducer G-proteins, and effector systems [76,88]. This multistep coupling mechanism and plurality of its components explains the variety of the responses and flexibility of regulation of the signal transduction system.

8. Two-site receptor model

The critical point in the mechanism of muscarinic receptor action is the differentiation between the agonist and antagonist molecules and the mechanism of action of partial agonists. Understanding of this problem was improved by the kinetic studies on ligand-receptor interaction, supplying evidence that agonists and antagonists occupy different binding sites on the muscarinic receptor. In a less direct way this is supported by several other observations on receptor structure and function. These two binding sites obviously possess different specificity patterns providing recognition of agonists and antagonists, respectively.

Furthermore, it can be proposed that these sites are related to different responses brought about by the binding of agonists and antagonists with the receptor. In other words, antagonists inhibit the receptor response not by preventing agonist binding to its target site, but by binding with a separate site that leads to another response of 'switching off' the signal transduction mechanism in the receptor molecule. This 'one site–single response' mechanism is easy to understand on the basis of the present knowledge about ligand-induced conformational changes of proteins.

In the case of muscarinic receptors, the discrimination between agonists and antagonists seems to be based chiefly on the quantitative differences in the interactions of ligands with the receptor. This means that muscarinic antagonists bind more effectively at the site which blocks the physiological response, but less effectively at the site which triggers off this response. As the former site is occupied at lower ligand concentration, the inhibitory effect prevails in the case of antagonists. On the other hand, muscarinic agonists show a preference for the site which governs the appearance of the response and bind at higher ligand concentrations at the antagonistic site. Therefore, the model predicts inhibition of muscarinic responses at high agonist concentrations.

The two-site model of the muscarinic receptor gives a simple explanation for the mechanism of action of partial agonists. It is obvious that between the typical sets of agonists and antagonists some compounds may possess a similar affinity against both types of binding sites. Therefore, these ligands simultaneously evoke and inhibit the physiological activity of the receptor.

References

1. Dale, H.H. (1914) J. Pharmacol. 6, 147–190.
2. Michelson, M.J. (1973) Comp. Pharmacol. 1, 191–227.
3. Järv, J. and Bartfai, T. (1988) In: The Cholinergic Synapse. Handbook of Experimental Pharmacology (Whittaker, V.P., Ed.), Chapter 10, pp. 315–345, Springer Verlag, Berlin, Heidelberg, New York, London, Paris, Tokyo.
4. Hoss, W., Messer, W.S., Monsma, F.J., Miller, M.D., Ellerbrok, B.R., Scranton, T., Ghodsi-Hovsepian, S., Price, M.A., Balan, S., Mazloum, Z. and Bohnett, M. (1990) Brain Res. 517, 195–201.
5. Kuhar, M.J. (1976) In: Biology of Cholinergic Function (Goldberg, A.M. and Hanin, I., Eds) pp. 3–27, Raven Press, New York.
6. Ringdahl, B. (1984) Pharmacol. Exp. Ther. 232, 67–73.
7. Ehlert, F.J. (1985) Mol. Pharmacol. 28, 410–421.
8. Leslie, E.B.A., Putney, J.M., Jr. and Sherman, J.M. (1976) J. Physiol. (London) 260, 351–370.
9. Michell, R.H. (1975) Biochem. Biophys. Acta 415, 81–147.
10. Bartfai, T., Study, R.F. and Greengard, P. (1977) In: Cholinergic Mechanisms and Psychopharmacology (Jenden, D.E., Ed.), pp. 285–295, Plenum, New York.
11. Onali, P., Olianas, M.C., Schwartz, J.P. and Costa, P. (1984) Adv. Biosci. 48, 163–191.
12. Adams, R.P., Brown, D.A. and Constanti, A. (1982) J. Physiol. (London) 332, 223–262.
13. Löffelholz, K. and Pappanino, A.J. (1985) Pharmacol. Rev. 37, 1–24.
14. Birnbaumer, L. (1990) Annu. Rev. Pharmacol. Toxicol. 30, 675–705.
15. Paton, W.D.M. and Rang, H.P. (1965) Proc. R. Soc. Lond. (Biol.) 163, 1–44.
16. Laduron, P.M., Verwimp, M. and Leysen, J.E. (1979) J. Neurochem. 32, 421–427.
17. Kloog, Y. and Sokolovsky, M. (1978) Brain Res. 144, 31–48.
18. Birdsall, N.J.M., Burgen, A.S.V., Hulme, E.C. and Wong, E.H.F. (1983) Br. J. Pharmacol. 80, 197–204.
19. Watson, M., Roeske, W.R. and Yamamura, H.I. (1982) Life Sci. 31, 2019–2023.

20. Hulme, E.C., Birdsall, N.J.M., Burgen, A.S.V. and Mehta, P. (1978) Mol. Pharmacol. 14, 737–750.
21. Yamamura, H.I. and Snyder, S.H. (1974) Proc. Natl Acad. Sci. USA, 71, 1725–1729.
22. Kloog, Y., Egozi, Y. and Sokolovsky, M. (1979) Mol. Pharmacol. 15, 545–558.
23. Eller, M., Järv, J. and Loodmaa, E. (1989) Org. React. (USSR) 26, 199–210.
24. Gibson, R.E., Rzeszotarski, W.J., Jagoda, E.M., Francis, B.E., Reba, R.C. and Eckelman, W.C. (1984) Life Sci. 34, 2287–2296.
25. Hrdina, P.D. (1986) In: Neuromethods (Boulton, A.A., Baker, G.B. and Hrdina, P.D., Eds), pp. 1–21, Humana Press, Clifton, NJ.
26. Eller, M. and Järv, J. (1988) Neurochem. Int. 13, 419–428.
27. Gurwitz, D., Kloog, Y. and Sokolovsky, M. (1984) Proc. Natl Acad. Sci. USA, 81, 3650–3654.
28. Birdsall, N.J.M., Hulme, E.C. and Burgen, A. (1980) Proc. R. Soc. Lond. Ser. B 207 (1166), 1–12.
29. Hedlund, B. and Bartfai, T. (1981) Naunyn-Schmiedebergs Arch. Pharmacol. 317, 126–130.
30. Ehlert, F.J., Dumont, Y., Roeske, W.R. and Yamamura, H.I. (1980) Life Sci., 26, 961–967.
31. Järv, J. (1991) Period Biol. 93, 197–200.
32. Taylor, P. (1980) In: The Pharmacological Basis of Therapeutics (Goodman and Gilman, Eds), pp. 91–99, Macmillan, New York.
33. Schmiedeberg, O. and Koppe, R. (1869) Das Muscarin, das giftige Alkaloid des Fligenpilzes (*Agaricus muscarius* L.), seine Darstellung, chemischen Eigenschoften, physiologischen Wirkungen, toxikologische Bedeutung und sein Verhaltnis zum Piltz vergiftungen im Allgemeinen, Leipzig.
34. Rumack, B.H. (1973) Pediatrics 52, 449–451.
35. Abramson, F.B., Barlow, R.B., Mustafa, M.G. and Stephenson, R.P. (1969) Br. J. Pharmacol. 37, 207–233.
36. Stephenson, R.P. (1956) Br. J. Pharmacol. 11, 379–393.
37. Järv, J. (1992) In: Selective Neurotoxicity. Handbook of Experimental Pharmacology (Herken, H. and Hucho, F., Eds), Springer Verlag, Berlin, Heidelberg, New York, London, Paris, Tokyo (in press).
38. Pratesi, P., Villa, L., Ferri, V., De Micheli, C., Grana, E., Silipo, C. and Vittoria, A. (1984) In: Highlights in Receptor Chemistry (Melchirre, C. and Gianella, M., Eds), pp. 225–249, Elsevier Science Publishers B.V., Amsterdam.
39. Maayani, S., Weinstein, H., Cohen, S. and Sokolovsky, M. (1973) Proc. Natl Acad. Sci. USA, 70, 3103–3107.
40. Angeli, P., Gianella, M. and Pigini, M. (1984) Eur. J. Med. Chem. 19, 495–500.
41. Dahlbom, R. (1981) In: Cholinergic Mechanisms: Phylogenetic Aspects, Central and Peripheral Synapses and Clinical Significance (Pepeu, G. and Ladinsky, H., Eds), pp. 621–638, Plenum Press, New York.
42. Triggle, D.J. and Triggle, C.R. (1976) In: Chemical Pharmacology of the Synapse, pp. 233–430, Academic Press, New York, London.
43. Adem, A., Asblom, A., Johansson, G., Mbugua, P.M. and Karlsson, E. (1988) Biochim. Biophys. Acta 968, 340–345.
44. Riker, W.F. and Wescoe, W.C. (1951) Ann. N.Y. Acad. Sci. 54, 373–394.
45. Hammer, R., Berrie, C.P., Birdsall, N.J.M., Burgen, A.S.V. and Hulme, E.C. (1980) Nature (London) 283, 90–92.
46. Ladinsky, H., Schiavi, G.B., Monferini, E. and Giraldo, E. (1990) In: Progress in Brain Research (Aquilonius, S.-M. and Gillberg, P.-G., Eds) Vol. 84, pp. 193–200, Elsevier Science Publishers B.V., Amsterdam, New York, Oxford.
47. Hammer, R. and Giachetti, A. (1982) Life Sci. 31, 2991–2998.
48. Eglen, R. and Whiting, R.L. (1986) J. Auton. Pharmacol. 5, 323–346.
49. Kubo, T., Fukuda, K., Mikami, A., Takahashi, H., Mishina, M., Haga, T., Ichiyama, A., Kangawa, K., Kojima, M., Matsuo, H., Hirose, T. and Numa, S. (1986) Nature (London) 323, 411–416.
50. Kubo, T., Maeda, A., Sugimoto, K., Akiba, I., Mikami, A., Takahashi, H., Haga, T., Ichiyama, A., Kangawa, K., Matsuo, H., Hirose, T. and Numa, S. (1986) FEBS Lett. 209, 367–372.

51. Bonner, T.I. (1989) TINS 12, 148–151.
52. Rinken, A., Langel, U. and Järv, J. (1987) Biokhimiya 52, 303–310.
53. Hurko, O. (1978) Arch. Biochem. Biophys. 190, 434–445.
54. Gavish, M. and Sokolovsky, M. (1982) Biochem. Biophys. Res. Commun. 109, 819–824.
55. Florio, V.A. and Sternweis, P.C. (1985) J. Biol. Chem. 260, 3477–3478.
56. Rinken, A. and Haga, T. (1993) Arch. Biochem. Biophys. 301, 158–164.
57. Rinken, A.A., Langel, U.L. and Järv, J.L. (1984) Biol. Membr. 1, 341–348.
58. Haga, K. and Haga, T. (1983) J. Biol. Chem. 258, 13575–13579.
59. Hulme, E.C., Wheatley, M., Curtis, C. and Birdsall, N.J.M. (1987) In: Muscarinic Cholinergic Mechanism. (Cohen, S. and Sokolovsky, M., Eds), pp. 192–211, Freund Publishing House, London.
60. Haga, K., Haga, T., Ichiyama, A., Katada, T., Kurose, H. and Ui, M. (1985) Nature (London) 316, 731–733.
61. Lieber, D., Harbon, S., Guillet, J.G., Andre, C. and Strosberg, A.D. (1984) Proc. Natl Acad. Sci. USA 81, 4331–4334.
62. Luetje, C.W., Brumwell, C., Norman, M.W.G., Peterson, G.L., Shimerlik, M.I. and Nathanson, N.M. (1987) Biochemistry 26, 6892–6896.
63. Peralta, E.G., Ashkenazi, A., Winslow, J.W., Smith, D.H., Ramachandran, J. and Capon, D.J. (1987) Eur. Mol. Biol. Organ. (EMBO) J. 6, 3923–3929.
64. Liao, C.-F., Themmen, A.P.N., Joho, R., Barberis, C., Birnbaumer, M. and Birnbaumer, L. (1989) J. Biol. Chem. 264, 7328–7337.
65. Kubo, T., Bujo, H., Akiba, I., Nakai, J., Mishina, M. and Numa, S. (1988) FEBS Lett. 241, 119–125.
66. Peterson, G.L., Rosenbaum, L.C., Broderick, D.J. and Schimerlik, M.I. (1986) Biochemistry 25, 3189–3202.
67. Herron, G.S. and Schimerlik, M.I. (1983) Neurochemistry 41, 1414–1420.
68. Uchijama, H., Ohara, K., Haga, K., Haga, T. and Ichiyama, A. (1990) J. Neurochem. 54, 1870–1881.
69. Ho, A.K.S., Ling, Q.-L., Duffield, R., Lam, P.H. and Wang, J.H. (1987) Biochem. Biophys. Res. Commun. 142, 911–918.
70. Järv, J. and Rinken, A. (1987) In: Receptors and Ion Channels (Ovchinnikov, Y.A. and Hucho, F., Eds), pp. 101–108, Walter De Gruyter & Co, Berlin, New York.
71. Kurtenbach, E., Curtis, C.A.M., Pedder, K.E., Aitken, A., Harris, A.C.M. and Hulme, E.C. (1990) J. Biol. Chem. 265, 13702–13708.
72. Hedlund, B. and Bartfai, T. (1979) Mol. Pharmacol. 15, 531–544.
73. Fraser, C.M., Wang, C.-D., Robinson, D.A., Gocayne, J.D. and Venter, J.C. (1990) Mol. Pharmacol. 36, 840–847.
74. Wang, J.-K., Yamamura, H.I., Wang, W. and Roeske, W.R. (1988) Biochem. Pharmacol. 37, 3787–3790.
75. Hulme, E.C., Birdsall, N.J.M. and Buckley, N.J. (1990) Annu. Rev. Pharmacol. Toxicol. 30, 633–673.
76. Haga, K., Haga, T. and Ichiyama, A. (1986) J. Biol. Chem. 261, 10133–10140.
77. Jencks, W.P. (1969) Catalysis in Chemistry and Enzymology, Ch. 11, McGraw-Hill Book Company, New York, St. Louis, San Francisco, London, Sidney, Toronto, Mexico, Panama.
78. Järv, J., Hedlund, B. and Bartfai, T. (1979) J. Biol. Chem. 254, 5595–5598.
79. Schimerlik, M.L. and Searles, R.P. (1980) Biochemistry 19, 3407–3413.
80. Eller, M. and Järv, J. (1989) Neurochem. Int. 15, 301–305.
81. Luthin, G.R. and Wolfe, B.B. (1984) Mol. Pharmacol. 26, 164–169.
82. Galper, J.B., Klein, N. and Catterall, W.A. (1977) J. Biol. Chem. 252, 8692–8699.
83. Järv, J., Sillard, R. and Bartfai, T. (1987) Proc. Acad. Sci. Estonia Chemistry 36, 172–180.
84. Schreiber, G., Henis, Y.I. and Sokolovsky, M. (1985) J. Biol. Chem. 260, 8795–8802.
85. Eller, M., Järv, J. and Palumaa, P. (1988) Org. React. 25, 372–386.
86. Eller, M., Järv, J. and Loodmaa, E. (1989) Org. React. 26, 199–210.
87. Järv, J., Hedlund, B. and Bartfai, T. (1980) J. Biol. Chem. 255, 2649–2651.

88. Wess, J., Bonner, T.I., Dorje, F. and Brann, M.R. (1990) Mol. Pharmacol. 38, 517–523.
89. Ashkenazi, A., Peralta, E.G., Winslow, J.W., Ramachandran, J. and Capon, D.J. (1989) TIPS, Suppl., 16–22.
90. Kaziro, Y., Itoh, H., Kozasa, T., Nakafuku, M. and Satoh, T. (1991) Annu. Rev. Biochem. 60, 349–400.
91. Szabo, G. and Otero, A.S. (1989) TIPS, Suppl., 46–49.
92. Levey, A.I., Storman, T.M. and Brann, M.R. (1990) FEBS Lett. 275, 65–69.
93. Levey, A.I., Kitt, C.A., Simonda, W.T., Price, L. and Brann, M.R. (1991) J. Neurosci. 11, 3218–3226.
94. Dörje, F., Levey, A.I. and Brann, M.R. (1991) Mol. Pharmacol. 40, 459–462.

F. Hucho (Ed.) *Neurotransmitter Receptors*
© 1993 Elsevier Science Publishers B.V. All rights reserved.

CHAPTER 8

Receptors for 5-hydroxytryptamine

DANIEL H. BOBKER and JOHN T. WILLIAMS

Vollum Institute, L474, Oregon Health Sciences University, 3181 SW Sam Jackson Park Rd., Portland, Oregon 97201 USA

1. Introduction

5-Hydroxytryptamine (5-HT, serotonin) is an indolealkylamine that is found in multiple tissues of vertebrates and invertebrates. As is true for most neurotransmitters, the last decade has seen a remarkable growth in the understanding of serotonergic systems. This chapter will attempt to review the biochemical, pharmacological, and physiological data that has accumulated. It is not possible, however, to be comprehensive and the reader will note that we have not included much discussion of the behavioral or clinical pharmacology. Instead, reference will be made to some excellent reviews in these fields.

2. History

In a series of papers in 1948–1949, Rapport and colleagues isolated and eventually crystallized a substance from serum that had potent vasoconstrictor properties. They termed it serotonin, and eventually identified its structure as 3-(β-aminoethyl)-5-hydroxyindole or 5-hydroxytryptamine (5-HT) [1–4]. During the same period, Er-

---

Abbreviations: 5-hydroxytryptamine (5-HT, serotonin); inhibitory postsynaptic potential (IPSP); slow excitatory postsynaptic potential (sEPSP); lysergic acid diethylamide (LSD); spiroperidol (spiperone); 5-hydroxytryptophan (5-HTP); monoamine oxidase (MAO); dorsal raphe (DR); methylenedioxymethamphetamine (MDMA); 8-hydroxy-2-(di-n-propylamino)tetralin (8-OH-DPAT); 5-carboxamidotryptamine (5-CT); 5-methoxytryptamine (5-MeOT); 1-(2-methoxyphenyl)4-[4-(2-phthalimido)butyl]piperazine(NAN-190); 1-(m-trifluoromethylphenyl)piperazine (TFMPP); 1-(3-chlorophenyl)piperazine (mCPP); 1-(2,5-dimethoxy-4-iodophenyl)-2-aminopropane (DOI); 1-(2,5,-dimethoxy-4-bromophenyl)-2-aminopropane (DOB); GR-38032F (odansetron); 1-phenyl-biguanide (PBG); 1-(m-chlorophenyl)-biguanide (mCBG); GTP-binding protein (G proteins); phosphatidylinositol (PI); phospholipase C (PLC); inositol-1,4,5-triphosphate (IP3); diacylglycerol (DAG); protein kinase C (PKC); choroid plexus (CP); dissociation constant (K_d); concentration causing 50% maximal effect (EC_{50}); hyperpolarization-activated current (I_h).

spamer identified a constrictor substance in intestinal chromaffin cells that he called enteramine. This was eventually found to be identical to 5-HT [5,6]. Shortly thereafter, 5-HT was identified in brain tissue [7] and in a remarkably inciteful paper, Woolley and Shaw speculated that 5-HT had a role in mental illness [8]. Brodie et al. [9] provided further evidence for the role of 5-HT in brain function by demonstrating that the effect of reserpine in causing sedation was closely related to its ability to deplete brain 5-HT. In 1957, Bogdanski et al. [10] determined the 5-HT content and 5-HT decarboxylase activity of multiple brain regions, giving further support for its role as a central transmitter substance. It was also at that time that Gaddum and Picarelli [11] proposed two distinct 5-HT receptors. In the 1960s it was shown that 5-HT had both excitatory and inhibitory actions on central neurons [12,13] and central 5-HT-containing neurons were identified [14]. Using extracellular recording techniques, Aghajanian et al. [15] demonstrated the effects of lysergic acid diethylamide (LSD) and serotonin at inhibiting raphe neurons. The first 5-HT postsynaptic potentials were demonstrated in *Aplysia* [16,17], followed by discovery of a 5-HT-mediated inhibitory postsynaptic potential (IPSP) in mammalian brain [18]. In the last 10–20 years, the field has seen a dramatic increase in the number of receptor types and the identification of their effects on cellular biochemistry and physiology. The most recent development has been the identification of the primary structure of the majority of the receptors.

3. Nomenclature

The classification of 5-HT receptors has undergone several revisions as knowledge has progressed from the pharmacologic to the molecular level. Gaddum and Picarelli [11] originally proposed the existence of two 5-HT receptors based on work on the guinea-pig ileum. One receptor mediated a direct contraction of smooth muscle and was antagonized by dibenzyline. It was thus termed the D-receptor. The other type was called the M-receptor because it was antagonized by morphine and caused muscle contraction by exciting cholinergic neurons [11].

This system stood until Peroutka and Snyder [19] presented evidence from radioligand-binding studies in 1979. Using [^{3}H]-labeled LSD, 5-HT, and spiroperidol (spiperone), they identified two LSD-binding sites. The first site had low nanomolar affinity for 5-HT and was termed the 5-HT$_1$ receptor. A second site had picomolar affinity for spiperone, but only low micromolar 5-HT affinity and was designated 5-HT$_2$. Subsequently, [^{3}H]5-HT label was found to be displaced in a biphasic manner by spiperone [20]. This suggested further subdivisions, eventually known as the 5-HT$_{1A}$ and 5-HT$_{1B}$ sites (type 1 designation still implied high affinity for 5-HT). Later, the choroid plexus was found to be enriched with high affinity, mesulergine-sensitive sites, defined as 5-HT$_{1C}$ receptors [21].

Apart from the pharmacologic data, two newer approaches have led to a reanalysis of the 5-HT nomenclature [22]. The first is information on the second messenger coupling of receptors. The 5-HT$_{1A}$, 5-HT$_{1B}$, and 5-HT$_{1D}$ receptors all effect adenylate cyclase via a G-protein. The 5-HT$_{1C}$ receptor, despite its high affinity for 5-HT, is more like the 5-HT$_2$ receptor in that both modulate phosphatidylinositide (PI) turnover via a distinct G protein. Other significant advances have come from molecular biology. The primary sequence of the 5-HT$_{1C}$ and 5-HT$_2$ receptors show a high degree of homology, whereas the 5-HT$_{1A}$ and 5-HT$_{1D}$ receptors form a separate subfamily. Thus, a more updated classification should be based on molecular structure and has been suggested [23] (see Table I).

4. Biochemistry of 5-HT synthesis, storage and neurotransmission

5-HT does not cross the blood-brain barrier, so neurons must synthesize their own. Dietary tryptophan is transported into the CNS by a specific carrier of large neutral amino acids. Tryptophan is taken up by neurons, then hydroxylated by tryptophan hydroxylase to 5-hydroxytryptophan (5-HTP). This is the rate-limiting step in the synthesis of 5-HT. The mechanism by which tryptophan hydroxylase is regulated has been the subject of much controversy. In addition to tryptophan, tryptophan hydroxylase requires molecular oxygen and tetrahydrobiopterin. Thus, changes in the concentration of any of these factors will effect the rate of 5-HT synthesis. This subject has been more fully reviewed elsewhere [24–26].

The next step in 5-HT biosynthesis is decarboxylation of 5-HTP to 5-HT by 5-HTP decarboxylase. After synthesis, 5-HT is packaged into vesicles, thus protecting it from cytosolic degradation by monoamine oxidase (MAO). These vesicles are similar to, but distinct from, those for catecholamines. Some differences include the lack of ATP colocalization and the presence of a specific serotonin-binding protein in 5-HT vesicles. The transport into either type of vesicle, however, is inhibited by reserpine [27].

Following vesicular storage, 5-HT can be synaptically released where it acts upon both pre- and postsynaptic receptors. Serotonergic transmission is terminated primarily by reuptake of the amine by a specific 5-HT transporter. This protein has been cloned [28,29], but a discussion of reuptake is beyond the scope of this chapter. After reuptake, the free indoleamine is either repackaged or metabolized by MAO and aldehyde dehydrogenase to 5-hydroxyindoleacetic acid. This metabolite is then excreted in the urine [4,26].

5. Anatomy of central 5-HT

5-HT is found throughout the animal (vertebrates and invertebrates) and plant kingdoms. In mammals, it is present in platelets, mast cells, enterochromafin cells, and

TABLE I
Classification of 5-HT-receptors

Receptor	Agonist	Antagonist[1]	Tissue[2]	Effect on potential	Ion conductance	Second messenger
5-HT$_{1A}$	8-OH-DPAT DP-5-CT	NAN-190 Spiperone Spiroxatrine	Hippocampus, dentate gyrus, septal nucleus, dorsal raphe, enteric neurons, spinal cord, lymphoid tissue	Hyperpolarization	g_K increase	G-protein[3] Inhibit AC
5-HT$_{1B/1D}$	RU 24969 TFMPP Sumatriptan	CYP SDZ 21-009	Basal ganglia Cingulate cortex, nerve terminals	?	?	Inhibit AC
5-HT$_{1C}$	(−)-α-Me-5-HT mCPP	Mesulergine Mianserin	Choroid plexus Hippocampus	?	?	PI turnover
5-HT$_2$	DOB, DOI LSD	Ketanserin Spiperone	Neocortex Olfactory bulb, claustrum, basal ganglia, aortic smooth muscle	Depolarization	g_K decrease	PI turnover
5-HT$_3$[4]	2-Me-5-HT PBG mCPBG	ICS 205-930 Ondansetron GR65630	Autonomic postgan. Sensory neurons Entorhinal cortex, amygdala, area postrema	Depolarization	$g_{Na,K}$ increase	Ligand-gated channel
5-HT$_4$	5-MeOT	ICS 205-930 (low affinity)	Enteric neurons Colliculi neurons, hippocampus	Depolarization	?	Activate AC
5-HT$_?$	?	?	Lateral geniculate, preps. hypoglossi, spinal motoneurons	Depolarization	$g_{Na,K}$ increase (I_h)	Activate AC

[1] Some of these compounds have been reported to be partial agonists, depending on the tissue, e.g., spiroxatrine, CYP, SDZ 21-009.
[2] Only some representative tissues are listed, i.e., those where autoradiographic labeling is dense or physiologic responses have been demonstrated.
[3] A G-protein inhibits AC, and also activates gK; however, the latter is not caused by changes in cAMP.
[4] Receptor subtypes have been described, but electrophysiologic effects are the same (see text).

8-OH-DPAT, 8-hydroxy-(2-N,N-dipropyl-amino)tetralin; DP-5-CT, N,N-dipropyl-5-carboxamidotryptamine; 1-(2-methoxyphenyl)4-[4-(2-phthalimido)butyl]piperazine (NAN-190); TFMPP, m-trifluoromethylphenylpiperazine; mCPP, 1-(3-chlorophenyl)piperazine; LSD, d-lysergic acid diethylamide; DOB, 1-(2,5-dimethoxy-4-bromophenyl)-2-aminopropane; DOI, 1-(2,5-dimethoxy-4-iodophenyl)-2-aminopropane; CYP, cyanopindolol; PBG, 1-phenyl-biguanide; mCPBG, 1-(m-chlorophenyl)-biguanide; 5-MeOT, 5-methoxytryptamine; AC, adenylate cyclase; PI, phosphatidyl inositol.

neurons. Central nervous system serotonin is synthesized predominantly by a group of raphe nuclei, designated B1–B9. The rostral group (B4–B9) contains the dorsal raphe (DR) and median raphe and gives rise to ascending projections to the cortex, striatum, thalamus, limbic system, and cerebellum. Projections from the caudal group (B1–B3) descend to regions of the spinal cord. These include lamina I (substantia gelatinosa), lamina IX (motoneurons), and intermediolateral gray (sympathetic preganglionic) [14,30,31].

The anatomy of the ascending groups deserves elaboration. Serotonergic fibers rise through the medial forebrain bundle and project widely. Regions that receive a particularly dense innervation include cortical layer I, visual cortex layer IV, striatum, hypothalamus, and septal area. Two types of ascending projection are seen. The first type of fiber has fine fusiform varicosities and is the most common. They come from the DR and are sensitive to methylenedioxymethamphetamine (MDMA), which causes fiber degeneration. The other fiber type appears as large beaded varicosities originating from median raphe and is MDMA-insensitive. Electron microscopy demonstrates that fine varicosities form few true synaptic connections, while up to 80% of beaded varicosities form synaptic specializations [31,32].

Another level of complexity is introduced when the distribution of 5-HT receptors is studied. Each receptor subtype has a unique distribution. 5-HT_{1A} receptors are located most densely in the limbic system, 5-HT_{1B} in striatum, 5-HT_{1C} in choroid plexus, 5-HT_2 in neocortex, and 5-HT_3 in area postrema (details in the respective receptor sections). In addition, the cortical layers have laminar differences in receptor subtype [33]. Thus, the serotonergic innervation of the brain, once felt to be a simple hormonal system, now appears more finely detailed and containing at least some direct synaptic connections.

6. 5-HT_{1A} receptor

6.1. Pharmacology

In 1983, Gozlan et al. published data from radioligand-binding studies using a new compound, 8-hydroxy-2-(di-n-propylamino)tetralin (8-OH-DPAT) [34]. The introduction of this 5-HT_{1A}-selective compound greatly accelerated an understanding of the receptor. In addition to 8-OH-DPAT, other potent agonists include 5-HT, 5-carboxamidotryptamine (5-CT), (+)-LSD, buspirone, and ipsapirone. Antagonist compounds with some degree of selectivity include spiperone, spiroxatrine, and more recently, 1-(2-methoxyphenyl)4-[4-(2-phthalimido)butyl]piperazine (NAN-190) [35–38].

6.2. Anatomy

The brain distribution of 5-HT_{1A} receptors is similar across species, including man [39]. In rat brain, regions that have high densities of 1A sites include limbic structures (dentate gyrus, septal nucleus, CA1 and CA3 hippocampal regions), lamina IV of motor and striate cortex, entorhinal cortex, DR, and lamina I of spinal gray [40].

The cellular location and function of the 5-HT_{1A} receptor has been a source of controversy. In the study by Gozlan et al., hippocampal [^{3}H]5-HT-binding sites were found to correlate well with [^{3}H]8-OH-DPAT sites. This contrasted with the striatum where, despite high levels of [^{3}H]5-HT binding, there were few [^{3}H]8-OH-DPAT sites. In addition, the binding sites in the hippocampus were found to be unaffected by serotonergic lesioning with 5,7-dihydroxytryptamine, while those in striatum were reduced in number by about 65%. This implied that the 8-OH-DPAT sites in the hippocampus were postsynaptic and those in striatum presynaptic. Furthermore, the striatal sites had the properties expected of 5-HT autoreceptors and were of a different type than those in the hippocampus [34]. Nevertheless, the concept of the 5-HT_{1A} site being the autoreceptor persisted because of its predominance in the DR nucleus. Using frontal cortex slices incubated with [^{3}H]5-HT, Middlemiss [41] demonstrated that 8-OH-DPAT was inactive in a functional assay of autoreceptor activity. These results prompted further investigation into the identity of the terminal 5-HT receptor, which was the 'non-1A' site found in striatum. Several studies demonstrated that the autoreceptor had the pharmacologic profile of the 5-HT_{1B} receptor (see Section 6.3) [42,43]. It is now a general consensus that the terminal autoreceptor (or heteroreceptor on non-5-HT terminals) is of the 5-HT_{1B} type (in rat and mouse; 5-HT_{1D} in other species), while the somatic receptor responsible for inhibition in the DR is the 5-HT_{1A} receptor.

6.3. Biochemistry

A consistently observed effect of 5-HT_{1A} activation has been modulation of adenylate cyclase activity. Both increases in basal cyclase [44,45] and inhibition of forskolin-stimulated cyclase activity have been reported [44,46]. These discrepancies may have resulted from tissues that had more than one type of closely related 5-HT receptor. This problem was surmounted by using mammalian cell lines that had no 5-HT receptors and transfecting them with the 5-HT_{1A} gene. The transfectants expressed the receptor, which was shown to only inhibit forskolin- or isoproterenol-stimulated adenylate cyclase. 5-HT had no stimulatory effect by itself [47]. The functional significance of the inhibitory effect is not clear, but it remains a useful assay for studying of 5-HT_{1A} pharmacology [48].

One consensus that did emerge from these studies was the rank order potency of several agonists. Thus, the pharmacologic profile that has emerged for the 5-HT_{1A} receptor is: 5CT > 8-OH-DPAT > LSD > 5-HT > RU 24969 > buspirone. The efficacy of these compounds varies depending on the assay, but 8-OH-DPAT, LSD, and buspirone were frequently found to be partial agonists [44–48].

6.4. Physiology

6.4.1. Potassium conductance

The identification of 5-HT in nerve terminals in brain and spinal cord stimulated interest in its role as a central transmitter [14]. Early studies identified both pure inhibitory [12,13] and pure excitatory [13,17] effects of 5-HT applied to cortical neurons. Inhibitory effects were also seen in DR neurons [15] and hippocampus [49,50]. With the introduction of intracellular recording techniques, it became possible to demonstrate a decreased cell input resistance in association with the 5-HT-induced hyperpolarization [51,52].

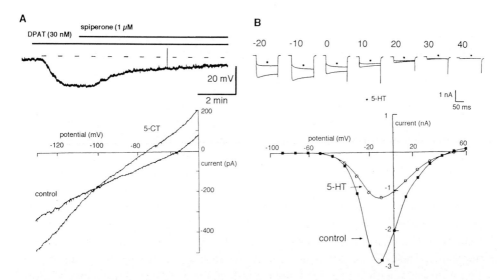

Fig. 1. 5-HT$_{1A}$-receptor activation increases potassium conductance and inhibits calcium conductance in dorsal raphe neurons. A, Intracellular recordings from dorsal raphe neurons in the slice. Top trace: membrane potential recording shows a hyperpolarization caused by 8-OH-DPAT, which is then reversed by spiperone. Bottom trace: two superimposed current/voltage (I/V) plots obtained from a slow depolarizing ramp potential going from −130 to −50 mV. In the presence of 5-CT the slope of the I/V plot is increased and the zero current level is shifted from −60 mV to about −75 mV. (Taken from Williams et al. [56] by courtesy of the Editors of *J. Neurosci.*) B. Whole cell recordings from acutely isolated adult rat dorsal raphe neurons. Top trace: superimposed current records of inward barium currents evoked from a holding potential of −100 mV to the potential indicated. The * above the trace indicates the current obtained in the presence of 5-HT (10 μM). The amplitude of the inward current and the time course of activation were decreased by 5-HT. Bottom trace: an I/V plot of the data shown above. 5-HT did not change the voltage-dependence of the barium current but caused about a 50% reduction of the current measured at −10 mV. (Taken from Pennington et al. [67] by courtesy of the Editors of *J. Neurosci.*).

The developments in 5-HT pharmacology soon made it apparent that the hyperpolarizing effect of 5-HT in the hippocampus resulted from 5-HT_{1A}-receptor activation. Colino and Halliwell [53] demonstrated three distinct potassium conductances that were modulated by 5-HT in CA1 hippocampal neurons. One of the responses was inhibition while the other two were excitations. 8-OH-DPAT had no agonist properties at concentrations up to $10\,\mu\text{M}$, but was an antagonist at 100–200 nM. The hyperpolarizing response to 5-HT was shown to be due to activation of a potassium conductance that was independent of external calcium and blocked by $BaCl_2$ (1–2 mM) [53]. The properties of this conductance were similar to those reported for $GABA_B$ [54], adenosine, α_2-adrenergic, and μ-opioid receptors [55]. All of these systems activate a current that is outward at resting potentials, thus causing a hyperpolarization. The slope of the current/voltage plots tends to flatten at potentials positive to the potassium reversal potential, demonstrating an inward rectification (Fig. 1A) [53,56]. A second paper extended this work by examining the regional distribution of responses along the dendritic axis and using more pharmacology. These authors found 8-OH-DPAT to be a weak partial agonist at $30\,\mu\text{M}$ and demonstrated antagonism by spiperone [57]. A later study found that the anxiolytic buspirone, a relatively selective 5-HT_{1A} compound, also caused hyperpolarization in the hippocampus [58].

6.4.2. Inhibitory postsynaptic potential

Further insight into the 5-HT_{1A} receptor came from studies in the DR nucleus. 5-HT, 5CT, and 8-OH-DPAT all act as full agonists, indicating a greater number of spare receptors than in hippocampus. Spiperone antagonized these compounds with a K_d of about 30 nM. Most important, however, was the identification of a serotonergic IPSP in DR. Yoshimura and Higashi [18] demonstrated a slow IPSP following focal electrical stimulation that was blocked by methysergide and enhanced by imipramine. Subsequently, Williams et al. [56,59] determined that the IPSP had voltage-dependent properties that were identical to those of applied 5-HT. The latter report went on to demonstrate the blockade of the IPSP by spiperone and prolongation by fluoxetine. The IPSP in DR is pertussis toxin-sensitive, implying G protein mediation. Consistent with such a transduction system, the IPSP has a long latency following stimulus (40–65 ms) and a long duration (1–2 s) [56,59].

Hyperpolarizations due to 5-HT_{1A} activation have been observed in several other regions [60–63]. In addition, a 5-HT_{1A}-mediated IPSP has been observed in the guinea pig prepositus hypoglossi, a medullary nucleus with a dense 5-HT terminal input but few cell bodies (Fig. 2A). This has provided an excellent opportunity to study 5-HT synaptic responses in a terminal field. In many respects, the 5-HT hyperpolarization and IPSP are similar to that reported in DR. One unique feature is that uptake inhibitors, such as cocaine and fluoxetine, do not cause membrane hyperpolarization as they do in DR [64,65]. This implies that the serotonergic neurons may have some type of somatic release of 5-HT that does not occur in peripheral fields. Another finding made in the prepositus was that presynaptic inhibition of the IPSP

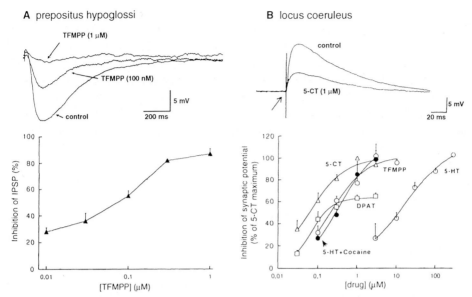

Fig. 2. Presynaptic inhibition of transmitter release by 5-HT receptors. A. Intracellular recording from a neuron in the nucleus prepositus hypoglossi of guinea pig. Top trace: electrical stimulation evoked a 5-HT-mediated IPSP through activation of 5-HT$_{1A}$ receptors. The amplitude of IPSP was depressed by TFMPP (100 nM, 1 μM) shown in three superimposed IPSPs. Bottom trace: concentration-response curve for TFMPP against the inhibition of the 5-HT-mediated IPSP. The maximum inhibition was about 90% and the EC$_{50}$ was near 100 nM. (Taken from Bobker and Williams [64] by courtesy of the Editors of *J. Physiol.*) B. Intracellular recording from a locus ceruleus neuron in rat. Top trace: electrical stimulation evoked the release of GABA and glutamate (presumably from different release sites), causing a fast synaptic potential. The amplitude of the synaptic potential was depressed by 5-CT through a presynaptic mechanism. This inhibition induced by 5-CT was the maximum effect (about 50% of control). Bottom trace: concentration-response curves for several agonists plotted as the percent of the maximum inhibition caused by 5-CT (1 μM). 8-OH-DPAT had a submaximal effect. This may have been due to its being a partial agonist, or because it affected only the glutamate-mediated synaptic potential (it had no effect on GABA transmission). The other agonists, being less selective, may act at both 5-HT$_{1A}$ and 5-HT$_{1B}$ receptors on different sets of terminals. (Taken from Bobker and Williams [94] by courtesy of the Editors of *J. Pharmacol. Exp. Ther.*).

was mediated in part by the 5-HT$_{1D}$ receptor [64]. This subject will be more fully discussed in Section 7.

The system of transduction between the 5-HT$_{1A}$ receptor and K channel is still incompletely understood. Despite the evidence that the 5-HT$_{1A}$ receptor is linked to adenylate cyclase, there has been no demonstration that this system mediates the increase in K conductance. In both DR and hippocampus, 8-bromo-cAMP and forskolin had no effects on the 5-HT-evoked hyperpolarization [54,56]. Thus, it remains to be determined whether a G protein activates the K conductance directly or if some other system is involved.

6.4.3. Calcium conductance

The 5-HT$_{1A}$ receptor can also modulate calcium currents. Using whole-cell recordings from acutely dissociated DR cells from adult rats, Pennington and Kelly [66] demonstrated T-, N- and L-type calcium currents. Application of 5-HT and 8-OH-DPAT produced a maximal 50% reduction in the N-type current, while having essentially no effect on the T-current (Fig. 1B). This action was selectively blocked by NAN-190. In addition, pertussis toxin blocked this effect [66,67], as it does the 5-HT-induced inhibition of adenylate cyclase and hyperpolarization. Thus, DR cells appear to couple the 5-HT$_{1A}$ receptor to multiple effectors via G-proteins. The net effect should be a significant inhibition of transmitter release.

7. 5-HT$_{1B}$ and 5-HT$_{1D}$ receptors

7.1. Pharmacology

The 5-HT$_{1B}$ and 5-HT$_{1D}$ receptors will be considered together in this section because of their marked similarities of pharmacology, distribution, and function. Pedigo et al. originally noted that spiperone inhibition of the [^{3}H]5-HT-binding curve was biphasic. From these results, he proposed two distinct receptors that became known as the 1A and 1B sites [20]. Cyanopindolol was identified as a suitable ligand for radiolabeling studies [68]. The 5-HT$_{1B}$ receptor has been identified in rat and mouse, while being absent from calf, pig, guinea pig, and man [39,69]. In 1987, Heuring and Peroutka [70] identified a receptor in bovine brain membranes related to the 1B site and named it the 5-HT$_{1D}$ receptor. This receptor is present in those species where there are no 5-HT$_{1B}$ receptors.

The 5-HT$_{1B}$ receptor, by definition as a type 1 receptor, has a relatively high affinity for 5-HT and 5-CT (< 10 nM). Other high affinity agonist compounds include RU-24969 and 1-(m-trifluoromethylphenyl)piperazine (TFMPP), while cyanopindolol, SDZ-21-009 and methiothepin are partial agonists or antagonists depending upon the preparation [68,71,72]. The 5-HT$_{1D}$ receptor has a similar profile with the difference that SDZ-21-009 and RU-24969 are not as potent ligands and the α_1-antagonist yohimbine has a relatively high potency [73–75]. As none of the above compounds are selective, the most widely used method of defining the receptor has been to show high 5-HT and cyanopindolol or RU-24969 affinity in the presence of spiperone and 8-OH-DPAT insensitivity [42,70,73,75]. Recently, sumatriptan has been developed as a relatively selective 5-HT$_{1B/1D}$ agonist and is being tested in clinical trials for its antimigraine effects [76,77].

7.2. Anatomy

5-HT$_{1B/1D}$ sites are widely distributed in the brain and have also been identified on pig coronary arteries [75] and mouse kidney [78]. Brain regions with the highest density include the basal ganglia (globus pallidus, substantia nigra pars reticulata), dorsal subiculum, cortex (especially cingulate), and lateral nucleus of cerebellum. Approximately 40% of 5-HT binding in raphe nuclei is of the 5-HT$_{1B/1D}$ type [40].

To assess the cellular localization of the receptor, Verge et al. [79] studied the distribution of 5-HT$_{1A}$ and 5-HT$_{1B}$ sites after raphe neuron lesioning with 5,7-dihydroxytryptamine. This caused the expected loss of 5-HT$_{1A}$ receptors in DR (~ 60%), but had relatively little effect on 5-HT$_{1B}$ receptors. The only exception was in substantia nigra, where 1B sites were reduced (~ 40%). They concluded that the autoreceptors located on 5-HT terminals, with the exception of substantia nigra, were neither of the 1A nor 1B type [79]. A study on guinea pig brain (1D instead of 1B) extended these findings. Waeber et al. [80] found that quinolinic acid lesions of the caudate led to a marked decrease in 5-HT$_{1D}$ binding in caudate and substantia nigra (~ 70%), while 6-hydroxydopamine lesions caused almost no change in binding. Three conclusions are possible from these studies: (1) the dopaminergic neurons of substantia nigra do not have appreciable numbers of 5-HT$_{1B/1D}$ receptors; (2) striatal neurons do have 5-HT$_{1B/1D}$ receptors and account for all the 1B/1D binding in the striatum and about two-thirds of the binding in nigra; (3) raphe terminals in nigra, but not striatum, have 5-HT$_{1B/1D}$ autoreceptors. The explanation for this latter point awaits further study.

In rat cingulate cortex, a combination of lesioning techniques was used to localize 5-HT$_{1A}$ and 5-HT$_{1B}$ receptors. Crino et al. [81] found that 5-HT$_{1A}$ receptors were located only postsynaptically on cortical neurons (i.e., they found none on nerve terminals). Unlike the above studies, however, they reported a significant reduction in 5-HT$_{1B}$ receptors following both raphe lesions and quinolinic acid lesions of the cortex. This indicated that 5-HT$_{1B}$ receptors were on both raphe terminals and cortical cell somata [81]. Thus, these studies suggest that the 5-HT$_{1B/1D}$ receptor is found both pre- and postsynaptically and it may be on a variety of cell types (i.e., serotonergic, GABAergic, etc.).

7.3. Biochemistry

The 5-HT$_{1B/1D}$ receptor is negatively coupled to adenylate cyclase. This has been demonstrated in rat and calf substantia nigra, human prefrontal cortex, and mouse kidney. The inhibition was generally weak, unless forskolin was added. Under these conditions, a typical maximal response was a 20–25% inhibition. This effect had an absolute requirement for guanyl, but not adenyl, nucleotides [78,82–85]. 5-HT also directly inhibited adenylate cyclase (i.e., without forskolin stimulation) in fibroblast cultures, resulting in mitogenesis [86].

7.4. Physiology

7.4.1. 5-HT autoreceptor

As is the case for most neurotransmitters, 5-HT release is under autoinhibitory control [87–90]. Thus, application of 5-HT extracellularly to 5-HT releasing terminals will inhibit release [89]. It was readily established that these autoreceptors were of the 5-HT$_1$ type [91]. As noted, early expectations were that the 5-HT$_{1A}$ receptor would be the autoreceptor. However, when 8-OH-DPAT was found to have no autoreceptor activity [41], Engel et al. [42] analyzed the effect of a series of compounds on [^{3}H]5-HT release from rat cortical slices. The conclusion was that autoreceptor activity correlated very well with 5-HT$_{1B}$ affinity and fairly well with 5-HT$_{1A}$ affinity. However, since 8-OH-DPAT and spiperone were inactive, the 5-HT$_{1B}$ receptor was the likely autoreceptor. Similar results were obtained in other regions [43,92] and in species were the 5-HT$_{1D}$ receptor was found [74].

Chaput et al. have studied autoreceptor function using an in-vivo model. By stimulating ascending 5-HT pathways, they inhibited CA3 hippocampal pyramidal neuron firing. The duration of this inhibition was increased by methiothepin, an autoreceptor antagonist. Furthermore, as they increased the frequency of stimulation, which should decrease 5-HT release by increasing feedback inhibition, the duration of the ascending inhibition decreased [93].

In the guinea-pig prepositus hypoglossi, the autoreceptor has been studied on a 5-HT synapse using intracellular recordings. Electrical stimulation near an impaled neuron caused a 5-HT-mediated IPSP. The IPSP was inhibited (~ 90%) by 5-HT$_{1D}$ agonists (Fig. 2A). When a series of stimuli were given in a train (> 0.1 Hz), a rapid rundown of the IPSP was observed. This rundown was partly reversed by blocking the autoreceptor [64]. Similar results have been obtained in rat DR (Fig. 3; Grudt and Williams, unpublished observations). This supports the concept that the 5-HT$_{1B/1D}$ receptor acts to regulate serotonin transmission by providing negative feedback control to 5-HT terminals [64,93].

7.4.2. Terminal heteroreceptor

In addition to functioning as an autoreceptor, there is both anatomical (reviewed above) and physiological evidence that the 5-HT$_{1B/1D}$ receptor modulates non-serotonergic neurotransmission. We have observed an inhibitory effect of 5-HT on both glutamatergic and GABAergic transmission in the rat locus ceruleus. 5-HT, 5-CT, and 8-OH-DPAT all caused a presynaptic inhibition of glutamatergic and GABAergic transmission (Fig. 2B). While the effect of 8-OH-DPAT was spiperone-sensitive, that of 5-CT was not. The pharmacologic evidence supported both 5-HT$_{1A}$ and 5-HT$_{1B}$ receptors as presynaptic neuromodulators [94]. These receptors would then be termed *heteroreceptors*, because they are modulating nonserotonergic transmission. This has also been observed at adrenergic and cholinergic terminals [95–97]. The cumulative evidence implies that 5-HT$_{1A}$ receptors are found on the somatodendritic

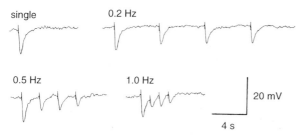

Fig. 3. Presynaptic depression of 5-HT-mediated IPSPs in rat dorsal raphe. Intracellular recordings of membrane potential from a neuron in the slice preparation. A single electrical stimulus caused an IPSP that was about 12 mV in amplitude. When stimuli were separated by an interval of 30 s or greater, the amplitude of this synaptic potential remained constant. When repetitive stimuli were applied at increasing frequencies (indicated above the trace), the amplitude of the second and subsequent IPSPs were depressed. When a single stimulus was applied 30 s following the repetitive stimuli, the amplitude of the IPSP had recovered (not shown). (Grudt and Williams, unpublished data; see also [64]).

tree, and therefore, the effect is to hyperpolarize cell somata. This would effectively shift some neurons beyond threshold with a resultant decrease in synaptic potential amplitude. On the other hand, 5-HT$_{1B/1D}$ receptors are on nerve terminals as well as somata, so they can inhibit transmission through direct actions on terminals. The precise mechanism awaits clarification.

8. 5-HT$_{1C}$ receptor

8.1. Pharmacology

Several investigators noted a high concentration of [^{3}H]5-HT labeling in the choroid plexus. The pharmacologic profile of this binding site was notable in that it was labeled by [^{3}H]5-HT, [^{3}H]LSD, and by the 5-HT$_2$ ligand, [^{3}H]mesulergine. The receptor was unique and dubbed the 5-HT$_{1C}$ receptor [21,98]. It is identified by a high affinity for 5-HT and LSD as agonists, and for methysergide, mianserin, and mesulergine as antagonists. Although there are no ideal ligands for this receptor, (−)α-methyl-5-HT and 1-(3-chlorophenyl)piperazine (mCPP) are somewhat selective agonists. It is insensitive to 8-OH-DPAT, SDZ-21-009, and cyanopindolol. Ketanserin must still often be used to differentiate this profile from that of the 5-HT$_2$ receptor (K_D ~ 1 nM for 5-HT$_2$; ~ 50 nM for 5-HT$_{1C}$) [75,99].

8.2. Anatomy

The distribution of the 5-HT$_{1C}$ receptor is relatively restricted (however, see Section

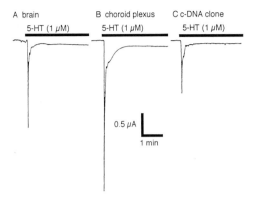

Fig. 4. 5-HT$_{1C}$ receptors evoke IP3-mediated calcium-sensitive chloride currents in mRNA-injected *Xenopus* oocytes. A. The inward current induced by 5-HT following injection of polyA+ mRNA extracted from rat whole brain. B. Following injection of polyA+ mRNA extracted from rat choroid plexus, a larger current is seen. C. Using this response in oocytes, a cDNA clone of the 5-HT$_{1C}$ receptor was isolated. Injection of mRNA obtained from this cDNA clone (from a mouse choroid plexus papilloma) also caused an inward current. (Taken from Lubbert et al. [104] by courtesy of the Editors of *Proc. Natl Acad. Sci. USA*).

12). Appreciable numbers of binding sites are found only on cells of the choroid plexus. Much lower concentrations of binding sites are found in olfactory bulb, hippocampus, amygdala, layer IV of cortex (frontal, sensory, striate) and claustrum [40].

8.3. Biochemistry and physiology

Having earlier found that 5-HT$_2$ receptors activate phosphatidylinositol (PI) turnover in cortex [100], Conn et al. [101] studied the response in choroid plexus (CP). They found that serotonin caused a 5-fold increase in [^{3}H]inositol 1-phosphate release from CP, whereas in the cortex only a 2.5-fold increase over basal release was found. In addition, the pharmacology in CP correlated well with the 5-HT$_{1C}$ receptor, while the cortical response appeared to be due to mixed receptor types. This data supported the contention that the CP possesses a relatively homogeneous and dense population of receptors that are coupled to the PI pathway. It has also been reported that the 5-HT$_{1C}$ receptor can stimulate guanylate cyclase activity [99].

Demonstration of the potent effect of 5-HT$_{1C}$-receptor activation on PI turnover strongly suggests this is a physiologically significant system. Determining this significance in the CP has been difficult, however, because 5-HT modulates blood flow to the CP, which then has its own effect on cerebrospinal fluid production [102]. An electrophysiological effect of 5-HT$_{1C}$-receptor activation was observed as early as 1984, when Gundersen et al. [103] reported on the novel technique of injecting mRNA into *Xenopus* oocytes in order to obtain functional expression of receptor proteins. Among their findings was that serotonin caused an oscillatory inward chloride current in oocytes injected with rat or human brain mRNA. Their current was blocked by methysergide but not ketanserin, consistent with 5-HT_{1C} pharmacology [103]. It was this system that was used to isolate mRNA and eventually clone the receptor (Fig. 4; also see Section 12) [99,104,105]. Whether a chloride current is relevant for CP cell functioning has not been determined.

9. 5-HT_2 receptors

9.1. Pharmacology

Originally defined as the high-affinity spiperone-binding site [19], the 5-HT_2 receptor now has a better defined pharmacology than most 5-HT receptors. The phenylalkylamine hallucinogens, 1-(2,5-dimethoxy-4-iodophenyl)-2-aminopropane (DOI) and 1-(2,5,-dimethoxy-4-bromophenyl)-2-aminopropane (DOB), are among the most selective and potent agonists at this site [106]. LSD is a potent, but nonselective agonist, while 5-HT has a relatively low potency ($K_d \sim 3\ \mu M$). The antagonist used most often to define the 5-HT_2 receptor is ketanserin, which is both potent and selective [69,107]. Spiperone can also be used as an antagonist, but it competes at 5-HT_{1A}, α_1-adrenergic, and D_2 dopamine receptors as well.

9.2. Anatomy

Autoradiographic analysis of [^{125}I]DOI and [^{3}H]ketanserin binding demonstrates the following high-density regions: olfactory bulb and tubercle, cerebral cortex (lamina I, IV, V), claustrum, caudate, putamen, nucleus accumbens, ventral dentate gyrus, and mammillary bodies [108,109]. Using in-situ hybridization to a 5-HT_2 receptor oligonucleotide, the distribution of 5-HT_2 receptor mRNA was roughly similar to that described by autoradiography. The area of greatest hybridization was to layer V of frontal cortex [110].

9.3. Biochemistry

9.3.1. PI turnover
The 5-HT_2 receptor, like the 5-HT_{1C}, α_1-adrenergic, M1 muscarinic receptors, can stimulate PI turnover (Fig. 5). The effect on this system is dependent on GTP, suggesting that a G protein mediates its actions [100]. This inference has been borne out by the molecular structure of the protein (see Section 12). The specific G-protein(s) has not been completely characterized and may be dependent on the response that is being measured. G-protein activation by the 5-HT_2 receptor causes activation of

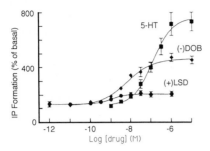

Fig. 5. Activation of 5-HT$_2$ receptors increases the hydrolysis of phosphoinositide. Concentration-response curve shows the increase in [^{3}H]IP formation induced by three 5-HT$_2$ agonists in choroid plexus. LSD was a very weak partial agonist. DOB had a greater efficacy, inducing an increase that was about 50% of that caused by 5-HT. (Taken from Barker et al. [113] by courtesy of the Editors of *Brain Res.*).

phospholipase C (PLC). This phosphodiesterase then hydrolyzes phosphatidyl-4,5-bisphosphate to inositol-1,4,5-trisphosphate (IP3) and diacylglycerol (DAG). These two compounds act as intracellular second messengers. IP3 can release calcium from intracellular stores, activating calcium-dependent events. DAG has at least two roles: (1) activation of protein kinase C (PKC) and (2) serving as an eicosanoid precursor [100,111–113]. The consequences of these pathways on cellular physiology have not been definitively determined (see Section 9.4.).

9.3.2. Desensitization

5-HT$_2$ receptors are generally found to demonstrate desensitization. This observation has been made in humans with hallucinogens [114], in behavioral paradigms measuring head-twitch responses [115,116], in electrophysiologic studies measuring depolarizations [117], in radioligand-binding assays [118], and biochemically measuring phosphoinositide breakdown [118]. The mechanism of desensitization has been studied in the clonal cell line P11 because of its high density of 5-HT$_2$ receptors. Exposure of P11 cells to 5-HT for as little as 20 min caused a progressive decrease in the ability of 5-HT to stimulate PI turnover. After 4 hours, the 5-HT response was totally abolished. Desensitization occurred with relatively low receptor occupancy, as low concentrations (100 nM) applied for 8 hours depressed the maximal response by about 50% (EC$_{50}$ ~ 600 nM, for acutely applied 5-HT). The increase in PI turnover seen with α_1-adrenoceptor agonists was unaffected by prior exposure to 5-HT, indicating that the 5-HT$_2$ desensitization was *homologous*, i.e., it did not cross-desensitize to other receptors. Associated with the loss in response to 5-HT, there was a 30% decrease in the number of 5-HT$_2$ receptors [118]. Such a decrease could be a mechanism to account for desensitization, although the events leading to decreased receptor number remain to be determined.

9.4. Physiology

9.4.1. Potassium conductance

An increase in excitability or a membrane depolarization has been reported in all sites where 5-HT$_2$-receptor activation has been studied. One of the first descriptions of a depolarizing action of 5-HT was in facial motoneurons. Associated with the depolarization was an increase in the input resistance measured with voltage recording and a decline in the membrane conductance under voltage clamp. These observations suggested the closure of a potassium channel [119,120]. It has now been shown that at least part of this depolarization results from activation of 5-HT$_2$ receptors since the depolarizations to 5-HT and α-methyl-5-HT were blocked by ketanserin, mianserin, and ritanserin [121,122]. Interestingly, application of the hallucinogens, LSD, mescaline, or psilocin, all caused a small and long-lasting depolarization and facilitated the excitation induced by both 5-HT and noradrenaline [120,123]. These results suggest two conclusions: (1) hallucinogens may be partial agonists at 5-HT$_2$ receptors because they evoke only weak depolarizations, and (2) the facilitation of 5-HT and noradrenaline by hallucinogens could be the result of actions at different receptor sites.

The depolarization of motoneurons is thought to be mediated through a G protein since the 5-HT-induced inward current was prolonged and became irreversible with intracellular injection of GTPγS. Injection of GDPβS, on the other hand, caused a reduction in the amplitude of the 5-HT current. In that study, superfusion with H-7 and sphingosine (PKC inhibitors) unexpectedly enhanced the inward current, whereas phorbol diacetate (PKC activators) reduced the 5-HT current, but had no effect on its own. These results suggest that the depolarization induced by 5-HT is not directly mediated by the activation of PKC, but may be modulated by the second messenger [123]. One possibility is that activated PKC is responsible for the desensitization of the 5-HT$_2$ receptor; such an action would account for the augmentation of the current observed with PKC inhibitors.

The ionic mechanism of 5-HT$_2$ depolarizations has been studied in the nucleus accumbens. North and Uchimura found that 85% of neurons in a slice preparation were depolarized by 5-HT (Fig. 6A). This 5-HT effect was blocked by ketanserin, mianserin, and spiperone, implicating the 5-HT$_2$ receptor. Depolarization was due to a decrease in an inwardly rectifying potassium conductance, i.e., a potassium conductance that was active at resting potential was inhibited. In addition, superfusion with BaCl$_2$ blocked the inward rectification of the resting membrane, caused a depolarization (or an inward current), and occluded the current induced by 5-HT [124]. A second paper [125] reported the effects of phorbol esters in nucleus accumbens. The PKC activators also decreased the same potassium conductance and prevented further depolarization by 5-HT. Thus, in accumbens, an increase in PI turnover may be the mediator of potassium conductance inhibition [125]. The discrepancy between

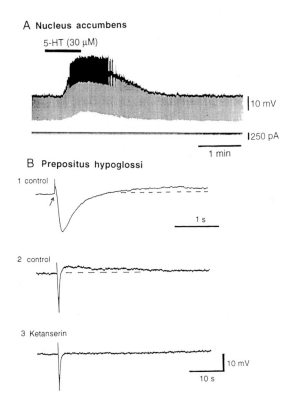

Fig. 6. 5-HT$_2$ receptors cause depolarizations. A. Intracellular recording of membrane potential from a neuron in the nucleus accumbens in a brain slice. Superfusion with 5-HT caused a depolarization of the membrane potential and evoked action potentials. The down-going deflections are electrotonic potentials caused by constant current pulses passed across the membrane (indicated in the bottom trace). During the 5-HT-induced depolarization, the amplitude of the electrotonic potentials increased suggesting that cell resistance was increasing, i.e., channels were being closed. These were found to be inwardly rectifying potassium channels (not shown). (Taken from North and Uchimura [124] by courtesy of the Editors of *J. Physiol.*). B. Intracellular recording from a neuron in guinea pig prepositus hypoglossi in a brain slice. 1. A single electrical stimulus was applied at the arrow. This evoked a series of synaptic potentials, beginning with a rapid depolarizing potential, followed by an IPSP that lasted for about 1 s and a slow EPSP. The beginning of the slow EPSP is illustrated by the depolarization from the resting potential indicated by the dashed line. 2. A longer recording illustrating the 30–40 s duration of the slow EPSP. 3. after superfusion with ketanserin (1 μM), the slow EPSP was abolished. (From Bobker and Williams, unpublished observations).

these results and those of Aghajanian [123] remains to be explained, but may have been related to the concentrations of phorbol esters used.

9.4.2. Excitatory postsynaptic potential
Ketanserin-sensitive depolarizations to be applied have also been observed in cortical [117] and brainstem neurons [126]. The neurons in prepositus hypoglossi are note-

worthy because an electrically evoked, ketanserin-sensitive, slow excitatory postsynaptic potential (sEPSP) has been observed following IPSP (Fig. 6B). This sEPSP is augmented by cocaine, mimicked by DOI, and blocked by ketanserin, supporting the contention that it is due to 5-HT$_2$ receptor activation (Bobker and Williams, unpublished observations, [65]). These neurons are always first hyperpolarized by 5-HT and then depolarized following the washout of 5-HT, before returning to resting potential. This is similar to what is seen with cortical neuronal 5-HT responses [117]. It may then be a common occurrence that both 5-HT$_{1A}$ and 5-HT$_2$ receptors are found on the same cell.

10. 5-HT$_3$ receptors

10.1. Pharmacology

The 5-HT$_3$ receptor is unique among monoamine receptors because it is a ligand-gated ion channel, more akin to the nicotinic acetylcholine receptor than to G-protein-coupled receptors. Selective compounds have been synthesized using either 5-HT or cocaine as the starting nucleus [127,128]. Widely used antagonists include ICS 205-930, GR67330 and GR38032F (ondansetron) [129]. Agents that block the channel include (+)-tubocurare, cocaine and MDL-72222. These channel blockers showed little or no selectivity for the 5-HT$_3$ channel over the nicotinic channel. A distinguishing feature is that the nicotinic blocker, hexamethonium, is inactive at the 5-HT$_3$ channel [130]. Selective agonists at this receptor include 2-methyl-5-HT, 1-phenyl-biguanide, and 1-(m-chlorophenyl)-biguanide, but most studies to date have used 5-HT. Some authors have also suggested 5-HT$_3$ receptor-subtypes based on varying antagonist affinities in different tissues. For instance, MDL-72222 is inactive in guinea pig ileum, while having potent effects elsewhere [129]. For the purpose of this review, however, all 5-HT$_3$ sites will be considered together because their physiologic properties do not vary significantly.

10.2. Anatomy

5-HT$_3$ receptors are widely distributed in the peripheral nervous system, while having a more restricted distribution in the central nervous system. 5-HT$_3$ receptors have been demonstrated on postganglionic autonomic neurons (sympathetic and parasympathetic), enteric neurons, and sensory neurons (skin, heart, nodose ganglion) [129]. The presence of 5-HT$_3$ receptors on cell lines derived from peripheral ganglion, such as NG108-15 and N1E-115 cells, has facilitated studies at the single cell level (see below).

Radiolabeled studies in the central nervous system have revealed binding sites in various part of the brain. Using [^{3}H]GR65630, Kilpatrick et al. [131] found specific binding in entorhinal cortex, amygdala, hippocampus, and nucleus accumbens. In a later study, two regions of highest binding were found in the area postrema and vagus nerve, both structures important in the emesis reflex [132]. These findings have led to the successful development of clinically potent antiemetics [133]. In human brain, the greatest binding is observed in the amygdala and hippocampus [132].

10.3. Physiology

10.3.1. Peripheral nervous system
Early work on 5-HT_3 receptors was carried out in peripheral ganglia. A complete characterization of the ionic conductance activated by 5-HT was done by Higashi and Nishi [134] using voltage clamp techniques from acutely isolated nodose ganglia. Iontophoretic application of 5-HT caused a rapid depolarization of the membrane potential. Under voltage clamp, an inward current associated with an increase in membrane conductance was observed (Fig. 7A). The reversal potential of the 5-HT current was about 0 mV, consistent with a mixed sodium/potassium current. These investigators made three other important observations. The first was that the Hill slope of the concentration-response curve was steep, implying the presence of multiple 5-HT binding sites on the receptor. Secondly, (+)-tubocurarine was found to be a non-competitive inhibitor of the 5-HT-mediated conductance, whereas hexamethonium was without effect. Finally, other known 5-HT-receptor antagonists were ineffective, indicating that the depolarization was mediated by a novel 5-HT receptor [134]. Similar findings have been made in rabbit ciliary ganglion [135], guinea pig submucous plexus [136,137], and rabbit superior cervical ganglion [138].

More recently, whole cell and single channel recordings were made from guinea pig ileum submucous plexus ganglion cells with patch clamp methods. Single channel cationic currents were evoked upon application of 5-HT onto outside-out membrane patches. These currents reversed polarity at about 0 mV and this reversal potential was more negative in reduced sodium concentrations. In addition, both ICS 205-930 and ondansetron blocked the action of 5-HT (Fig.7B). The novel observation was that this 5-HT-activated channel remained functional in the absence of G-proteins, i.e., in excised patches [139]. This indicates that the 5-HT_3 receptor, like the nicotinic receptor, is a receptor-channel complex. It is a distinct protein, however, as evidenced by its pharmacology. Similar results have been obtained from guinea pig celiac ganglion neurons in culture (Gerzanich, unpublished results).

The biophysical properties of the 5-HT_3 receptor channel have been studied in the most detail using cell lines such as NG108-15 [190], N18 [141], NCB-20 [142], N1E-115 [142,143]. Most of these reports are in general agreement that the 5-HT-gated channel is permeable to both sodium and potassium. It also appears that under certain conditions calcium may permeate the channel. This is an important issue since the terminals of primary afferent neurons are known to have 5-HT_3 receptors that may modulate transmitter release and calcium entry could be a mechanism of modu-

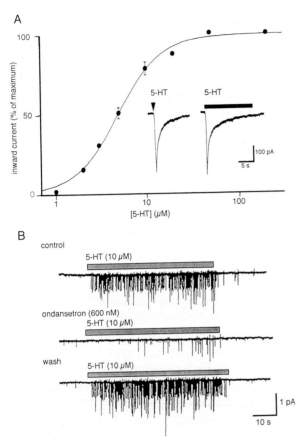

Fig. 7. 5-HT$_3$ receptors increase a cationic nonselective conductance. A. Concentration-response curve to 5-HT measuring the peak current induced by rapid application of 5-HT onto rat glioma X mouse neuroblastoma hybrid cells. The EC$_{50}$ was 3.3 μM and the Hill coefficient was 1.8. The insets are inward currents induced by 5-HT using whole cell recording configuration. At the arrow a brief application was applied, whereas in the next trace 5-HT was applied during the period indicated by the bar. This experiment indicates that the peak current was rapidly followed by desensitization. Yakel et al. [146] by courtesy of the Editors of *J. Physiol.*). B. Single channel currents induced by 5-HT recorded from cultured celiac ganglion cells using the excised outside-out patch configuration. In each trace 5-HT was applied during the period indicated by the bar above the trace. Preincubation of the patch with ondansetron blocked the increase in channel activity induced by 5-HT. (From Gerzanich, V., unpublished observations).

lation. There is also consensus that the Hill coefficient is near 2, indicating two 5-HT-binding sites on the receptor. There are differences, however, in the observed current/voltage relationships of the 5-HT-activated current; in some cases they are linear [134,139,143], while others have reported inward rectification [140,144]. In addition, voltage-dependent cationic blockade of the current has been reported from a cloned receptor (see Section 12) [145].

10.3.2. Central nervous system

There have been only a few reports of 5-HT$_3$-mediated actions on single neurons in the central nervous system. The first was in cultured hippocampal and striatal neurons, where the response was similar in all respects to that found in peripheral ganglia and cell lines [63]. One problem encountered in these experiments was the low percentage of neurons that had 5-HT$_3$ depolarizations (10% in hippocampus, 6% in striatum). This has prompted some investigators to conduct studies on cell lines with a higher yield of responsive cells [140,146].

An indirect effect of 5-HT$_3$ receptors has been observed in 85% of CA1 pyramidal hippocampal neurons in the slice preparation. Superfusion with 5-HT and 2-methyl-5-HT caused a large increase in the frequency of spontaneous GABA-mediated synaptic potentials. This was antagonized by ICS 205-930 and blocked by tetrodotoxin, indicating that the synaptic potentials were caused by a 5-HT$_3$-mediated excitation of GABA interneurons. The amplitude of the GABA-mediated synaptic potentials was not affected by 5-HT, suggesting that there was no direct action at the terminal [147].

10.3.3. Synaptic potentials

Most recently, a 5-HT-mediated fast synaptic potential was found in the lateral nucleus of the amygdala in rat brain. In about 40% of neurons, electrical stimulation evoked a synaptic potential that was insensitive to blockade of glutamate, GABA$_A$- and nicotinic receptors. The peak amplitude was between 5–15 mV from the resting membrane potential. Both the synaptic potential and synaptic current reversed polarity at about 0 mV. Most importantly, antagonism by ICS 205-930, GR-67330 and GR-38032F at low concentrations was demonstrated. Blockade of 5-HT reuptake with fluoxetine increased the amplitude of the synaptic potential but curiously did not prolong the duration [148].

10.3.4. Desensitization

It was apparent from the earliest studies of the 5-HT$_3$ receptor that 5-HT application had to be rapid in order to avoid desensitization. This was accomplished by Higashi and Nishi [134] in the study on nodose ganglia with the use of iontophoretic application. Although the amplitude of the responses was reproducible and directly related to the amount of ejection current, this method of application had the disadvantage of delivering the drug at unknown concentrations and under nonequilibrium conditions. Using cells in culture, the rapid application (tenths of ms) of agonists with 'fast-flow' pipettes has allowed drug delivery at known concentrations and at equilibrium [143,146]. These studies indicate that the apparent dissociation constant for 5-HT was 2–4 mM. The rate of desensitization was dependent on the 5-HT concentration, with a near maximal rate observed at 5 mM ($t_{1/2}$ about 2 s). The voltage-dependence of desensitization varied among preparations, but it was generally found that the rate of desensitization was slowed at depolarized potentials [146].

The rate of desensitization was found to decrease with the duration of whole cell recording. The decline in the rate of desensitization was greatly reduced with the inclusion of nonhydolyzable ATP analogs in the recording pipette. Conversely, the rate of desensitization could be increased with agents expected to increase cAMP levels in the cell [146]. A complete understanding of the mechanisms involved requires further investigation and may be significantly aided by the molecular identification of the components of the 5-HT_3 receptor.

11. 5-HT receptors that stimulate adenylate cyclase

11.1. 5-HT_4 receptor

As previously noted, the 5-HT_{1A} receptor has been reported to both stimulate and inhibit adenylate cyclase, depending on conditions. Shenker et al.[44] were among those to recognize that stimulation of cyclase appeared to be due to the actions of two distinct receptors. A high affinity site was sensitive to 8-OH-DPAT and spiperone, identifying it as the 5-HT_{1A} receptor. A low affinity site was insensitive to those compounds, as well as 5-HT_2 and 5-HT_3 ligands. Dumuis et al. [149] confirmed these results and extended the pharmacology. In mouse embryo colliculi neurons, 5-HT caused a 2.5-fold increase in basal cAMP production with an EC_{50} of 109 nM. 5-Methoxytryptamine (5-MeOT) was an equipotent agonist, while 5-CT was relatively weak. The only antagonist that was identified is ICS 205-930, although its affinity for this site is much lower than for 5-HT_3 receptors [149]. The receptor with this pharmacology has been named 5-HT_4.

Functional effects of 5-HT-induced adenylate cyclase stimulation have been observed in several tissues. Like the pharmacological studies, not all of these responses are due to the same receptor. At least two groups have reported physiological responses due to activation of a receptor corresponding to the 5-HT_4 site. Andrade and Chaput [150], like other groups, found that 5-HT inhibited the action potential afterhyperpolarization in CA1 hippocampal neurons. In addition, after blocking the 5-HT_{1A} receptor, 5-HT caused a small depolarization. Both the depolarization and the effect on the afterhyperpolarization had the pharmacologic profile of a 5-HT_4 receptor, although 5-MeOT was not particularly potent [150]. In the guinea pig ileum, several 5-HT receptors appear to mediate smooth muscle contraction, either by direct effects (5-HT_2) or by causing acetylcholine release. One of these receptors was highly sensitive to 5-HT and 5-MeOT, while being insensitive to the 5-HT_3 agonist 2-methyl-5-HT. ICS 205-930 was a low affinity antagonist ($pA_2 = 6.4$), consistent with the 5-HT_4 receptor [151].

11.2. Modulation of h-current (I_h)

Three laboratories have reported a 5-HT-evoked depolarization mediated by augmentation of the cationic current, I_h (a time-dependent, *h*yperpolarization-activated current). We have observed a slow depolarizing response to 5-HT in a subset of prepositus hypoglossi neurons [152,153]. Under voltage clamp, hyperpolarizing steps from -50 mV (in control) revealed a time- and voltage-dependent inward current with properties identical to those of I_h described in other tissues (also known as I_f and I_q) [154]. 5-HT augmented this inward current, thus accounting for the depolarization. Extracellular CsCl blocked I_h, the 5-HT current, and the depolarization induced by 5-HT. In contrast, $BaCl_2$ had no effect on the 5-HT current. Two other laboratories have reported the same finding in guinea pig thalamic neurons and rat spinal motoneurons [155–157].

In all three of the above preparations, forskolin, 3-isobutyl-1-methylxanthine and 8-bromo-cAMP mimicked the 5-HT depolarization, suggesting that 5-HT augmented I_h via activation of adenylate cyclase. The identity of the receptor mediating the response was not determined, as it did not correspond to known subtypes of monoamine receptors. Further work will be needed to determine if this receptor is related to the $5-HT_4$ receptor or is a novel subtype.

12. Molecular biology

12.1. G-protein-coupled receptors

12.1.1. Modulation of PI turnover: $5-HT_{1C}$, $5-HT_2$

All 5-HT receptors, with the exception of the $5-HT_3$ receptor, belong to the family of G-protein-coupled receptors. Members of this family consist of a single polypeptide chain and have seven putative membrane-spanning regions. The amino terminus of these receptors is extracellular, while the carboxy terminus is intracellular. The greatest homology between members is within the putative membrane-spanning regions, while nonspanning regions show little homology and may be important for desensitization mechanisms [158]. Among the G protein-coupled receptors, there appear to be two subgroups: (1) receptors that are strongly coupled to PI turnover and (2) those that modulate adenylate cyclase. The former group includes the $5-HT_{1C}$, $5-HT_2$, α_1-adrenergic and M1 muscarinic receptors, while those in the latter include $5-HT_{1A}$, $5-HT_{1B/1D}$, α_2- and β-adrenergic, dopamine, M2 muscarinic, and opioid receptors.

The first 5-HT receptor to be cloned was the $5-HT_{1C}$ receptor. This was by virtue of its high density in choroid plexus and the development of a functional assay in *Xenopus* oocytes [103]. Injection of oocytes with choroid plexus mRNA leads to expression of the $5-HT_{1C}$ receptor. Application of 5-HT then causes an increase in intracellular Ca^{2+} via PI turnover. This activates a Ca^{2+}-dependent chloride current. This response was used by Julius et al. [159] and Lubbert et al. [104,105] to isolate a full length clone of the $5-HT_{1C}$ receptor. In-situ hybridization demonstrated a wider distribution of $5-HT_{1C}$ receptors than anticipated from autoradiographic studies. In addition to choroid plexus, significant levels of receptor expression were seen in hippocampus, medial geniculate, posterior thalamus, substantia nigra pars compacta, and raphe nuclei [160].

Using oligonucleotides directed against portions of the 5-HT_{1C} receptor, Pritchett et al. [161] cloned the 5-HT_2 receptor. Their clone demonstrated 51% homology overall, and 80% homology in the transmembrane region, with the 5-HT_{1C} receptor. Northern blot analysis showed expression predominantly in cerebral cortex, with low levels in hypothalamus, spinal cord, and olfactory bulb [162]. Peripheral expression was also seen on aortic smooth muscle cells [163].

12.1.2. Modulation of adenylate cyclase: 5-HT_{1A}, 5-$HT_{1B/1D}$

The 5-HT_{1A} receptor was cloned using a full length β_2-adrenoceptor as a probe under low stringency conditions. The clone had high affinity for 8-OH-DPAT, ipsapirone, 5-HT and spiroxatrine, consistent with the 5-HT_{1A} receptor [38]. An interesting result of work on the cloned receptor has been the study of its effect on adenylate cyclase. In mammalian cell lines that express the 5-HT_{1A} receptor, serotonin has been shown only to inhibit adenylate cyclase activity after stimulation with an agent such as forskolin. No direct stimulation of cAMP synthesis by serotonin, as reported in tissue culture, has been demonstrated [164]. The discrepancy between these findings and those in tissue culture remains to be explained, but it has been suggested that cyclase stimulation may be mediated in some preparations by the 5-HT_4 receptor [158], as discussed above. Another unexpected result came from Northern analysis, which demonstrated high levels of 5-HT_{1A} receptor mRNA in lymphoid tissue, such as lymph node, spleen, and thymus. This suggests a possible role for 5-HT as an immunomodulator [38].

The human 5-HT_{1D} receptor has been cloned using hybridization to an unidentified G protein-coupled canine receptor. The clone was 43% identical to the 5-HT_{1A} receptor. Its pharmacological profile was generally in agreement with that reported from tissue membranes, although notable exceptions included 8-OH-DPAT (affinity higher than expected), methiothepin (higher affinity), and sumatriptan (lower affintiy) [75,165]. The authors also demonstrated inhibition of forskolin-stimulated cAMP accumulation by the cloned receptor, suggesting G_i coupling [165].

12.2. Ligand-gated channels: 5-HT_3

The 5-HT_3 receptor has been recently cloned, again using *Xenopus* oocytes. A neuroblastoma expression library was screened for 5-HT-gated currents. The 487 amino acid protein had approximately 25% homology with the α subunit of the nicotinic acetylcholine receptor, the $\beta1$ subunit of the $GABA_A$ receptor, and the 48K subunit of the glycine receptor. Inward currents generated by the receptor were sensitive to blockade by curare, ICS 205-930, and MDL72222, but not to methysergide. Electrophysiological characterization of the channel showed a Hill coefficient of about 1.8, a reversal potential near 0 mV, and rapid desensitization with continuous 5-HT exposure. An unexpected result was that current-voltage relations generated from the cloned receptor showed negative-slope conductance at potentials more negative than −50 mV. This effect was augmented by increasing the Ca^{2+} or Mg^{2+} concentration [145]. This type of voltage-dependent, divalent cation channel blockade is also observed with the NMDA glutamate receptor (Mg^{2+} only), and may indicate similar types of voltage regulation of the two channels.

13. Conclusion

Since its discovery, 5-HT research has undergone two periods of intense activity. An early era where 5-HT was identified as a neurotransmitter, and the past decade with its rapid expansion in understanding 5-HT receptor subtypes. Knowledge of subtype pharmacology has permitted analysis of regional anatomy, differential biochemical and physiologic effects, and behavioral phenomenona [166]. Perhaps the most exciting result of these advances is in the development of drugs [167,168] with new clinical applications, such as ondansetron, sumatriptan, buspirone and fluoxetine.

References

1. Rapport, M., Green, A. and Page, I. (1948) Science 108, 329–330.
2. Rapport, M., Green, A. and Page, I. (1948) J. Biol. Chem. 176, 1243–1250.
3. Rapport, M. (1949) J. Biol. Chem. 180, 961–969.
4. Garrison, J. (1990) In: The Pharmacological Basis of Therapeutics (Gilman, A., Rall, T., Nies, A. and Taylor, P., Eds.), Ch. 23, pp. 575–599, Pergamon Press, New York.
5. Erspamer, V. and Boretti, G. (1951) Arch. Int. Pharmacodyn. 88, 296–332.
6. Erspamer, V. and Asero, B. (1952) Nature 169, 800–801.
7. Twarog, B. and Page, I. (1953) Am. J. Physiol. 175, 157–161.
8. Woolley, D. and Shaw, E. (1954) Proc. Natl Acad. Sci. 40, 228–231.
9. Brodie, D., Pletscher, A. and Shore, P. (1955) Science 122, 968.
10. Bogdanski, D., Weissbach, H. and Udenfriend, S. (1957) J. Neurochem. 1, 272–278.
11. Gaddum, J. and Picarelli, Z. (1957) Br. J. Pharmacol. 12, 323–328.
12. Krnjevic, K. and Phillis, J. (1963) J. Physiol. 165, 274–304.
13. Roberts, M. and Straughan, D. (1967) J. Physiol. 193, 269–294.
14. Dahlstrom, A. and Fuxe, K. (1964) Acta Physiol. Scand. 62, Suppl. 232, 1–55.
15. Aghajanian, G., Haigler, H. and Bloom, F. (1972) Life Sci. 11, 615–622.
16. Cottrell, G. (1970) Nature 225, 1060–1062.
17. Gerschenfeld, H. and Paupardin-Tritsch, D. (1974) J. Physiol. 243, 457–481.
18. Yoshimura, M. and Higashi, H. (1985) Neurosci. Lett. 53, 69–74.
19. Peroutka, S. and Snyder, S. (1979) Mol. Pharmacol. 16, 687–699.
20. Pedigo, N., Yamamura, H. and Nelson, D. (1981) J. Neurochem. 36, 220–226.
21. Pazos, A., Hoyer, D. and Palacios, J. (1984) Eur. J. Pharmacol. 106, 539–546.
22. Bradley, P., Engel, G., Feniuk, W., Fozard, J., Humphrey, P., Middlemiss, D., Mylecharane, E., Richardson, B. and Saxena, P. (1986) Neuropharmacology 25, 563–576.

23. Pierce, P. and Peroutka, S. (1989) Semin. Neurosci. 1, 145–153.
24. Lovenberg, W., and Kuhn, D. (1982) In: Serotonin in Biological Psychiatry, (Ho, B., Ed.), pp. 73–82. Raven Press, New York.
25. Lookingland, K., Shannon, N., Chapin, D. and Moore, K. (1986) J. Neurochem. 47, 205–212.
26. Cooper, J., Bloom, F. and Roth, R. (1986) In: The Biochemical Basis of Neuropharmacology, 5th edn., Ch. 11, pp. 315–351. Oxford University Press, New York.
27. Tamir, H. and Gershon, M. (1990) Ann. N.Y. Acad. Sci. 600, 53–66.
28. Blakley, R., Berson, H., Fremeau, R., Caron, M., Peek, M., Prince, H. and Bradley, C. (1991) Nature 354, 66–70.
29. Hoffman, B., Mezey, E. and Brownstein, M. (1991) Science 254, 579–580.
30. Steinbusch, H. (1981) Neuroscience 6, 557–618.
31. Tork, I. (1990) Ann. N.Y. Acad. Sci. 600, 9–35.
32. Papadopoulos, G., Parnavelas, J. and Buijs, R. (1987) J. Neurocytol. 16, 883–892.
33. Goldman-Rakic, P., Lidow, M. and Gallager, D. (1990) J. Neurosci. 10, 2125–2138.
34. Gozlan, H., El Mestikawy, S., Pichat, L., Glowinski, J. and Hamon, M. (1983) Nature 305, 140–142.
35. Middlemiss, D. and Fozard, J. (1983) Eur. J. Pharmacol. 90, 151–153.
36. Peroutka, S. (1986) J. Neurochem. 47, 529–540.
37. Glennon, R., Naiman, N., Pierson, M., Titeler, M., Lyon, R. and Weisberg, E. (1988) Eur. J. Pharmacol. 154, 339–341.
38. Fargin, A., Raymond, J., Lohse, N., Kobilka, B., Caron, N. and Lefkowitz, R. (1988) Nature 335, 358–360.
39. Hoyer, D., Pazos, A., Probst, A. and Palacios, J. (1986) Brain Res. 376, 85–96.
40. Pazos, A. and Palacios, J. (1985) Brain Res. 346, 205–230.
41. Middlemiss, D. (1984) Naunyn-Schmiedebergs Arch. Pharmacol. 327, 18–22.
42. Engel, G., Gothert, M., Hoyer, D., Schlicker, E. and Hillenbrand, K. (1986) Naunyn-Schmiedebergs Arch. Pharmacol. 332, 1–7.
43. Maura, G., Roccatagliata, E. and Raiteri, M. (1986) Naunyn-Schmiedebergs Arch. Pharmacol. 334, 323–326.
44. Shenker, A., Maayani, S., Weinstein, H. and Green, J. (1985) Eur. J. Pharmacol. 109, 427–429.
45. Markstein, R., Hoyer, D. and Engel, G. (1986) Naunyn-Schmiedebergs Arch. Pharmacol. 333, 335–341.
46. De Vivo, M. and Maayani, S. (1986) J. Pharmacol. Exp. Ther. 238, 248–253.
47. Fargin, A., Raymond, J., Regan, J., Cotecchia, S., Lefkowitz, R. and Caron, M. (1989) J. Biol. Chem. 25, 14848–14852.
48. Dumuis, A., Sebben, M. and Bockaert, J. (1987) Mol. Pharmacol. 33, 178–186.
49. Segal, M. (1980) J. Physiol. 303, 423–439.
50. Jahnsen, H. (1980) Brain Res 197, 83–94.
51. Aghajanian, G. and VanderMaelen, C. (1982) Brain Res. 238, 463–469.
52. Aghajanian, G. and Lakoski, J. (1984) Brain Res. 305, 181–185.
53. Colino, A. and Halliwell, J. (1987) Nature 328, 73–77.
54. Andrade, R., Malenka, C. and Nicoll, R. (1986) Science 234, 1261–1265.
55. North, R., Williams, J., Surprenant, A. and Christie, M. (1987) Proc. Natl Acad. Sci. USA 84, 5487–5491.
56. Williams, J., Colmers, W. and Pan, Z. (1988) J. Neurosci. 8, 3499–3506.
57. Andrade, R. and Nicoll, R. (1987) J. Physiol. 394, 99–124.
58. Andrade, R. and Nicoll, R. (1987) Naunyn-Schmiedebergs Arch. Pharmacol. 336, 5–10.
59. Pan, Z., Colmers, W. and Williams, J. (1989) J. Neurophysiol. 62, 481–486.
60. Ireland, S. (1987) Br. J. Pharmacol. 92, 407–416.
61. Joels, M., Shinnick-Gallagher, P. and Gallagher, J. (1987) Brain Res. 417, 99–107.
62. Galligan, J., Suprenant, A., Tonini, M. and North, R. (1988) Am. J. Physiol. 255, G603–611.

63. Yakel, J., Trussell, L. and Jackson, M. (1988) J. Neurosci. 8, 1273–1285.
64. Bobker, D. and Williams, J. (1990) J. Physiol. 422, 447–462.
65. Bobker, D. and Williams, J. (1991) J. Neurosci. 11, 2151–2156.
66. Pennington, N. and Kelly, J. (1990) Neuron 4, 751–758.
67. Pennington, N., Kelly, J. and Fox, A. (1991) J. Neurosci. 11, 3594–3609.
68. Hoyer, D., Engel, G. and Kalkman, H. (1985) Eur. J. Pharmacol. 118, 1–12.
69. Heuring, R., Schlegel, J. and Peroutka, S. (1986) Eur. J. Pharmacol. 122, 279–282.
70. Heuring, R. and Peroutka, S. (1987) J. Neurosci. 7, 894–903.
71. Sills, N., Wolfe, B. and Frazer, A. (1984) J. Pharmacol. Exp. Ther. 231, 480–486.
72. Offord, S., Ordway, G. and Frazer, A. (1988) J. Pharmacol. Exp. Ther. 244, 144–153.
73. Hoyer, D., Waeber, C., Pazos, A., Probst, A. and Palacios, J. (1988) Neurosci. Lett. 85, 357–362.
74. Schlicker, E., Fink, K., Gothert, N., Hoyer, D., Molderings, G., Roschke, I. and Schoeffter, P. (1989) Naunyn-Schmiedebergs Arch. Pharmacol. 340, 445–451.
75. Schoeffter, P. and Hoyer, D. (1990) J. Pharmacol. Exp. Ther. 252, 387–395.
76. Peroutka, S. and McCarthy, B. (1989) Eur. J. Pharmacol. 163, 133–136.
77. Ferrari, M. (1991) N. Engl. J. Med. 325, 316–321.
78. Ciaranello, R., Tan, G. and Dean, R. (1990) J. Pharmacol. Exp. Ther. 252, 1347–1351.
79. Verge, D., Deval, G., Marcinkiewicz, M., Paty, A., El Mestikawy, S., Gozlan, H. and Hamon, N. (1986) J. Neurosci. 6, 3474–3482.
80. Waeber, C., Zhang, L. and Palacios, J. (1990) Brain Res. 528, 197–206.
81. Crino, P., Vogt, V., Volicer, L. and Wiley, R. (1990) J. Pharmacol. Exp. Ther. 252, 651–656.
82. Bouhelal, R., Smounya, L. and Bockaert, J. (1988) Eur. J. Pharmacol. 151, 189–196.
83. Schoeffter, P., Waeber, C., Palacios, J. and Hoyer, D. (1988) Naunyn-Schmiedebergs Arch. Pharmacol. 337, 602–608.
84. Hoyer, D. and Schoeffter, P. (1988) Eur. J. Pharmacol. 147, 145–147.
85. Herrick-Davis, K., Titeler, M., Leonhardt, S., Struble, R. and Place, D. (1988) J. Neurochem. 51, 1906–1912.
86. Seuwen, K., Magmaldo, I. and Pouyssegur, J. (1988) Nature 335, 254–256.
87. Farnebo, L. and Hamberger, B. (1974) J. Pharm. Pharmacol. 26, 642–644.
88. Cerrito, F. and Raiteri, M. (1979) Eur. J. Pharmacol. 57, 427–430.
89. Gothert, M. (1980) Naunyn-Schmiedebergs Arch. Pharmacol. 314, 223–230.
90. Starke, K. (1981) Annu. Rev. Pharmacol. Toxicol. 21, 7–30.
91. Martin, L. and Sanders-Bush, E. (1982) Naunyn-Schmiedebergs Arch. Pharmacol. 321, 165–170.
92. Sharp, T., Bramwell, S. and Grahame-Smith, D. (1989) Br. J. Pharmacol. 96, 283–290.
93. Chaput, Y., Blier, P. and De Montigny, C. (1986) J. Neurosci. 6, 2796–2801.
94. Bobker, D. and Williams, J. (1989) J. Pharmacol. Exp. Ther. 215, 37–43.
95. Kannisto, P., Owman, C., Schmidt, G. and Sjoberg, N. (1987) Br. J. Pharmacol. 92, 487–497.
96. Bianchi, C., Siniscalchi, A. and Baeni, L. (1989) Br. J. Pharmacol. 97, 213–221.
97. Molderings, G., Werner, K., Likungu, J. and Gothert, M. (1990) Naunyn-Schmiedebergs Arch. Pharmacol. 342, 371–377.
98. Young, W. and Kuhar, M. (1980) Eur. J. Pharmacol. 62, 237–240.
99. Hartig, P., Hoffman, B., Kaufman, M. and Hirata, F. (1990) Ann. N.Y. Acad. Sci. 600, 149–167.
100. Conn, P. and Sanders-Bush, E. (1984) Neuropharmacology 23, 993–996.
101. Conn, P., Sanders-Bush, E., Hoffman, B. and Hartig, P. (1986) Proc. Natl Acad. Sci. USA 83, 4086–4088.
102. Faraci, F., Mayhan, W., and Heistad, D. (1989) Brain Res. 478, 121–126.
103. Gundersen, C., Miledi, R. and Parker, I. (1984) Nature 308, 421–424.
104. Lubbert, H., Hoffman, B., Snutch, T., VanDyke, T., Levine, A., Hartig, P., Lester, H. and Davidson, N. (1987) Proc. Natl Acad. Sci. USA 84, 4332–4336.
105. Lubbert, H., Snutch, T., Daskal, N., Lester, H. and Davidson, N. (1987) J. Neurosci. 7, 1159–1165.

106. Frazer, A., Maayani, S. and Wolfe, B. (1990) Annu. Rev. Pharmacol. Toxicol. 30, 307–348.
107. Glennon, R., Titeler, M. and McKenney, J. (1984) Life Sci. 35, 2505–2511.
108. Pazos, A., Cortes, R. and Palacios, J. (1985) Eur. J. Pharmacol. 346, 231–249.
109. Appel, N., Mitchell, W., Garlick, R., Glennon, R., Teitler, M. and De Souza, E. (1990) J. Pharmacol. Exp. Ther. 255, 843–857.
110. Mengod, G., Pompeiano, M., Martinze-Mir, M. and Palacios, J. (1990) Brain Res. 524, 139–143.
111. Berridge, M. and Irvine, R. (1989) Nature 341, 197–205.
112. Sanders-Bush, E., Tsutsumi, M. and Burris, K. (1990) Ann. N.Y. Acad Sci. 600, 224–235.
113. Barker, E., Burris, K. and Sanders-Bush, E. (1991) Brain Res. 552, 330–332.
114. Cholden, L., Kurland, A. and Savage, C. (1955) J. Nerv. Ment. Dis. 122, 211–221.
115. Eison, A., Eison, M., Yocca, F. and Gianutsos, G. (1989) Life Sci. 44, 1419–1427.
116. Leysen, J. and Pauwels, P. (1990) Ann. N.Y. Acad. Sci. 600, 183–193.
117. Araneda, R. and Andrade, R. (1991) Neuroscience 40, 399–412.
118. Ivins, K. and Molinoff, P. (1991) J. Pharmacol. Exp. Ther. 259, 423–429.
119. VanderMaelen, C. and Aghajanian, G. (1980) Nature 287, 346–347.
120. McCall, R. and Aghajanian, G. (1980) Life Sci. 26, 1149–1156.
121. Sheldon, P. and Aghajanian, G. (1990) Brain Res. 506, 62–69.
122. Anwyl, R. (1990) Prog. Neurobiol. 35, 451–468.
123. Aghajanian, G. (1990) Brain Res. 524, 171–174.
124. North, R. and Uchimura, N. (1989) J. Physiol. 417, 1–12.
125. Uchimura, N. and North, R. (1990) J. Physiol. 422, 369–380.
126. Davies, M., Wilkinson, L. and Roberts, M. (1988) Br. J. Pharmacol. 94, 483–491.
127. Fozard, J. (1984) Naunyn-Schmiedebergs Arch. Pharmacol. 326, 36–44.
128. Richardson, B., Engel, G., Donatsch, P. and Stadler, P. (1985) Nature 316, 126–131.
129. Richardson, B. and Engel, G. (1986) Trends Neurosci. 9, 424–428.
130. Vanner, S. and Surprenant, A. (1990) Br. J. Pharmacol. 99, 840–844.
131. Kilpatrick, G., Jones, B. and Tyers, M. (1987) Nature 330, 746–748.
132. Kilpatrick, G., Jones, B. and Tyers, M. (1989) Eur. J. Pharmacol. 159, 157–164.
133. Cubeddu, L., Hoffman, I., Fuenmayor, N. and Finn, A. (1990) N. Engl. J. Med. 322, 810–816.
134. Higashi, H. and Nishi, S. (1982) J. Physiol. 323, 543–567.
135. Tatsumi, H. and Katayama, Y. (1987) J. Auton. Nerv. Syst. 20, 137–145.
136. Neild, T. (1981) Gen. Pharmacol. 12, 281–284.
137. Surprenant, A. and Christ, J. (1988) Neuroscience 24, 283–295.
138. Wallis, D. and North, R. (1978) Neuropharmacology 17, 1023–1028.
139. Derkach, V., Surprenant, A. and North, R. (1989) Nature 339, 706–709.
140. Yakel, J. and Jackson, M. (1988) Neuron 1, 615–621.
141. Yang, J. (1990) J. Gen. Physiol. 96, 1177–1198.
142. Lambert, J., Peters, J., Hales, T. and Dempster, J. (1989) Br. J. Pharmacol. 97, 27–40.
143. Neijt, H., Plomp, J. and Vijverberg, H. (1988) J. Physiol. 411, 257–269.
144. Robertson, B. and Bevan, S. (1991) Br. J. Pharmacol. 102, 272–276.
145. Maricq, M., Peterson, A., Brake, A., Myers, R. and Julius, D. (1991) Science 254, 432–437.
146. Yakel, J., Shao, X. and Jackson, M. (1991) J. Physiol. 436, 293–308.
147. Ropert, N. and Guy, N. (1991) J. Physiol. 441, 121–136.
148. Sugita, S., Shen, K. and North, R. (1991) Neuron (in press).
149. Dumuis, A., Bouhelal, R., Sebben, N., Corey, R. and Bockaert, J. (1988) Mol. Pharmacol. 34, 880–887.
150. Andrade, R. and Chaput, Y. (1991) J. Pharmacol. Exp. Ther. 257, 930–937.
151. Craig, D. and Clarke, D. (1990) J. Pharmacol. Exp. Ther. 2, 1378–1386.
152. Bobker, D. and Williams, J. (1989) Neuron 2, 1535–1540.
153. Bobker, D. and Williams, J. (1990) Trends Neurosci. 13, 169–173.

154. DiFrancesco, D. (1985) Prog. Biophys. Mol. Biol. 46, 163–183.
155. Pape, H-C. and McCormick, D. (1989) Nature 340, 715–718.
156. Takahashi, T. and Berger, A. (1990) J. Physiol. 423, 63–76.
157. McCormick, D. and Pape, H-C. (1990) J. Physiol. 431, 319–342.
158. Julius, D. (1991) Annu. Rev. Neurosci. 14, 335–360.
159. Julius, D., MacDermott, A., Axel, R. and Jessell, T. (1988) Science 241, 558–564.
160. Molineaux, S., Jessell, T., Axel, R. and Julius, D. (1989) Proc. Natl Acad. Sci. USA 86, 6793–7997.
161. Pritchett, D., Bach, A., Wozny, N., Taleb, O., Daltoso, R., Shih, J. and Seeburg, P. (1988) Eur. Mol. Biol. Organ. (EMBO) J. 7, 4135–4140.
162. Julius, D., Huang, K., Livelli, T., Axel, R. and Jessell, T. (1990) Proc. Natl Acad. Sci. USA 87, 928–932.
163. Kavanaugh, W., Williams, L., Ives, H. and Coughlin, S. (1988) Mol. Endocrinol. 2, 599–605.
164. Albert, P., Zhou, Q.-Y., Van Tol, H., Bunzow, J. and Civelli, O. (1990) J. Biol. Chem. 265, 5825–5832.
165. Hamblin, M. and Metcalf, M. (1991) Mol. Pharmacol. 40, 143–148.
166. Tricklebank, M. (1985) Trends Pharmacol. Sci. 6, 403–407.
167. Angus, J. (1989) Trends Pharmacol. Sci. 10, 89–90.
168. Editor (1989) Lancet ii, 717–719.

F. Hucho (Ed.) *Neurotransmitter Receptors*
© 1993 Elsevier Science Publishers B.V. All rights reserved.

CHAPTER 9

Dopamine receptors

PHILIP G. STRANGE

Biological Laboratory, The University, Canterbury, Kent, CT2 7NJ, U.K.

1. Introduction

Many important functions are mediated by the neurotransmitter dopamine both in the brain and in the periphery. Systems employing dopamine as neurotransmitter have been associated with certain brain disorders such as parkinsonism and schizophrenia [1]. There has, therefore, been much interest over the recent years in receptors for dopamine and these have been studied intensively using physiological, pharmacological, biochemical, and more recently molecular biological techniques [2]. Dopamine receptors are also important targets for drugs, adding to the interest. In this Chapter the present state of knowledge on the characteristics, structure and function of dopamine receptors will be outlined.

2. Functions and distribution of dopamine receptors

The physiological functions attributed to dopamine are mediated via the binding of dopamine to dopamine receptors. In the brain the principal functions mediated by dopamine are the control of movement and the control of certain aspects of behaviour, e.g., motivation, cognitive function, aspects of emotion. In the pituitary gland dopamine controls the release of prolactin and αMSH and in the cardiovascular system it is important in the regulation of blood pressure.

Dopamine receptors have therefore been identified functionally in the tissues related to these effects, e.g., pituitary gland, certain blood vessels. In the brain dopamine receptors have been detected in electrophysiological studies in the caudate nucleus and putamen where it is thought the control of movement is elicited and in certain limbic brain areas, e.g., nucleus accumbens and cerebral cortical areas, where the effects of dopamine on behaviour are thought to be localised.

It is now clear that a family of dopamine receptors exists which may mediate different aspects of the functions of dopamine. In the next section the development of the concept of multiple dopamine receptors will be covered.

3. Dopamine receptor subtypes: definitions and overall properties

3.1. Subtypes defined from physiological, pharmacological, and biochemical studies

Dopamine receptors were studied in the 1960s and 1970s using physiological techniques to define the functions mediated by the receptors in different tissues. The pharmacological profiles of these effects were defined with an ever-growing series of dopamine agonists and antagonists. The introduction of biochemical techniques for studying receptors, particularly ligand-binding assays and effector, e.g. adenylyl cyclase, responses, expanded knowledge in the field. It gradually became apparent that the concept of a single dopamine receptor was insufficient to define the effects of dopamine and in 1979 Kebabian and Calne [3] proposed that there were two distinct subtypes of dopamine receptor, D_1 and D_2. The properties of D_1 and D_2 receptors are outlined in Table I and from this suggestion a large body of work was then performed examining the properties of these receptor subtypes. Some of this will be summarised below.

Although there were a number of attempts to expand the number of dopamine receptor subtypes, none of these stood the tests of time and experimentation in other laboratories until molecular biological techniques were applied (see next section).

3.2. Subtypes defined using molecular biology techniques

Molecular biology techniques were applied to receptors in the 1980s (Chapter 3). It soon became apparent that for many receptor systems the number of distinct receptor isoforms defined from different cloned genes was greater than had been suspected from physiological, pharmacological, and biochemical studies. This is strikingly illustrated for the dopamine receptors.

In 1988 the first cloned dopamine receptor sequence was published [4] for a D_2 receptor and since then several additional sequences have appeared [5] (Table II). Each cloned sequence corresponds to a protein of about 400 amino acids and M_r in the region of 50,000. When analysed for hydropathy, the amino acid sequences all show a pattern of seven stretches of hydrophobic amino acids which are thought to form membrane spanning α-helices. This pattern has been generally observed for receptors that signal via G-proteins and it is now thought that these receptors contain an extracellular amino terminus with attached oligosaccharide chains, seven transmembrane spanning α-helices which bundle together in the membrane to form the ligand-binding site, and an intracellular carboxy terminus with a site for palmitoylation. In addition to these conserved features at the gross structural level, there are certain amino acid residues found in each dopamine receptor sequence that may be functionally important. In particular there are an aspartic acid residue in the third putative transmembrane region and serine residues in the fifth putative transmem-

TABLE I

Dopamine receptor subtypes defined from physiological, pharmacological, and biochemical studies

	D_1	D_2
Pharmacological characteristics		
Selective antagonists	SCH23390	(−)-Sulpiride
		YM09151-2
Selective agonists	SKF38393	Quinpirole
		N-0437
Specific radioligands	[^{3}H]SCH23390*	[^{3}H]YM09151-2
		[^{3}H]Spiperone**
Physiological functions	Aspects of motor function (brain), cardiovascular function	Aspects of motor function and behaviour (brain), control of prolactin and αMSH secretion from pituitary, cardiovascular function
Biochemical responses	Adenylyl cyclase ↑	Adenylyl cyclase ↓
	phospholipase C ↑	K^+ channel ↑
		Ca^{2+} channel ↓
Localisation	Caudate nucleus, putamen, nucleus accumbens, olfactory tubercle, cerebral cortex (brain), cardiovascular system	Caudate nucleus, putamen, nucleus accumbens, olfactory tubercle, cerebral cortex (brain) anterior and neurointermediate lobes of pituitary gland, cardiovascular system
Size of receptor protein (daltons)		
Affinity labelling	72,000	85–150,000 (major species of 94,000, 118,000, 140,000)
Purification	–	92–95,000 (brain)
		120,000 (pituitary)

With the advent of molecular biological studies (Table II), these subtypes should be termed D_1-like and D_2-like receptors. The localisation data are from functional and ligand-binding studies on dispersed tissues and tissue slices.
*[^{3}H]SCH23390 can also bind to 5-HT$_2$ receptors if present [8];
**[^{3}H]Spiperone can also bind to 5-HT$_{1A}$, 5-HT$_2$ receptors, and α_1-adrenoceptors if present [7].

brane region that may be important in interacting with the ligand dopamine (see Section 4.5).

Although the different cloned sequences show considerable similarities, there are also differences. Some of these differences are outlined in Table II. The features highlighted, such as the presence of introns in certain genes and the relative sizes of the third intracellular loop and carboxy terminus, suggest that the cloned sequences can

TABLE II
Dopamine receptor subtypes from molecular biological studies

The properties of the principal dopamine receptor subtypes identified by gene cloning are shown. They are divided into 'D$_1$-like' and 'D$_2$-like' groups to reflect amino acid homology, functional similarity, structural similarity, and pharmacological properties. This grouping conforms with a previous classification based on pharmacological and biochemical properties (Table I). Under amino acids h and r refer to human and rat sequences, respectively. D$_{2(short)}$ and D$_{2(long)}$ refer to different alternatively spliced forms of the D$_2$ receptor gene as outlined in the text. Under pharmacological characteristics the figures are the dissociation constants (nM) for selected ligands taken from the published data, some are for rat, some human. The figures for dopamine are in the presence of Gpp(NH)p. D$_{2(short)}$ and D$_{2(long)}$ do not differ appreciably pharmacologically where antagonist affinities are concerned. The homology values are for the transmembrane-spanning regions, a dash indicates that the information has not been published. The localisations shown are the principal ones known at present from in-situ hybridisation and use of the polymerase chain reaction and in some cases have not been examined exhaustively. For general reviews, see Strange [5,6].

	'D$_1$-like'		'D$_2$-like'		
	D$_1$	D$_5$	D$_{2(short)/(long)}$	D$_3$	D$_4$
Amino acids	446(h,r)	477(h)	414/443(h)	400(h)	387(h)
		475(r)	415/444(r)	446(r)	368(r)
Pharmacological characteristics (Kd, nM)	SCH23390 (0.35) Dopamine (2340)	SCH23390 (0.30) Dopamine (228)	Spiperone (0.05) Raclopride (1.8) Clozapine (56) Dopamine (1705)	Spiperone (0.61) Raclopride (3.5) Clozapine (180) Dopamine (27)	Spiperone (0.05) Raclopride (237) Clozapine (9) Dopamine (450)
Homology					
with D$_1$ receptor	100	80	44	–	–
with D$_{2(short)}$ receptor	44	–	100	75	53
Localisation	Caudate/putamen, nucleus accumbens, olfactory tubercle, frontal cortex	Caudate/putamen, nucleus accumbens, olfactory tubercle, hippocampus, hypothalamus, frontal cortex (all low)	Caudate/putamen, nucleus accumbens, olfactory tubercle, cerebral cortex (low)	Nucleus accumbens, olfactory tubercle, islands of Calleja, cerebral cortex (low)	Frontal cortex, midbrain, amygdala, medulla (all low), cardiovascular system, retina

	'D₁-like'		'D₂-like'		
	D₁	D₅	D₂(short/long)	D₃	D₄
Response	Adenylyl cyclase ↑	Adenylyl cyclase ↑	Adenylyl cyclase ↓	?	?
Introns in gene	None	None	Yes	Yes	Yes
Organisation of amino acid sequence					
putative third intracellular loop	Short	Short	Long	Long	Long
Carboxyl terminal tail	Long	Long	Short	Short	Short
Reference (examples)	[50]	[51,52]	[4]	[23]	[53,54]

be grouped structurally into two subfamilies (D_1, D_5 and $D_{2(short)}$, $D_{2(long)}$, D_3, D_4). When the sequences are expressed in a suitable cell system and the pharmacological properties and biochemical responses of the receptor are determined, this subgrouping is also apparent and it seems that subfamilies of D_1-like (D_1, D_5) and D_2-like ($D_{2(short)}$, $D_{2(long)}$, D_3, D_4) receptors can be defined both in structural and functional terms. Broadly, the D_1-like receptors show high affinities for compounds such as SCH23390, a typical D_1 receptor antagonist. D_2-like receptors show high affinities for compounds such as haloperidol and spiperone, typically used to define D_2 receptors. Within each subfamily, however, pharmacological differences are also apparent and some data on the affinities of key compounds are given in Table II. This suggests that in the future it should be possible to design more selective compounds for the different receptor isoforms.

The existence of the short and long forms of the D_2 isoform deserves some comment. These are derived by alternative splicing of a common gene and differ only by a 29 amino acid insertion in the third intracellular loop region of the long form. This part of the receptor is not thought to be directly involved in forming the ligand-binding site and so far it seems that $D_{2(short)}$ and $D_{2(long)}$ do not display differences in their pharmacological properties at least where antagonist binding is studied (see also Section 4.3 and 4.4). Since the D_3 and D_4 receptor genes also contain introns, there is also the potential for alternatively spliced variants of these receptor isoforms. Typically, for the G-protein-linked receptors, diversity of receptors is apparent through the existence of different genes. The alternatively spliced variants of D_2 receptors are a rare example in this receptor family of another means of achieving diversity.

The different receptor subtypes also have different localisation patterns. This has been probed mainly using in-situ hybridisation and the polymerase chain reaction and some of the key data are given in Table II. It seems that in the brain where this has been examined the mRNA for the D_4 and D_5 subtypes is not present at high levels relative to that for other subtypes. Therefore except in certain specific brain regions these receptors may play rather minor roles. The distribution of the D_3 receptor suggests that it may have an important role in the control of aspects of behaviour [6]. Within the striatum (caudate nucleus, putamen), the region of the brain where dopamine is thought to elicit control of motor function, the principal receptor subtypes are D_1, $D_{2(short)}$, and $D_{2(long)}$. This means that in biochemical studies, if striatal tissue is used, the principal receptor subtypes under study are quite restricted despite the apparently daunting number of possible dopamine receptor isoforms (Table II). The D_4 receptor may represent a peripheral D_2-like receptor, as it is found at high levels in the cardiovascular system (Table II).

In subsequent sections when I refer to a specific receptor subtype e.g. D_1 I shall be referring to the receptor defined by molecular biology. Where the subtype has not been defined by molecular biology but the broad pharmacological classification (as in Table I) is clear, then I shall refer to D_1-like or D_2-like receptors.

4. Biochemical characterisation of dopamine receptors

4.1. Detection of dopamine receptors by ligand-binding assays

Ligand-binding assays have contributed greatly to the definition of dopamine receptor subtypes and some typically used radioligands are given in Table I. Sometimes, however, the results of these assays have been confusing or the interpretations incorrect. This has mostly been due to cross-reactivity of radioligands with other receptor sites. For example [^{3}H]spiperone, a very popular ligand for labelling D_2-like receptors, also will label 5-HT_{1A} and 5-HT_2 (serotonin) receptors, α_1-adrenoceptors and spirodecanone-binding sites where present [7].

Similarly [^{3}H]SCH23390, popularly used for labelling D_1-like receptors, will also label 5-HT_2 receptors [8]. Despite these problems if care is taken with the definitions of specific radioligand binding and certain cross-reacting sites are occluded, then either radioligand may be used with confidence. Newer more selective ligands are also becoming available, e.g. [^{3}H]YM09151-2 for D_2-like receptors. Recent studies in cell lines expressing single cloned receptor genes at high levels also circumvent the problems of cross-reactivity.

4.2. Characterisation of dopamine receptor proteins

Affinity-labelling studies have given values for the molecular sizes of D_1-like and D_2-like receptors (Table I). In the case of the D_2-like receptors from brain and the pituitary gland, photoaffinity labelling identified several species with different molecular weights. These seem to be variants at the glycosylation level, since deglycosylation yields a common core protein of 40–44,000 Da [9]. This is in the correct size range predicted for the deglycosylated species from gene cloning.

Purification of D_1-like and D_2-like receptors has been attempted using affinity chromatography [10–14]. In the case of D_1-like receptors only partial purification has been reported, whereas for D_2-like receptors full purification has been achieved. The purified D_2-like receptors from brain have a molecular size of 92–95,000 Da [11,12,14] and this has been found in several brain regions known to possess D_2-like receptors from ligand binding. From the neurointermediate lobe of the pituitary gland, a 95,000 Da species was also found [14], whereas in the anterior pituitary species of 120–140,000 Da were purified [13,14].

Comparison of the relative distributions of the D_2-like receptor species identified by photoaffinity labelling and by purification with distributions of the D_2-like receptor isoforms (Table II) suggests that the major differences in molecular size are due to differential glycosylation of similar core proteins. Some disparity in the observed sizes may also reflect uncertainties in the analytical techniques [15]. It is not possible at present to determine which receptor genes are responsible for the core protein, but in some studies it can be inferred that the major species expressed and purified from brain and pituitary [14] are the $D_{2(short)}$ and $D_{2(long)}$ genes. More refined analytical techniques, such as the use of specific antibodies will be required to assign these definitively.

4.3. Coupling of dopamine receptors to G-proteins

The structures of the dopamine receptors predicted from the amino acid sequence showing seven putative transmembrane-spanning regions place the dopamine receptors firmly in the family of G-protein-linked receptors. Evidence was, however, obtained for this previously from biochemical studies. Principally, this came from studies of the binding of agonists to D_1-like and D_2-like receptors where the affinities of agonists were reduced in the presence of added GTP or nonhydrolysable analogues of GTP [16,17]. Thus, GTP reduces the binding of a radiolabelled agonist such as [^{3}H]*N*-propylnorapomorphine to D_2-like receptors [17] or shifts the dopamine agonist competition curve versus a radiolabelled antagonist to lower affinity at either D_1-like or D_2-like receptors [2,16,17].

Agonist competition curves at both D_1-like and D_2-like receptors are best described by models of two binding sites of higher and lower affinity. The effect of GTP is sometimes to convert the two binding sites into a single set of binding sites for the agonist. This behaviour has been described in terms of a model where the receptor (R) associates with a G-protein [18]. This R/G complex has a higher affinity for agonists than R and GTP destabilises R/G leading to its conversion to R and G (Fig. 1).

Not all laboratories have reported a simple conversion of higher and lower affinity agonist-binding sites into lower affinity agonist-binding sites by GTP for D_2-like receptors. Partial conversions and effects on the affinities of both classes of site have been described [19,20]. There also seem to be tissue-specific differences in the effects seen with some brain regions having apparently GTP-insensitive agonist binding [19,20].

Phosphorylation of preparations containing D_2-like receptors by protein kinase A [21] or protein kinase C [22] has been reported to affect agonist binding to the receptors. The D_2-like receptors in phosphorylated preparations show reductions in agonist affinity which are similar to the effects of GTP. In neither case has it been shown definitively that it is the receptor that is being phosphorylated rather than the associated G-protein in order to cause the effect. Nevertheless, there are potential sites on D_2 receptor sequences for phosphorylation by these kinases and roughly

$$R + G \rightleftharpoons R \cdot G$$

lower affinity for agonists   higher affinity for agonists

Fig. 1. Model for receptor/G-protein interaction.

stoichiometric phosphorylation of purified D_2-like receptor by protein kinase A has been shown [21]. Thus, differences in the phosphorylation states of receptor or G-proteins from different tissues and in different preparations could be one reason why agonist binding shows different sensitivities to GTP in different reports.

Differences between the effects of GTP on agonist binding in different tissues could also reflect different populations of the cloned dopamine receptor subtypes or different populations of G-proteins. GTP effects on agonist binding to cloned receptor subtypes have been examined after expression in animal cells. For the members of the D_2-like subfamily of receptor isoforms, $D_{2(short)}$, $D_{2(long)}$, and D_4 show GTP-sensitive agonist binding when expressed in animal cells, whereas D_3 is notably much less sensitive to the effects of GTP [23]. Also, $D_{2(short)}$ and $D_{2(long)}$ show small differences in the sensitivity of agonist binding to GTP [24], which may reflect the presence of the peptide insertion in $D_{2(long)}$ in the region of the receptor thought to interact with G-proteins, the third intracellular loop.

Attempts have been made to determine the specific G-protein subtype associated with D_2-like receptors using reconstitution of R/G interaction and by attempting to purify R/G complexes. This has provided evidence for the specific interaction of D_2-like receptors with G_o and G_{i2} [25–27] (see Section 4.4).

4.4. Cellular responses linked to dopamine receptor activation

The initial subdivision of dopamine receptors into D_1 and D_2 (now D_1-like and D_2-like) subclasses rested heavily on the observation that the former subclass was linked to stimulation of adenylyl cyclase, whereas the latter was not [3]. It soon became apparent that this simple division of effector function was an oversimplification and several effector systems have now been found to be associated with each broad subclass of dopamine receptor (Table I). With the discovery of the different cloned receptor isoforms, a key question is then whether these isoforms are individually responsible for different effector responses. For example within the D_2-like subfamily, there are $D_{2(short)}$, $D_{2(long)}$, D_3, and D_4 receptors and at least three effector responses (adenylyl cyclase inhibition, K^+ channel stimulation, Ca^{2+} channel inhibition). Can isoforms and effector responses be matched?

At present the answer to this question is not available and the limited information on the linkage of isoforms to effectors is given in Table II. In terms of linkage to effectors, the case of $D_{2(short)}$ and $D_{2(long)}$ is of interest as these differ in the region of the receptor thought to be involved in triggering effector responses via coupling to G-proteins. There are minor indications that $D_{2(short)}$ couples to inhibition of adenylyl cyclase more effectively than $D_{2(long)}$ [28,29], although this needs further investigation.

There is also some information on the cellular responses caused by dopamine receptor activation. In the case of D_1-like receptors, it is assumed that the activation of adenylyl cyclase leads to activation of protein kinase A. A prominent substrate for protein kinase A, whose phosphorylation state is increased by D_1-like receptor activation, is DARPP-32 first identified by Greengard's laboratory [2]. DARPP-32 is a 32,000 dalton soluble protein found in the cytosol of dopamine neurones. When phosphorylated it is a potent inhibitor of protein phosphatase-I, so that D_1-like receptor activation may lead to changes in the phosphorylation state of a number of proteins.

Drugs that affect D_1-like and D_2-like receptors have been shown to alter the expression of Fos-like proteins in brain [30–32]. Fos proteins are the gene products of c-fos, one of the so-called immediate early genes whose expression is thought to alter transcription of other genes. It may be, therefore, that alterations of c-fos expression provide a link between the activation of dopamine receptors and the longer term changes in neuronal function observed after treatment of animals with receptor-directed drugs (Section 4.6).

4.5. Mechanism of ligand binding to dopamine receptors

It is important to understand the mechanism of ligand binding to receptors as this will help in the design of drugs directed at specific receptor isoforms. It can be assumed that a ligand binds to a receptor at its binding site via a series of interactions with amino acid side chains. The forces contributing to the binding energy will be due to electrostatic, hydrogen-bonded, hydrophobic, and Van der Waal's interactions. In the case of dopamine receptors, all ligands are cationic so that electrostatic interaction with a negatively charged amino acid side chain is to be expected. Antagonists often have several aromatic moieties so hydrophobic interactions are anticipated in that case.

The ligand-binding site of the dopamine receptors is thought to be formed from the bundling of the seven putative transmembrane regions [5]. Some clues to the amino acids involved in ligand binding may therefore be obtained from amino acids in the transmembrane regions that are conserved in the dopamine receptors. Such considerations highlight aspartic acid residues in the second and third transmembrane regions and serine residues in the fifth transmembrane region. These are conserved in all the dopamine receptors and may be involved in electrostatic interaction (aspartic acids) and hydrogen-bond interaction (serine) with the catechol moiety of the dopamine molecule.

Chemical modification studies have given some information in agreement with these ideas. For both D_1-like and D_2-like receptors chemical modification of carboxyl groups on the receptors abolishes ligand binding [33,34], supporting the likely importance of aspartic acid residues. Chemical modification has also given some evidence for the participation of thiol groups in D_1-like receptors from brain and D_2-like receptors from the pituitary gland but not brain [34,35]. The significance of this is not clear at present.

The pH dependence of ligand binding can also give information on the amino acid residues involved in ligand binding. For D_1-like and D_2-like receptors from brain, antagonist binding depends on groups of pKa 6.9 and 5.5, respectively [33,34] and these pKa values presumably correspond to ionisation of the aspartic acid residues involved in electrostatic interaction with the ligand.

For D_2-like receptors the binding of certain ligands, e.g. substituted benzamides such as sulpiride, was found to be more sensitive to changes in pH than the majority of antagonists. Whereas the binding of most antagonists seemed to depend on a group of pKa about 5.5, for substituted benzamides the apparent pKa was nearer 7 [36]. This suggested that the substituted benzamide drugs bound differently to D_2-like receptors compared with other ligands but did not identify the amino acids involved.

In order to identify the particular amino acids involved, site-specific mutagenesis of amino acid residues is required. For the dopamine receptors only limited information is available. Mutation of the aspartic acid in the second transmembrane region has been performed for the $D_{2(short)}$ receptor [37]. In the altered receptor substituted benzamide binding was less sensitive to changes of pH and the stimulation of binding of these drugs by Na^+ was also reduced [37]. This suggests that the amino acid, pKa about 7, whose ionisation state affects specifically the binding of substituted benzamides, is the aspartic acid residue in the second transmembrane region. This residue does not, however, interact directly with drugs [38] and its ionisation state probably regulates the conformation of the receptor (see below). The aspartic residue that interacts directly with the cationic drugs is probably the aspartic acid residue in the third transmembrane region. Evidence in support of this idea comes from work on other receptors for cationic amines. Both β-adrenergic and muscarinic acetylcholine receptors have the conserved aspartic acid residue in the third transmembrane region and mutation of this residue severely attenuates ligand binding [39,40].

Therefore, at present for the D_2 receptor it can be assumed the ligands (dopamine agonists and antagonists of the substituted benzamide group and other chemical classes) bind to the receptor partly via an electrostatic interaction with an aspartic acid residue in the third transmembrane-spanning region. This residue is buried about a third of the distance into the membrane from the extracellular surface suggesting that this is the region where ligands bind. Dopamine itself also probably interacts with two serine residues via hydrogen bonds to the catechol ring. These serine residues are also found about a third of the distance into the membrane along the fifth transmembrane-spanning region. Mutation of these serine residues for other catecholamine receptors inhibits ligand binding [40]. For antagonist ligands hydrophobic interactions are likely to be important where the ligands have bulky aromatic moieties. For the substituted benzamide class of antagonist where binding is affected by both pH and Na^+, it seems that the conformation of the receptor is affected by the ionisation state of the aspartic acid residue in the second transmembrane-spanning region. It may therefore be that the receptor can exist in two conformations and in one of these this aspartic acid is ionised. Whereas both conformations have high affinity for most antagonists, only the conformation with this ionised aspartate has a

$$R \underset{Na^+}{\overset{H^+}{\rightleftharpoons}} R'$$

Asp-COO⁻ Asp-COOH

high affinity for   high affinity for most
all antagonists   antagonists, low affinity
                           for substituted benzamides

Fig. 2. Regulation of affinity of D_2-like receptors by H^+ and Na^+. Two conformations of the receptor protein are shown with different ionisation states of the aspartic acid residue (pKa about 7) in the second putative transmembrane region.

high affinity for the substituted benzamides. The equilibrium between the two conformations is also then affected by Na^+ as shown in Figure. 2.

4.6. Regulation of dopamine receptors

Both acute and chronic regulation of G-protein-linked receptors by agonists has been studied most extensively for the β-adrenoceptor. Phosphorylation of the receptor, particularly in the third intracellular loop and the carboxy terminal tail is important in acute regulation (desensitisation) and may be important for chronic regulation (downregulation). The details of these processes are summarised in [41] and elsewhere in this book (Chapter 4).

A comparable detailed molecular dissection of the processes involved in agonist regulation has not been performed for dopamine receptors. Nevertheless, there are changes in the number of dopamine receptors in certain disorders where dopamine neuronal systems are perturbed and when animals or patients are treated with dopamine receptor-directed drugs (agonists and antagonists) [2] so that understanding the basis of the regulation process will be important.

Examination of the published sequences for dopamine receptor isoforms shows that there are consensus sequences for phosphorylation by different protein kinases in the intracellular regions of the sequences, particularly the third intracellular loop. For D_2-like receptors, roughly stoichiometric phosphorylation of purified receptor from brain has been demonstrated by protein kinase A [21]. As outlined above, a likely consequence of this is to affect agonist binding to the receptor. Therefore, the potential exists for dopamine receptors for regulatory events based on phosphorylations similar to those described for β-adrenoceptors.

Preliminary descriptive studies have been performed on the regulation of dopamine receptors in cell lines expressing the receptors following treatment with agonists. For D_1-like receptors expressed in neural cell lines [42,43], agonist treatment leads to loss of the dopamine stimulation of adenylyl cyclase and a somewhat slower reduction in total receptor number. For D_2-like receptors, agonist treatment leads to a reduction in the ability of dopamine to inhibit adenylyl cyclase in the cells [44,45]. The number of D_2-like receptors at the cell surface is also reduced, but the total number of receptors associated with the cell is unchanged. The desensitisation of the effector system appears to precede the receptor loss from the cell surface. Thus, in both cases although there may be differences in detail, it seems that there may be an initial agonist-induced desensitisation of the agonist effector response and somewhat later receptors may be lost from the cell surface and in the case of D_1-like receptors degraded, presumably using the cellular proteolytic machinery. Related changes in receptor numbers and responses may occur when patients are treated with dopamine agonist drugs. If a dopamine antagonist drug is used, the system may also be perturbed, leading to changes in the opposite direction.

There is much to be learned here about the exact molecular details of the processes involved and the importance of early and late events. Another way this is being investigated is by examining changes in mRNA for receptor isoforms in experimental animals that have undergone manipulation to change receptor number. In animals treated with the D_2-like receptor antagonist, haloperidol, an increased number of D_2-like receptors is seen in the brains (striatum) of the animals in response to the interruption of receptor-agonist interaction. Although not all reports agree, it seems that this increase in receptor number is not accompanied by a major increase in receptor mRNA [46–48]. In contrast, if the dopamine neurones innervating the striatum are lesioned, an increased number of D_2-like receptors is seen and this is accompanied by a large increase in receptor mRNA [49]. Superficially, it might be expected that the effects of haloperidol (blocking dopamine transmission) and denervation (removing dopamine transmission) would be similar. It may be that the effective degree of dopamine depletion is important here. A greater effective depletion is achieved by denervation than with drug blockade and only with a high degree of effective depletion do levels of receptor mRNA increase. If this is true, it suggests that there may be at least two phases to the response to interruption of dopamine transmission. The first is to alter the level of the receptor protein at the cell surface, perhaps by altering the rate of receptor removal from the membrane. This is then followed by alterations of mRNA levels which depend on other mechanisms in the cell. Alternatively, interruption of dopamine transmission by drug blockade and by denervation may be fundamentally different processes, for example the loss of nerve terminals could remove some important trophic factor.

Therefore, dopamine receptor regulation may be rather complex and temporally different events may occur. In response to an increased level of agonist, a rapid desensitisation of receptor-linked responses occurs and may be followed subsequently by increased internalisation of receptors so that receptor number at the cell surface is reduced (downregulation). Internalisation is a reflection of perturbations in the normal receptor cycling processes to and from the cell membrane (see [1] for review) and these may also be perturbed by interference with dopamine transmission by drugs or by denervation. An increased level of agonist stimulation increases the rate of receptor removal from the membrane so that the steady-state receptor number is reduced, while an antagonist blocks agonist access to the receptor, the removal rate is reduced, and the steady-state receptor level increases. Persistent and severe interruption of dopamine transmission can also lead to changes in transcription so that receptor mRNA levels change.

Acknowledgement

I thank Sue Davies for preparing the manuscript.

References

1. Strange, P.G. (1992) Brain Biochemistry and Brain Disorders, Oxford University Press.
2. Creese, I. and Fraser, C.M. (Eds.) (1987) Dopamine Receptors, A.R. Liss, N.Y.
3. Kebabian, J.W. and Calne, D.B. (1979) Nature 277, 93–96.
4. Bunzow, J.R., Van Tol, H.H.M., Grandy, D.K., Albert, P., Salon, J., Christie, M., Machida, C.A., Neve, K.A. and Civelli, O. Nature 336, 783–787.
5. Strange, P.G. (1991) Curr. Op. Biotechnol. 2, 269–277.
6. Strange, P.G. (1991) Trends. Neurosci. 14, 43–45.
7. Strange, P.G. (1987) Neurochem. Int. 10, 27–33.
8. Bischoff, S., Heinrich, M., Sonntag, J.M. and Krauss, J. (1986) Eur. J. Pharmacol. 129, 367–370.
9. Jarvie, K.R., Niznik, H.B. and Seeman, P. (1988) Mol. Pharmacol. 34, 91–93.
10. Gingrich, J.A., Amlaiky, N., Senogles, S.E., Chang, W.K., McQuade, R.D., Berger, J.D. and Caron, M.G. (1988) Biochemistry 27, 3907–3912.
11. Williamson, R.A., Worrall, S., Chazot, P.L. and Strange, P.G. (1988) Eur. Mol. Biol. Organ. (EMBO) J. 7, 4129–4133.
12. Elazar, Z., Kanety, H., David, C. and Fuchs, S. (1988) Biochem. Biophys. Res. Commun. 156, 602–609.
13. Senogles, S.E., Amlaiky, N., Falardeau, P. and Caron, M.G. (1988) J. Biol. Chem. 263, 18996–19002.
14. Chazot, P.L. and Strange, P.G. (1992) Neurochem. Int. 21, 159–169.
15. Clagget-Dame, M. and McElvy, J.F. (1989) Arch. Biochem. Biophys. 274, 145–154.
16. Hess, E.J., Battaglia, G., Norman, A.B., Iorio, L.C. and Creese, I. (1986) Eur. J. Pharmacol. 121, 31–38.
17. Sibley, D.R., De Lean, A. and Creese, I. (1982) J. Biol. Chem. 257, 6351–6361.
18. Wreggett, K.A. and De Lean, A. (1984) Mol. Pharmacol. 26, 214–227.
19. Leonard, M.N., Macey, C.A. and Strange, P.G. (1987) Biochem. J. 248, 595–602.
20. De Keyser, J., De Backer, J., Convents, A., Ebinger, G. and Vauquelin, G. (1985) J. Neurochem. 45, 497–979.
21. Elazar, Z. and Fuchs, S. (1991) J. Neurochem. 56, 75–80.
22. Roque, P., Zwiller, J., Malviya, A.N. and Vincendon, G. (1990) Biochem. Int. 22, 575–582.
23. Sokoloff, P., Giros, B., Martres, M.P., Bouthenet, M.L. and Schwartz, J.C. (1990) Nature 347, 146–151.
24. Giros, B., Sokoloff, P., Martres, M.P., Rios, J.F., Emorine, L.J. and Schwartz, J.C. (1989) Nature 342, 923–926.
25. Senogles, S.E., Spiegel, A.M., Padrell, E., Iyengar, R. and Caron, M.G. (1990) J. Biol. Chem. 265, 4507–4514.

26. Ohara, K., Haga, K., Berstein, G., Haga, T., Ichiyama, A. and Ohara, A. (1988) Mol. Pharmacol. 33, 290–296.
27. Elazar, Z., Siegel, G. and Fuchs, S. (1989) Eur. Mol. Biol. Organ (EMBO) J. 8, 2353–2357.
28. Dal Toso, R., Sommer, B., Ewert, M., Herb, A., Prichett, D.B., Bach, A., Shivers, B.D. and Seeburg, P.H. (1989) Eur. Mol. Biol. Organ. (EMBO) J. 8, 4025–4034.
29. Montmayeur, J.P. and Borelli, E. (1991) Proc. Natl Acad. Sci. USA 88, 3135–3139.
30. Miller, J.C. (1990) J. Neurochem. 54, 1453–1455.
31. Graybiel, A.M., Moratella, R. and Robertson, H.A. (1990) Proc. Natl Acad. Sci. USA 87, 6912–6916.
32. Young, S.T., Porrino, L.J. and Idarola, M.J. (1991) Proc. Natl Acad. Sci. USA 88, 1291–1295.
33. Hollis, C.M. and Strange, P.G. (1992) Biochem. Pharmacol. 44, 325–334.
34. Williamson, R.A. and Strange, P.G. (1990) J. Neurochem. 55, 1357–1365.
35. Chazot, P.L. and Strange, P.G. (1992) Biochem. J. 281, 377–380.
36. Presland, J. and Strange, P.G. (1991) Biochem. Pharmacol. 41, R9–R12.
37. Neve, K.A., Cox, B.A., Henningsen, R.A., Spannoyanis, A. and Neve, R.L. (1991) Mol. Pharmacol. 39, 733–739.
38. Neve, K.A. (1991) Mol. Pharmacol. 39, 570–578.
39. Fraser, C.M., Wang, C., Robinson, D.A., Gocayne, J.D. and Venter, J.C. (1989) Mol. Pharmacol. 36 840–847.
40. Dixon, R.A.F., Sigal, I.S. and Strader, C.D. (1988) Cold Spring Harbor Symp. Quant. Biol. 53, 487–497.
41. Hausdorff, W.P., Caron, M.G. and Lefkowitz, R.J. (1990) FASEB J. 4, 2881–2889.
42. Balmforth, A.J., Warburton, P. and Ball, S.G. (1990) J. Neurochem. 55, 2111–2116.
43. Barton, A.C. and Sibley, D.R. (1990) Mol. Pharmacol. 38, 531–541.
44. Barton, A.C., Black, L.E. and Sibley, D.R. (1991) Mol. Pharmacol. 39, 650–658.
45. Bates, M.D., Senogles, S.E., Bunzow, J.R., Liggett, S.B., Civelli, O. and Caron, M.G. (1991) Mol. Pharmacol. 39, 55–63.
46. Van Tol, H.H.M., Riva, M., Civelli, O. and Creese, I. (1990) Neurosci. Lett. 111, 303–308.
47. Le Moine, C., Normand, E., Guitteny, A.F., Fouque, B., Teoule, R. and Bloch, B. (1990) Proc. Natl Acad. Sci. USA 87, 230–234.
48. Goss, J.R., Kelly, A.B., Johnson, S.A. and Morgan, D.G. (1991) Life Sci. 48, 1015–1022.
49. Neve, K.A., Neve, R.L., Fidel, S., Janowsky, A. and Higgins, G.A. (1991) Proc. Natl Acad. Sci. USA 88, 2802–2806.
50. Dearry, A., Gingrich, J.A., Falardeau, P., Fremeau, R.T., Bates, M.D. and Caron, M.G. (1990) Nature 347, 72–76.
51. Tiberi, M., Jarvie, K.R., Silvia, C., Falardeau, P., Gingrich, J.A., Godinot, N., Bertrand, L., Yang Feng, T.C., Fremeau, R. and Caron, M.G. (1991) Proc. Natl Acad. Sci. USA 88, 7491–7495.
52. Sunahara, R.K., Guan, H.C., O'Dowd, B.F., Seeman, P., Laurier, L.G., Ng, G., George, S.R., Torchia, J., Van Tol, H.H.M. and Niznik, H.B. (1991) Nature 350, 614–619.
53. Van Tol, H.H.M., Bunzow, J.R., Guan, H.C., Sunahara, R.K., Seeman, P., Niznik, H.B. and Civelli, O. (1991) Nature 350, 610–614.
54. O'Malley, K.L., Harmon, S., Tang, L. and Todd, R.D. (1992) New Biol. 4, 137–146.

F. Hucho (Ed.) *Neurotransmitter Receptors*
© 1993 Elsevier Science Publishers B.V. All rights reserved.

CHAPTER 10

Glutamate receptors

GRAHAM E. FAGG[1] and ALAN C. FOSTER[2]

[1]*CIBA-GEIGY Ltd., Pharmaceutical Division, CH-4002 Basel, Switzerland;* [2]*Merck Sharp & Dohme Research Laboratories, Terlings Park, Harlow, Essex CM20 2QR, UK.*

1. Introduction

The concept that L-glutamate serves specific physiological and pathological functions in the central nervous system (CNS) originated from experimental observations made more than 30 years ago. In a series of studies in the 1950s, L-glutamate was shown to increase neuronal firing when ionophoresed onto cat spinal neurons, to induce convulsions when applied intracerebrally in dogs, and to destroy retinal neurons when injected systemically in the immature mouse (for recent reviews, see [1,2]). Although the full implications of these findings were not realized at that time (and, indeed, it was considered unlikely that L-glutamate would serve a specific transmitter role in addition to its widespread role in intermediary cellular metabolism), research during the 1970s and 1980s led to the now well-accepted view that acidic amino acids

Abbreviations: ACPD, (1S,3R)-1-amino-cyclopentane-1,3-dicarboxylic acid; AMPA, α-amino-3-hydroxy-5-methyl-4-isoxazole propionic acid; AP3, 2-amino-3-phosphonopropionic acid; AP4, 2-amino-4-phosphonobutanoic acid; AP5, 2-amino-5-phosphonopentanoic acid; AP7, 2-amino-7-phosphonoheptanoic acid; CGP 37849, DL-2-amino-4-methyl-5-phosphono-3-pentenoic acid; CGP 39551, DL-2-amino-4-methyl-5-phosphono-3-pentenoic acid-1-ethylester; CGP 39653, DL-2-amino-4-propyl-5-phosphono-3-pentenoic acid; CGP 40116, D-2-amino-4-methyl-5-phosphono-3-pentenoic acid; CGS 19755, 1-(*cis*-2-carboxypiperidine-4-yl)-propyl-1-phosphonic acid; CNS, central nervous system; CNQX, 6-cyano-7-nitroquinoxaline-2,3-dione; CPP, 3-(2-carboxypiperazin-4-yl)-propyl-1-phosphonic acid; CPPene, 3-(2-carboxypiperazin-4-yl)-propenyl-1-phosphonic acid; DαAA, D-α-aminoadipic acid; DNQX, 6,7-dinitroquinoxaline-2,3-dione; DOPA, dihydroxyphenylalanine; EDRF, endothelial derived relaxing factor; GABA, γ-aminobutyric acid; GDEE, glutamic acid diethylester; GYKI-52466, 1-(4-aminophenyl)-4-methyl-7,8-methylenedioxy-5H-2,3-benzodiazepine; HA-966, 3-amino-1-hydroxypyrrolid-2-one; IP_3, inositol-1,4,5-trisphosphate; L-689,560, 4-*trans*-2-carboxy-5,7-dichloro-4-phenylaminocarbonylamino-1,2,3,4-tetrahydroquinoline; L-687,414, 3(R)-amino-1-hydroxy-4(R)-methylpyrrolidin-2-one; LTP, long-term potentiation; MDL-29951, 4,6-dichloro-2-carboxyindole-3-propionic acid; MDL-100748, 4-carboxymethylamino-5,7-dichloroquinoline-2-carboxylic acid; MK-801, (+)-5-methyl-10,11-dihydroxy-5H-dibenzo(a,d)cyclohepten-5,10-imine (dizocilpine); MPTP, N-methyl-4-phenyl-1,2,5,6-tetrahydropyridine; NBQX, 2,3-dihydroxy-6-nitro-7-sulphamoyl-benzo(F)-quinoxaline; NMDA, N-methyl-D-aspartic acid; NO, nitric oxide; PCP, phencyclidine; PI, phosphatidyl inositol; SKF-10047, N-allyl-normetazocine; SL-82.0715, α-(4-chlorophenyl)-4-[(4-fluorophenyl)methyl]-1-piperidine ethanol

are major excitatory synaptic transmitters in the CNS [3]. Studies during the 1980s, in particular, served to consolidate hypotheses regarding the types of membrane receptors mediating the actions of these substances, and demonstrated that acidic amino acid receptor mechanisms are involved in phenomena ranging from neuronal development and synaptic plasticity to epilepsy and ischaemic neurodegeneration [1,2].

The aim of this Chapter is to overview current understanding of acidic amino acid receptor subtypes and the mechanisms and functions which they subserve. We have not attempted an exhaustive review of the literature, and the reader is referred to recent articles and volumes for a comprehensive coverage of this topic [1,2,4–7].

1.1. Glutamate receptor subtypes: evolution of the classification scheme

Five categories of acidic amino acid receptors have been described based on pharmacological studies (Table I). Evidence that more than a single receptor type might exist arose during the 1970s from observations that neurons in the dorsal and ventral horns of the cat spinal cord were differentially sensitive to the excitatory effects of acidic amino acid analogues; L-glutamate and kainate were shown preferentially to stimulate dorsal horn interneurons, and L-aspartate and N-methyl-D-aspartate (NMDA) to activate Renshaw cells. The discovery of substances (D-α-aminoadipate [DαAA], L-glutamic acid diethylester [GDEE], and Mg^{2+}) able to antagonize the excitatory actions of acidic amino acids led to the first clear subdivision of excitatory neuronal responses. Rank ordering of excitants on the basis of their susceptibility to blockade by DαAA or by Mg^{2+} yielded the sequence NMDA > L-homocysteate > L-aspartate > L-glutamate > quisqualate, and the reverse for antagonism by GDEE; kainate was unusual, in that its excitatory action was little affected by any of the antagonists. Three receptor types were thus defined; these were characterized as (1) NMDA-preferring and DαAA-sensitive, (2) quisqualate-preferring and GDEE-sensitive, and (3) kainate-preferring and DαAA- and GDEE-insensitive, and became widely known as the NMDA, quisqualate, and kainate receptor subtypes, respectively [8,9]. More recently, the 'quisqualate' receptor has been renamed the 'AMPA' receptor, based on the nonselectivity of quisqualate and the greater selectivity of AMPA (α-amino-3-hydroxy-5-methylisoxazole-4-propionic acid) as an agonist for this receptor subtype [10] (Table I).

This preliminary receptor classification was confirmed during the 1980s by studies using a variety of experimental approaches. In addition, evidence for two further glutamate receptor subtypes and for molecular heterogeneity within subtypes has been obtained. One of the two 'new' receptor subtypes (termed the L-AP4 receptor) was defined on the basis of its sensitivity to low concentrations of the glutamate analogue, L-2-amino-4-phosphonobutanoate (L-AP4), and appears to be located presynaptically on subpopulations of glutamate-releasing nerve terminals [6]. The second (referred to as the Glu_G receptor) is linked via a GTP-binding protein ('G-pro-

TABLE I
Glutamate receptor subtypes: summary of pharmacological properties

	NMDA	AMPA	Kainate	L-AP4	Glu$_G$
Most useful agonists	NMDA trans-2,3-PDA L-glutamate	AMPA Quisqualate L-glutamate	Domoate Kainate	L-AP4 L-phosphoserine	trans-ACPD Quisqualate L-glutamate
Most useful antagonists	CGP40116 D-CPPene CGS19755 D-AP5	NBQX GYKI 52466			(AP3) (AP4)
Allosteric co-agonists or partial agonists	Glycine D-cycloserine L-678,414	(Aniracetam)			
Allosteric site antagonists	L-689,560 MDL100748 7-Chlorokynurenate				
Channel blockers	MK 801 Mg^{2+}				
Radiolabeled agonists	L-glutamate Glycine	AMPA L-glutamate	Kainate L-glutamate		
Radiolabeled antagonists	CGP39653 CGS19755 L-689,560 5,7-Dichloro-kynurenate MK 801	CNQX			
Location and mechanism*	Postsynaptic; voltage-dependent $Na^+/K^+/Ca^{2+}$ channel	Postsynaptic; Na^+/K^+ channel	May be pre- and postsynaptic	Presynaptic; G-protein; regulation of Ca^{2+} currents	Pre and postsynaptic; G-protein; membrane and intracellular responses

*Some types of glutamate receptors are found in glial cells in addition to the neuronal locations described in this chapter.

tein') to intracellular second-messenger systems in the postsynaptic neuron [11]. Molecular biological studies conducted within the past few years have confirmed the existence of all five glutamate receptor subtypes and, in addition, have begun to identify multiple isoforms with distinct functional properties (Section 4).

2. Function and occurrence

2.1. NMDA receptor

The NMDA receptor is the best characterized of the the acidic amino acid receptor subtypes. Functionally, electrophysiological studies have shown that the receptor gates a high conductance (~ 50 pS) cation channel in the subsynaptic membrane of the postsynaptic neuron. Critical advances for understanding NMDA receptor channel mechanisms were (1) the discovery that aspartate and related amino acids evoke a voltage-dependent conductance in cultured neurons [12], (2) the demonstration that this is due to a voltage-dependent blockade of the open NMDA receptor channel by Mg^{2+} [13], and (3) the finding that the NMDA receptor channel is permeable to Ca^{2+} (in addition to Na^+ and K^+) [14]. Prior to this time, electrophysiological studies had shown that the actions of glutamate and its analogues variably resulted from increases or decreases in membrane conductance [5]. The finding that some amino acids (e.g., aspartate) elicited a region of 'negative slope conductance' in the current-voltage relationship (due to the voltage-dependent channel block by Mg^{2+}) whereas others (e.g., quisqualate) did not [4,12,13], helped not only to resolve the earlier mixed results, but also provided a basis to explain later observations on the involvement of the NMDA receptor in phenomena such as long-term potentiation (LTP) [5], epileptiform activity [15], and neuronal degeneration [16,17].

The Ca^{2+} permeability of the NMDA receptor channel, shown initially in 1986 by means of voltage-clamp and fluorescent indicators to measure intracellular Ca^{2+} concentrations [14], also has helped to explain observations on the physiological and pathological roles of the NMDA receptor. For example, entry of Ca^{2+} through the NMDA receptor channel (rather than through voltage-dependent Ca^{2+} channels) has been shown to underlie LTP [18] and excitotoxic neurodegeneration in culture [19]. Presumably, Ca^{2+} entering via the NMDA receptor channel activates intradendritic mechanisms which are not immediately accessible to cytoplasmic Ca^{2+} originating via other routes, and which are critical for the cellular events subserved by the NMDA receptor. Ca^{2+} imaging studies following NMDA receptor stimulation suggest that such events may be highly localized to activated dendritic spines [20]. A recent investigation indicates that activation of a protein tyrosine kinase, followed by phosphorylation of a putative microtubule-associated protein (MAP2) kinase, may be one of the early consequences of Ca^{2+} entry via the NMDA receptor channel [21]. Increased formation of arachidonic acid [22] and synthesis of nitric oxide (NO; Section 5.1) have also been proposed as biochemical sequelae of NMDA receptor stimulation [22]. Clarification of the cascade of intracellular events following NMDA receptor activation will be a key step towards understanding long-term synaptic changes induced by this receptor.

Investigations using NMDA receptor antagonists (Section 3.1) were instrumental in defining the functions of the NMDA receptor in physiological and pathological events in the CNS. For example, it is now clear that the NMDA receptor does not play a major role in single fast monosynaptic transmission events (e.g., at the Schaffer collateral-CA1 hippocampal pyramidal neuron synapse or at the primary afferent-spinal motoneuron junction), but it does participate in postsynaptic responses generated as a result of repetitive afferent firing or convergent excitatory input (e.g., high frequency stimulation of the Schaffer collateral-CA1 synapse leading to LTP or polysynaptic activation of spinal motoneurons) [4,5]. In vivo, the NMDA receptor has been shown to play a role in neuronal development, in LTP (note that both metabotropic receptor and nitric oxide-mediated mechanisms may also contribute to LTP; Sections 2.5 and 5.1), in some forms of learning, in epileptiform activity, and in ischaemic/traumatic brain injury [1,2,5,6]. The participation of the receptor in such functions can be relatively easily understood within the framework of the voltage-dependence of the NMDA receptor conductance and the Ca^{2+} permeability of the receptor channel (see above).

Experiments involving the binding of radiolabeled receptor ligands have shown that the NMDA receptor is localized subcellularly in the postsynaptic density [23] and is present in high density in the cerebral cortex, hippocampus, striatum, septum, and amygdala [6]. Autoradiographic studies using radioligands which bind to different components of the NMDA receptor give the same overall distribution pattern, but small differences are apparent when using, for example, agonist and antagonist radioligands. This has been interpreted as evidence for NMDA receptor heterogeneity [6], much in the way that multiple isoforms of other transmitter receptors have been found to exist. Confirmation of such heterogeneity is emerging from ongoing molecular biological studies (Section 4.1).

2.2. AMPA receptor

AMPA receptor agonists activate channels of conductance intermediate in size between those elicited by NMDA and kainate [4,5]. They reverse close to 0 mV, show little voltage-dependence, and are permeable to Na^+ and K^+, and not to Ca^{2+} (but see Section 4.2). The response to quisqualate rapidly desensitizes, and this can partially be blocked by the lectin, concanavalin A [24].

Autoradiographic studies have shown that AMPA receptors are widely distributed in the CNS, with high levels in the cortex, hippocampus, lateral septum, striatum, and the molecular layer of the cerebellum [6]. With the exception of the cerebellum, this distribution corresponds closely to that of NMDA receptors, and has led to the suggestion that these two receptor subtypes act in concert to activate the postsynaptic neuron. Electrophysiological findings that fast synaptic responses are blocked primarily by non-NMDA receptor antagonists, and that NMDA receptor responses in the same pathways are evoked only under depolarizing conditions (owing to the voltage-dependence of the NMDA receptor conductance; Section 2.1) support this view [4–6]. Thus, in the absence of other excitatory activity, AMPA receptors may mediate fast depolarizing responses at most excitatory synapses in the CNS.

2.3. Kainate receptor

Kainate activates neuronal membrane channels which are distinguishable from those associated with NMDA and AMPA receptors on the basis of their conductance and desensitization properties [4,5]. Like AMPA receptors, kainate-activated channels are primarily permeable to Na^+ and K^+ (a kainate-activated Ca^{2+} conductance has been observed in cultured neurons [25] but, from the experiments reported, it is not clear whether this represented a kainate or an AMPA receptor response).

One of the first functions to be ascribed to the kainate receptor was pathological, that is, induction of a neurodegenerative response in discrete populations of brain neurons. Owing to the lack of selective antagonists, a clear synaptic function of kainate receptors has not been ascertained. However, specific high-affinity ^{3}H-kainate binding sites with a pharmacological profile similar to that of the kainate receptor mediating neuronal death have been identified, and these have been shown autoradiographically to be localized in those regions susceptible to the neurotoxic actions of this amino acid [6]. Furthermore, such regional localization is distinct from that shown by NMDA and AMPA receptors, indicating their separate identity. Thus, distinct 'kainate receptors' appear to be involved in some neuropathological events mediated by excitatory amino acids in the CNS.

The difficulty in clearly identifying synaptic responses mediated by the kainate receptor, the low magnitude of the kainate-evoked membrane conductance, together with observations indicating that kainate can also activate the AMPA receptor, have led to questions about its physiological significance as a postsynaptic excitatory amino acid receptor [26]. In addition, some data point towards a presynaptic role [27]. Resolution of these issues is likely to follow the ongoing molecular biological characterization of kainate receptors (Section 4.3), and additionally will require the development of selective antagonists to dissect out the role of this receptor in complex physiological events.

2.4. L-AP4 receptor

In contrast to the NMDA, AMPA, and kainate receptor subtypes, which were defined through the actions of exogenous excitatory amino acids, the L-AP4 receptor was identified on the basis of the potent antagonist properties of this glutamate analogue at subpopulations of excitatory synapses in the hippocampus, spinal cord, and olfactory cortex [6]. Despite the physiological significance of this site, it has proved difficult to elucidate the precise mechanism or membrane-binding site through which L-AP4 exerts its antagonist action. Electrophysiological studies have demonstrated that, at the low micromolar concentrations at which it blocks excitatory synaptic responses, L-AP4 does not block the depolarizing actions of glutamate, NMDA, quisqualate, or kainate; indeed, in the retina, L-AP4 appears to mimic the action of the natural transmitter (glutamate?) at ON-bipolar cells [6].

Following early observations that L-AP4 had no effect on the responses evoked by exogenous excitatory amino acids, it was hypothesized that it may act at a presynaptic autoreceptor to inhibit transmitter release. Support for this idea has arisen from several studies, in particular from quantal analyses showing that L-AP4 has no effect on miniature EPSPs at concentrations that clearly block evoked EPSPs [28,29]. Recent work using whole cell recording techniques in conjunction with cultured olfactory bulb neurons indicates that L-AP4 may act at a G-protein-coupled receptor which directly inhibits high-threshold Ca^{2+} currents and thereby suppresses transmitter release [30]. Although the actions of L-AP4 appear inconsistent with it acting at Glu_G receptors [30], a direct G-protein-mediated inhibition of Ca^{2+} currents by Glu_G receptors has been described [31] and several observations indicate that ACPD ((1S,3R)-1-amino-cyclopentane-1,3-dicarboxylic acid, a Glu_G agonist, Section 3.4) acts presynaptically to reduce excitatory transmission [32–34]. The precise relationship between the L-AP4 and Glu_G receptor types is emerging from molecular biological studies of glutamate receptors which currently are in progress (Section 4.4).

2.5. Glu_G receptor

Glu_G receptors are the only acidic amino acid receptors that have been shown to be directly linked to intracellular second-messenger systems. Evidence for the existence of such receptors arose in 1985 from studies showing that quisqualate and glutamate increased the turnover of phosphatidyl inositol (PI) in cultured mouse striatal neurons [35]. In the same year, experiments using voltage-clamped *Xenopus* oocytes injected with rat brain mRNA showed that these amino acids elicited oscillatory current responses typical of IP_3-induced fluctuations in intracellular Ca^{2+} (the response measured in oocytes is a Ca^{2+}-dependent Cl^- current), and moreover that the response was inhibited by pertussis toxin, thereby pointing to a G-protein-linked receptor-transduction system [36]. These experiments thus demonstrated that quisqualate and glutamate increase PI turnover by direct activation of an acidic amino acid receptor and not indirectly via depolarization or release of other transmitter substances.

A number of intracellular events has been described following Glu_G receptor activation, and it is likely that this multiplicity of responses at least partially reflects heterogeneity within this receptor class (Section 4.4). Measurements of PI turnover, for example, indicate that the Glu_G receptor linked to phospholipase C is present in several regions of the brain [11,37]. Another response, elicited by combined activation of Glu_G and AMPA receptors, is an increase in arachidonic acid, mediated by activation of phospholipase A_2 [11]. Molecular biology-based studies have led to the discovery of Glu_G receptors both positively and negatively coupled to adenylate cyclase (Section 4.4; see also [38,39]), while electrophysiological investigations point to direct Glu_G receptor-G-protein-mediated inhibition of both voltage-dependent Ca^{2+} and K^+ channels [31,40]. Studies involving cloned Glu_G receptors, as well as the development of subtype-selective agonists and antagonists, will be required to clarify the precise cellular functions subserved by individual receptor subtypes.

Although few specific pharmacological tools for the Glu_G receptor have been described, sufficient drugs are available (Section 3.4) to permit investigations of the role of Glu_G receptors in brain physiology and pathology. Using such agents, activation of Glu_G receptors has been shown to block the slow afterhyperpolarizing potential, suppress accomodation, and elicit repetitive cellular firing in hippocampal neurons [11,34,37,40]. These effects probably underlie the increased epileptiform activity in the form of augmented afterpotential duration and bursting activity, observed following application of ACPD to cortical slices (note that the frequency of epileptiform bursting is, however, reduced) [41]. A role in LTP has also received much study [37,42,43], although a unifying hypothesis to account for the functions and relative contributions of Glu_G, NMDA (Section 2.1), and NO (Section 5.1) mechanisms in different pathways still requires elaboration. Glu_G receptors additionally have been proposed to participate in ischemic neurodegeneration [16], although both positive and negative results have been obtained [44–48].

3. Pharmacology and toxicology

3.1. NMDA receptor

At least five distinct sites of pharmacological or regulatory significance have been identified on the NMDA receptor (Fig. 1) [7,27]. On the extracellular surface of the receptor complex are the transmitter recognition site and the glycine co-agonist site, two sites which are essential for activation of the NMDA receptor. Within the channel is a site(s) at which Mg^{2+} and channel-blocking drugs can bind, depending on the neuronal membrane potential. Two other sites, the locations of which are unclear, were described as binding sites for Zn^{2+} and polyamines. The binding site for Zn^{2+} has been proposed to lie within the mouth of the ion channel, where it may be involved in regulating cation selectivity and fluxes, whereas that for polyamines might be intra- or extracellular.

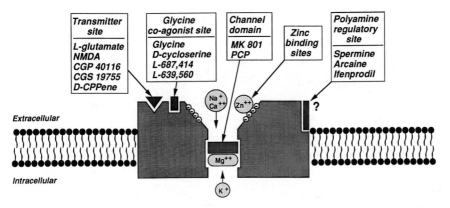

Fig. 1. Model of the NMDA receptor complex, illustrating the major sites of pharmacological and regulatory significance. Modified from refs. 27 and 78.

3.1.1. Transmitter recognition site

The transmitter-binding site is the most extensively studied of the pharmacological domains on the NMDA receptor. Structure-activity investigations at this site, led for many years by Watkins et al. [10], resulted in the discovery of agonists and antagonists which were critical for the discrimination of receptor subtypes (Section 1.1), and which played a key role in elucidating many of the physiological and pathological events subserved by the NMDA receptor. Early substances with high selectivity for the NMDA receptor included the agonists, NMDA and trans-2,3-piperidine dicarboxylic acid, and the antagonists, D-αAA, D-2-amino-5-phosphonopentanoic acid (D-AP5), and D-2-amino-7-phosphonoheptanoic acid (D-AP7) (Table I). At this point in time, NMDA is still the most widely used selective agonist, and the D-enantiomers of CGP37849 (i.e., CGP40116 [49]) and 3-(2-carboxypiperazin-4-yl)-propenyl-1-phosphonic acid (CPPene) [50] (two unsaturated phosphono amino acids derived by optimization of AP5 and AP7, respectively; Fig. 2) are the most potent and selective antagonists. A recent investigation, involving studies of the on- and off-rates of competitive NMDA antagonists using patch-clamp techniques, indicates that the high potency of CGP37849 is due to a combination of rapid association and slow dissociation; molecules with greater flexibility (e.g., AP5 and AP7) show lower potency due to their rapid dissociation rates, and those which are conformationally restricted through the presence of ring structures (e.g., 3-(2-carboxypiperazin-4-yl)-propyl-1-phosphonic acid (CPP) or CGS19755) show lower potency due to a slow rate of association [51]. Thus, CGP37849 appears to represent a balance between conformational restriction and flexibility which is optimal for high antagonist potency at the NMDA receptor.

Studies of the structure-activity relationships of NMDA receptor ligands have led to proposals that the binding sites for agonists and antagonists are not identical. It

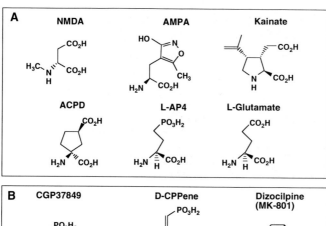

Fig. 2. Chemical structures of some acidic amino acid receptor agonists and antagonists. A. Agonists; B. NMDA receptor antagonists; and C. AMPA receptor antagonists.

has been postulated that, whereas the 1-carboxyl and α-amino groups of agonists interact with the same charged residues on the receptor surface as do those of antagonists, the ω-acidic terminals of agonists and antagonists bind to distinct sites. This hypothesis initially arose from difficulties in reconciling the different carbon chain lengths, enantiomeric preferences, and ω-terminal properties of agonists and antagonists, and gained support from molecular modeling studies showing that conformationally restricted antagonists such as CGS19755 cannot fold sufficiently to allow superimposition of their ionizable groups with those of NMDA [10,52]. This picture of the NMDA receptor recognition site has attained another level of complexity by studies showing that the 5-carbon (e.g., AP5, CGS19755) and 7-carbon (e.g., AP7, CPP) competitive antagonists have different allosteric interactions with the glycine co-agonist site [53]. Whether this indicates that the 5-carbon and 7-carbon competitive antagonists will show differing pharmacological properties at the synaptic level or in vivo is unknown.

3.1.2. Glycine co-agonist site

The exciting discovery that glycine plays a role in regulating the activity of the NMDA receptor originated from Ascher's laboratory in the late 1980s, and was based on patch-clamp studies of NMDA-evoked responses in cultured neurons [54]. Although initial work suggested that glycine acts allosterically to increase the frequency of NMDA-induced channel opening, much in the way that benzodiazepines modulate $GABA_A$ receptor activity, recent investigations indicate that glycine is more appropriately defined as a 'co-agonist' at the NMDA receptor (i.e., NMDA receptor responses cannot be elicited in the absence of this amino acid) [55].

The physiological/pharmacological significance of the glycine site has been the subject of much debate. One theory, based on the relative concentrations of glycine in cerebrospinal fluid and the K_d of the binding site, is that the site normally may be saturated and hence of little physiological import. However, a number of in-vivo studies have shown that glycine-site agonists or partial agonists increase seizure activity and improve learning performance (as predicted for compounds which facilitate NMDA receptor activity [7,27]), suggesting that glycine levels in vivo may be subsaturating. In the opposite direction, a question of current interest, given that glycine is essential for NMDA receptor activity, is whether antagonists at this site will exhibit pharmacological properties similar to or distinct from those of competitive NMDA receptor antagonists.

Currently available agonists or partial agonists at the co-agonist site include glycine itself and simple cyclic analogues such as D-cycloserine, (+)HA966, and L-687,414 (a low efficacy partial agonist with in-vivo activity [56,57]; Fig. 2); antagonists include kynurenic acid and derivatives, quinoline and quinoxaline analogues such as DNQX and MDL-100748, and indole carboxylic acid analogues like MDL-29951 [58]. Currently, the most potent and selective glycine antagonist described is L-689,560 (low nM K_d in membrane-binding assays [59]). Of these substances, L-687,414 has been shown to be anticonvulsant [60] and neuroprotective [61] when administered systemically to rodents, indicating a potential therapeutic role for drugs targeted at the glycine co-agonist site.

3.1.3. Open channel blockers

The identification of substances that antagonize the NMDA receptor by means of an open-channel blocking mechanism stemmed from the discovery by Lodge and his collaborators that dissociative anaesthetics and σ-opioids selectively blocked

NMDA-evoked neuronal responses recorded in vitro and in vivo [7]. Subsequent radioligand binding and electrophysiological investigations support the idea that blockade by such substances is of the uncompetitive type and is use- and voltage-dependent. It was further demonstrated that recovery of NMDA receptor responses from blockade, as well as dissociation of radiolabeled blockers from membrane preparations in vitro, require receptor activation, suggesting that blockers of this type bind and can become trapped inside the NMDA receptor-gated ion channel [62; see also 7,27].

Open-channel blockers of this type include phencyclidine (PCP), ketamine, SKF-10047 (N-allyl-normetazocine), and dextrorphan. The most potent and selective is MK-801 (dizocilpine [7]; Fig. 2). Agonists at this site have not been described and, given the channel location of the active site, it seems unlikely that any will be.

PCP is a major drug of abuse in the USA, producing a psychosis somewhat akin to that observed in schizophrenic patients. Psychotomimetic episodes also have been observed following ketamine anaesthesia. This has prompted two speculations. The first is that schizophrenia may involve a component of excitatory amino acid dysfunction (specifically involving decreased NMDA receptor activation) and that drugs which enhance NMDA receptor activity may exhibit antipsychotic properties. The second is that NMDA receptor blockers of this type, or indeed any type, may elicit psychotic episodes in man. Studies in rats indicate that the interoceptive cues elicited by PCP and MK-801 are mediated by NMDA receptor blockade [63], although CGS19755 and CGP40116 (the D-isomer of CGP37849) do not generalize to MK-801 or ketamine cues in primates [64; J.H. Woods, personal communication]. In any case, the nature of the cue in animals, as well as its cognitive or mechanistic relationship to psychosis in man, is unknown. Resolution of this question will await clinical trials of selective NMDA receptor antagonists in man.

3.1.4. Polyamine site
In a 1988 study of the pharmacological modulation of ^{3}H-MK-801 binding, polyamines were observed to increase the level of binding induced by glutamate and glycine [65]. Subsequent studies have shown that amines such as spermine and spermidine act as agonists at this site, and that putrescine and arcaine are antagonists [66]. In functional studies, spermidine has been shown to enhance NMDA-induced whole-cell currents in vitro (note that this has not been observed by all investigators [7]), and to potentiate NMDA-induced seizures in vivo. Ifenprodil and SL-82.0715, drugs which show neuroprotective properties in vivo, have also been shown to interact with the polyamine site on the NMDA receptor, albeit noncompetitively; unusually (by comparison with all other types of NMDA antagonists), these agents display weaker anticonvulsant than neuroprotective effects [67]. Given that drugs of this type exert multiple effects in the CNS, the question whether NMDA receptor blockade alone underlies their cerebroprotective properties in vivo is an issue which requires clarification.

3.2. AMPA receptor

Recent years have seen significant advances in the characterization of the AMPA receptor. As in the case of the NMDA receptor complex, such progress has largely followed the discovery of selective pharmacological tools, firstly the agonist, AMPA, which superceded quisqualate by virtue of its greater selectivity, and more recently the quinoxaline antagonists, 6-cyano-7-nitroquinoxaline-2,3-dione (CNQX) and 2,3-dihydroxy-6-nitro-7-sulphamoyl-benzo(F)-quinoxaline (NBQX) [10,68] (Fig. 2). In addition, several agents have been shown to act in other ways to modulate the activity of the AMPA receptor. For example, GYKI-52466, a 2,3-benzodiazepine, has been reported to be a noncompetitive antagonist of AMPA-mediated responses [69]. In addition, barbiturates have been shown to antagonize, and Zn^{2+} and the nootropic drug, aniracetam, to potentiate AMPA receptor responses [27]; the precise mode of action of these agents has not been fully defined, although aniracetam has been reported to reduce receptor desensitization [70].

As mentioned above (Section 2.2), studies in vitro indicate that AMPA receptor antagonists block fast monosynaptic responses widely throughout the CNS. However, most antagonists of this type show limited penetration of the blood-brain barrier, and their pharmacological properties in vivo have not been well characterized. Nevertheless, NBQX and GYKI-52466 have been reported to exhibit short-acting anticonvulsant properties following systemic administration to rodents and baboons [69] and to be neuroprotective in rodent models of global and focal ischaemia [68,71]. These observations certainly will increase the interest in searching for novel AMPA receptor antagonists or modulators with improved selectivity and pharmacokinetic properties.

3.3. Kainate and L-AP4 receptors

No selective antagonists for these receptors have been identified, although CNQX, DNQX and NBQX exhibit μM K_i values for ^{3}H-kainate-binding sites [10], and CNQX antagonizes responses elicited via a cloned kainate receptor protein (GluR6, Section 4.3).

3.4. Glu_G receptor

Pharmacologically, the Glu_G receptor is distinct from other acidic amino acid receptors, in that quisqualate, ibotenate, and glutamate are agonists, and NMDA, AMPA, and kainate are not [11,72]. Recently, another glutamate analogue, ACPD (Fig. 2), has been shown to increase PI turnover in hippocampal slices with greater selectivity than quisqualate, ibotenate, or glutamate, and members of the family of 2-(carboxycyclopropyl)glycines have been reported to exert selective agonist effects on cloned mGluR subtypes [73]. 2-Amino-3-phosphonopropionic acid (AP3) and AP4 have been reported to show weak antagonist activity in some Glu_G receptor systems, whereas the NMDA and AMPA receptor blockers, AP5 and CNQX, are ineffective. Inconsistencies in the agonist/antagonist profiles observed between different preparations (e.g., current responses in *Xenopus* oocytes versus PI turnover in tissue slices), and between native and cloned receptors (Section 4.4), together with the multiplicity of responses elicited by Glu_G agonists (Section 2.5), have been interpreted as indicating heterogeneity within this receptor class [34,74]. Evidence for the existence of such heterogeneity has been provided by recent molecular biological work (Section 4.4), and it is to be expected that potent and selective pharmacological tools to examine the roles of Glu_G receptor subtypes soon will follow.

Studies employing the few selective agents available for Glu_G receptors are pointing towards a number of disease states in which these receptors appear to be involved (although discrepancies between groups preclude definitive conclusions at present). Using in-vitro and in-vivo experimental systems, activation of Glu_G receptors has been shown both to augment [45,48] and to block excitotoxic or ischaemic neuronal damage [44,46,47] and to exert a dual action on epileptiform activity [41]. Subtype-selective agonists and antagonists will be required to discriminate the multiple actions of Glu_G receptors in such disease models.

3.5. Therapeutic utility

Preclinical investigations using a variety of animal species indicate that excitatory amino acid receptor antagonists may be of primary value for the treatment of epilepsy and ischaemic neurodegeneration (e.g., stroke and head trauma). More limited, but nevertheless solid data, point to therapeutic possibilities in Parkinson's disease and spasticity, while a smaller number of studies suggest anxiety, schizophrenia, and chronic neurodegenerative disorders as clinical targets [1,2].

3.5.1. Epilepsy
As indicated above, the only excitatory amino acid receptor for which a range of antagonists has been developed is the NMDA receptor, and therefore it is this receptor that has formed the focus of most therapeutically directed investigations to date. NMDA antagonists of all mechanistic types (Section 3.1) show anticonvulsant properties; moreover, such compounds are active in a wide range of animal models of epilepsy, suggesting that they may be therapeutically active against a high proportion of seizure types in man [75]. At higher doses, these agents induce muscle relaxation. An issue of some importance, therefore, is whether the therapeutic index (ratio of ED_{50}s producing muscle relaxant to anticonvulsant effects) is large enough to permit chronic usage in man. Studies in animals using noncompetitive antagonists (e.g., MK-801, PCP; therapeutic index close to 1 [75]) indicate that chronic use of these agents in ambulant patients will probably not be feasible. Competitive antagonists and glycine-site antagonists, on the other hand, show a distinct window between

doses eliciting wanted and unwanted effects. For example, CGP37849 and CGP39551 show therapeutic indices of 3-6 and 3-11, respectively, over the 24-hour period following oral administration to rats [75,76; J.P. Stables, personal communication; and unpublished observations], and L-687,414 exhibits a therapeutic index of 2-5 following systemic injection [60]. These values are comparable to those of established anticonvulsant agents such as valproate and carbamazepine.

3.5.2. Cerebral ischaemia

In addition to their anticonvulsant properties, all types of NMDA receptor antagonists display neuroprotective properties in animal models of brain ischaemia. Although early studies indicated that such compounds could prevent the diffuse neuronal death resulting from global brain ischaemia (i.e., a heart failure situation), more recent studies have shown that NMDA antagonists are most effective in reducing penumbral damage following focal cerebral arterial occlusion (as occurs in stroke) [17,77]. The basis for the differences between these models has not entirely been elucidated, although it is clear that in both circumstances a failure of energy-dependent processes leads to a rapid increase in the concentrations of glutamate and aspartate extracellularly, and hence hyperactivation of excitatory amino acid receptor mechanisms (note that a similar situation may pertain following head trauma, another target for treatment with excitatory amino acid blockers) [77]. Following focal brain ischaemia in animals, uncompetitive (e.g., MK-801, dextromethorphan) and competitive NMDA receptor antagonists (CGP40116, D-CPPene), as well as glycine-site antagonists (L-687,414) and polyamine-site antagonists (ifenprodil, SL-82.0715) markedly reduce the volume of infarcted brain tissue [61,67,77,78]. Investigations to date indicate that competitive, uncompetitive, and glycine-site antagonists are equally efficacious in this regard, although currently available uncompetitive blockers (e.g., MK-801) are the most potent. It remains to be ascertained whether drugs of these types will induce psychotic symptoms in man (Section 3.1.3) and whether this will influence the choice of agent for clinical development.

Recent work has shown that the AMPA receptor antagonists, NBQX and GYKI-52466, also show anticonvulsant and cerebroprotective properties in animals [68,69,71]. As anticonvulsants, neither compound is optimal in terms of duration of action (short) or therapeutic index. However, they may have advantages over NMDA receptor blockers as cerebroprotectants, since they are effective following both global and focal brain ischaemia. Unfortunately, it is difficult to draw firm conclusions about the potential therapeutic merits of AMPA receptor antagonists based on these two compounds, since their selectivity at therapeutic doses in vivo is unclear. However, these observations certainly justify further medicinal chemistry research efforts to develop compounds of greater potency and selectivity.

3.5.3. Parkinson's disease

In addition to their ability to block acute cerebral damage of the type that follows brain ischaemia or head trauma, both NMDA and AMPA receptor antagonists may be of value for the treatment of chronic neurodegenerative conditions, including Parkinson's [79] and Alzheimer's diseases [80] and AIDS-related dementia [81]. Currently, evidence in favour of a therapeutic role in Parkinson's disease is most well-developed. In this disease, a complex set of changes in basal ganglia circuitry is proposed to follow degeneration of dopaminergic neurons in the substantia nigra pars compacta, with increased activity of glutamatergic efferents from the subthalamic nucleus playing a key role in reducing the activity of thalamocortical neurons and hence in induction of parkinsonian symptoms. In rodent models mimicking some features of the disease, NMDA receptor antagonists (CPP, MK-801) and AMPA receptor blockers (NBQX), at doses which themselves elicit little or no behavioural activity, markedly potentiate the stimulant effects of L-DOPA or dopaminergic agonists [82–84]. Preliminary investigations using N-methyl-4-phenyl-1,2,5,6-tetrahydropyridine (MPTP)-treated monkeys indicate that NBQX also acts synergistically with L-DOPA to alleviate parkinsonian-like symptoms in this model [83]. These findings, together with observations that therapeutically relevant doses of many antiparkinsonian drugs block NMDA receptors [79], have led to the proposal that glutamate receptor antagonists may be valuable adjuncts for management of Parkinson's disease.

3.5.4. Dementia
Although studies of the therapeutic value of pharmacologically modulating brain glutamate receptors have focused primarily on antagonists, some observations suggest that agents which augment NMDA receptor activity might be used to improve cognitive function. This hypothesis is based on the idea that LTP may represent a neurophysiological substrate of some forms of learning, that NMDA receptor activation is required for the induction of LTP [85], and, more recently, that D-cycloserine, a partial agonist at the glycine co-agonist site, improves learning performance in animals [86]. As more pharmacological tools become available, it will be worthwhile to determine whether other glycine agonists, or agents which act via alternative mechanisms to facilitate the induction of LTP (e.g., $GluG$ receptor agonists, Section 2.5), exhibit similar properties. Clinical studies to assess the 'nootropic' effects of D-cycloserine in man may assist evaluation of the therapeutic potential of such approaches.

4. *Molecular biology*

Amongst established neurotransmitter receptors, those for glutamate were the last to yield to the power of molecular biology. The first glutamate receptor clone was reported in late 1989 [87] and, since then, members of most of the receptor subtypes described in previous sections have been cloned (Fig. 3). The information gained

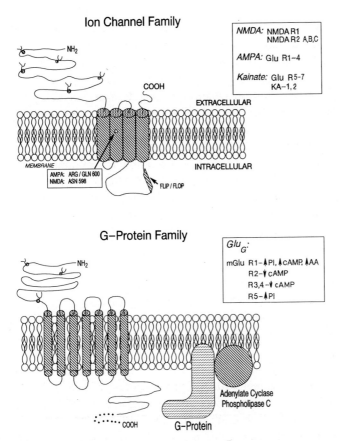

Fig. 3. Models of the two families of acidic amino acid receptors, incorporating recent information from molecular biological studies. Boxes indicate the receptors and subunits which have been cloned to date and, for the identified Glu_G subtypes, the intracellular response. For both families, the large, presumed extracellular, N-terminal domains contain several potential N-glycosylation sites. In the case of the ion channel family, 'ARG/GLN 600' denotes the single variable amino acid in the second transmembrane segment of GluR1–4 which regulates Ca^{2+} permeability (ASN 598 being the equivalent amino acid in NMDAR1 and 2), and 'FLIP/FLOP', the alternately-spliced region in the cytoplasmic loop of the AMPA receptor proteins GluR1–4. For the G-protein family, the C-terminal truncation site of mGluR1α (yielding mGluR1β) is indicated.

from these investigations has provided valuable new insights into the structural and functional heterogeneity of the glutamate receptor family and, in time, will almost certainly facilitate the identification of novel and more selective pharmacological agents for research and clinical study.

4.1. NMDA receptor

The NMDA receptor was the most recent addition to those glutamate receptors whose subunit cDNAs have been cloned and, like the others (see next sections), was initially identified by expression cloning in *Xenopus* oocytes [88]. The NMDA receptor proteins discovered so far fit into what is now becoming an expected pattern for glutamate receptors, being 938–1456 amino acids long, with a large N-terminal segment, and the expected 4 putative membrane-spanning domains. Somewhat surprisingly (given that channel-linked receptors described to date are hetero-oligomeric, with distinct receptor properties being conferred by separate protein subunits), the first clone to be identified, NMDAR1 [88], encodes all the key pharmacological features of native NMDA receptors; this includes co-activation by glutamate and glycine, antagonism by mechanistically distinct blockers (AP5, MK-801, Zn^{2+}), Ca^{2+} permeability of the ion channel, and voltage-dependent blockade by Mg^{2+}. The effects of polyamines have yet to be reported (note that it has proven difficult to demonstrate consistent effects of polyamines on native NMDA receptors using electrophysiological techniques [7]).

Although NMDAR1 is sufficient to account for the pharmacological properties of the NMDA receptor, homomeric channels formed by expression of this protein alone have conductances which are considerably smaller than those of native NMDA receptors, suggesting that other components may be required for a fully functional receptor. Recently, a second subunit, NMDAR2, has been identified by homology with conserved NMDAR1/AMPA receptor subunit sequences [89,90]. NMDAR2 alone does not form functional glutamate-responsive channels but, when co-expressed with NMDAR1, yields a ligand-gated channel with the pharmacological properties of NMDAR1, and with a conductance substantially greater than that of NMDAR1. Although NMDAR2 shares low overall sequence homology with NMDAR1, its regional expression in the brain, ability to potentiate the current responses of NMDAR1, and sequence similarities with NMDAR1 in the channel-lining domain all indicate that NMDAR2 is a structural component of the NMDA receptor. Whether additional distinct NMDA receptor subunits will be found, by analogy with the hetero-oligomeric nicotinic acetylcholine and $GABA_A$ receptors, remains to be determined.

Recent studies have identified multiple forms of the NMDA receptor with differing structural and functional characteristics and distributions within the brain. NMDAR1, for example, exists as seven alternately spliced forms, resulting from an inserted sequence in the *N*-terminal domain and two deletions in the C-terminus region. A truncated form, possessing the *N*-terminus but without the four putative transmembrane regions, also has been found [91]. NMDAR2 exists as three variants, A, B, and C, which show differential regional expression in the rat brain [90,92]. In-situ hybridization reveals that, like NMDAR1, NMDAR2A has a widespread expression in the CNS, consistent with the known localization of NMDA receptors. However, NMDAR2B is found only in forebrain structures such as cerebral cortex and hippocampus, whereas NMDAR2C expression is restricted to cerebellum. Co-expression of NMDAR2A or 2C with NMDAR1 gives rise to receptor complexes displaying differences in desensitization rates to glutamate and glycine and in the sensitivity to voltage-dependent block of the receptor channel by Mg^{2+}. It thus appears that the regional and functional heterogeneity of the NMDA receptor suggested by classical pharmacological studies (Section 2.1) may result from differential regional expression of protein subunits which comprise a hetero-oligomeric NMDA receptor complex.

The second membrane-spanning domain of ligand-gated channel receptors is believed to form the lining of the ion channel (Fig. 3) and to regulate the cation selectivity observed between members of this receptor family. In AMPA receptor subunits, the presence of a glutamine or arginine residue at one specific location (position 600) of this transmembrane segment is a crucial determinant of cation permeability (Section 4.2). In NMDAR1 and 2, the equivalent amino acid (position 598) is asparagine, and site-directed mutagenesis has confirmed that this single residue endows both the high Ca^{2+} permeability and the voltage-dependent blockade of this receptor channel by Mg^{2+} [93] and by MK-801 [94]. Further studies of this kind should provide detailed information about the structural basis underlying the unique properties of NMDA receptors.

At the same time that Nakanishi's group cloned NMDAR1, a cDNA was identified that encoded a glutamate-binding protein isolated as part of an NMDA receptor complex by conventional protein chemistry techniques [95]. This protein (57 kD) bears no similarity to any of the glutamate receptor subunits cloned to date and its function remains to be elucidated.

4.2. AMPA receptor

The first member of the glutamate receptor family to be cloned (GluR1) was identifed by functional expression in *Xenopus* oocytes and, although it is now clear that the pharmacology is that of a true AMPA receptor, this protein was originally designated a kainate receptor [87]. Three further cDNA clones, coding for proteins termed GluR2, 3, and 4, were discovered by homology screening [96,97]. Human forms of GluR1–4 recently have been cloned and the chromosomal locations of their genes have been mapped; unusually for hetero-oligomeric channel receptors, all four genes are located on different chromosomes [98].

cDNA sequencing indicates that GluR1–4 are 862–889 amino acids in length (approximately 100 kD), and hence are much larger than other members of the ligand-gated ion-channel family of receptors (e.g., nicotinic acetylcholine, $GABA_A$) with which they share only a small degree of sequence homology. Nevertheless, hydropathy plots suggest the presence of the 4 transmembrane-spanning regions characteristic of ion-channel receptors, plus a large, presumably extracellular, N-terminal domain. When expressed individually in *Xenopus* oocytes, GluR1–4 form homomeric ion channels which have the expected pharmacology of the AMPA receptor, although larger currents are evoked when two or more of the proteins are co-expressed, suggesting that (like the nicotinic and $GABA_A$ receptors) native receptors may have a hetero-oligomeric structure [97].

Experimental support for this postulate has arisen from two recent studies: (1) isolation of AMPA receptors from rat brain using antibodies directed towards the C-terminal portions of GluR1–4 has revealed that the native receptor exists as a pentameric structure comprising combinations of the GluR1–4 subunits [99] and (2) elegant work involving whole cell patch clamping of identified neurons in culture followed by analysis of their mRNA (amplified using the polymerase chain reaction) indicates that specific combinations of GluR1–4 subunits may be associated with distinct functional responses [100].

Further cDNA cloning has shown that each of the GluR1–4 proteins has an alternately spliced variant involving a 38 amino acid segment in the putative intracellular loop between the 3rd and 4th transmembrane domains (Fig. 3); the two isoforms of each protein have been designated 'flip' and 'flop' by Sommer et al. [101]. The variable region contains no known consensus sequences for protein kinases or other intracellular regulators but appears to be involved in regulating the functional properties of the receptor, since, when expressed in *Xenopus* oocytes, the flip and flop forms exhibit different electrophysiological characteristics. In-situ hybridization in the rat brain has revealed differential regional expression of GluR1–4 and their respective flip and flop forms. In general, the patterns of distribution of GluR1–3 are similar, with high expression in areas known to possess ^{3}H-AMPA-binding sites, such as the cerebral cortex, hippocampus, striatum, and cerebellum. In contrast, GluR4 is expressed to a lower extent and is restricted mainly to the cerebellum. Interestingly, the flip and flop forms of each protein show clear evidence of differential expression. For example, in the hippocampus, GluR2-flip is located primarily in pyramidal neurons of areas CA2-4, whereas GluR2-flop is expressed in CA1 neurons and in dentate gyrus granule cells [101]. Whether these observations indicate that excitatory synaptic transmission may be modulated in subpopulations of neurons by changing the expression of alternately spliced variants of AMPA receptor subunits remains to be seen; however, the hypothesis that gene regulation of this type might conceivably underlie some forms of synaptic plasticity (e.g., LTP) is sufficiently intriguing to warrant further investigation.

Further evidence for heterogeneity at the structure-function level has originated from studies of the Ca^{2+} permeability of AMPA receptor subtypes. When expressed alone, GluR1, 3, or 4 form channels which are permeable to Ca^{2+}, whereas GluR2 does not. On co-expression, the properties of GluR2 appear to dominate those of the other receptor subunits, leading to Ca^{2+}-impermeable channels [102]. This property of GluR2 has been traced to a single amino acid (Arg^{600} in GluR2, instead of Gln^{600} in GluR1, 3, and 4; Fig. 3) in the second putative transmembrane region (which may line the ion channel), and has been confirmed by site-directed mutagenesis [103]. Unexpectedly, the genomic DNA sequence for GluR2 encodes Gln at residue 600, and the single base change which leads to Arg appears to arise by a novel mechanism of

RNA editing [104,105]. If regulation of GluR2 expression plays a role in determining the Ca^{2+} permeability of AMPA receptor channel complexes in vivo, this may well have far-reaching implications for the physiology and pathology of glutamate receptor systems.

4.3. Kainate receptor

Screening rat cDNA libraries for clones homologous to GluR1–4 resulted in the identification of five related proteins, GluR5, 6, and 7, and KA1 and 2, which show properties expected of kainate receptors [106–109]. In-situ hybridization studies indicate that the regional expression of these proteins is consistent with that predicted on the basis of ^{3}H-kainate receptor autoradiography, although slight differences are apparent between the five mRNA species. When expressed alone in *Xenopus* oocytes, GluR5 is only weakly responsive to glutamate, and GluR6 forms channels which can be activated by kainate but not by AMPA; KA1 and 2 and GluR7 have not been shown to form functional channels, although radioligand-binding studies using cell lines expressing the proteins reveal pharmacological profiles consistent with those of high and low affinity ^{3}H-kainate-binding sites in brain membranes. Hence, molecular biological approaches have identified three classes of 'non-NMDA' receptor proteins which appear to provide a molecular basis for the ^{3}H-AMPA- and ^{3}H-kainate-binding sites identified in brain membrane preparations: (1) GluR1–4, which bind AMPA with high affinity (Section 4.2); (2) GluR5–7, which show preferential but low affinity binding of kainate (i.e., K_d in the 50 nM range); and (3) KA1 and 2, which bind kainate with high affinity (K_d of about 5 nM). Surprisingly, however, expression studies have shown that, although GluR5 or GluR6 alone are unresponsive to AMPA, combinations of either with KA2 endows AMPA sensitivity [110]. Whether other subunit combinations with unexpected pharmacological properties will be found, and whether these are of relevance to the physiology and pathology of the CNS, remains to be elucidated.

Two additional kainate-binding proteins have been identified in frog and chicken brain [111,112]. These are smaller (about 50 kD) than the AMPA and kainate receptor proteins described above, possess an amino acid sequence consistent with their comprising four membrane-spanning domains, but do not form functional receptor channels. Their function is unclear.

4.4. Glu_G receptor

The first member of the G-protein-linked receptors for glutamate (mGluR1α) was isolated by expression cloning from rat cerebellum [113]. The protein contains 1199 amino acids (about 133 kD) and therefore is much larger than other members of the family of G-protein receptors activated by neurotransmitters (e.g., muscarinic acetylcholine, dopamine). The amino acid sequence contains the typical 7 putative membrane-spanning domains and, as in the case of the cloned AMPA and kainate receptors, a large N-terminal segment (Fig. 3). A truncated form of mGluR1 (mGluR1β), lacking the C-terminal 22 amino acids, has also been identified, although its functional properties have not yet been determined [114]. In-situ hybridization reveals a widespread distribution of mGluR1 throughout the CNS, notable features of which are a high level of expression in cerebellar Purkinje cells and in hippocampal pyramidal and granule neurons. Interestingly, although mGluR1 (and the Glu_G receptor) was identifed on the basis of its ability to stimulate PI turnover [113], recent studies have shown that, in transfected CHO cells, activation of the same receptor protein with the same pharmacological selectivity also may increase cAMP formation and arachidonic acid release [115]. Activation of these three signal transduction pathways appears to be mediated by distinct G-proteins.

Recently, Nakanishi's group have identified four additional mGluR clones by homology screening [114,116]. These proteins (mGluR2–5) are 872–1171 amino acids in length and share substantial sequence homology with mGluR1. Surprisingly, although mGluR5 increases PI turnover with an agonist selectivity similar to that of mGluR1 [116], mGluR2, 3, and 4 do not, but are negatively coupled to adenylate cyclase [114,114a]. This is an unexpected second messenger response for a glutamate receptor, and would not have been predicted from earlier studies of acidic amino acid receptors in intact brain tissue preparations. Very recently, however, ACPD has been shown to inhibit cAMP formation in brain slices [38,39]. In addition, mGluR4 is uniquely sensitive to activation by L-AP4 and shows a regional expression consistent with that of the L-AP4 receptor [114a]. Thus, inhibition of adenylate cyclase, or direct G-protein-mediated coupling to Ca^{2+} channels, might underlie presynaptic autoreceptor functions (Sections 2.4 and 2.5).

It should be noted that native Glu_G receptors linked to phospholipase C show pharmacological properties not entirely in accord with those of mGluR1 and mGluR5, in that (at the former) ACPD is more potent as an agonist and AP3 is an effective antagonist (Section 3.4). This may suggest that the properties of the cloned proteins are modified post-translationally or, alternatively, that mGluR1 and 5 are not the predominant Glu_G receptors mediating PI turnover in the CNS.

5. Miscellaneous effects

5.1. Nitric oxide as a mediator

Nitric oxide (NO), described initially in physiological terms as a relaxant of vascular smooth muscle ('endothelium-derived relaxing factor', or EDRF), has recently been proposed as a 'tertiary messenger' following activation of excitatory amino acid receptors in the CNS [117]. Entry of Ca^{2+} through NMDA receptor channels (and perhaps through voltage-dependent Ca^{2+} channels following AMPA or kainate receptor-mediated depolarization) activates NO synthetase (via a Ca^{2+}-calmodulin complex), which catalyses the formation of NO from arginine. The primary target of NO in the CNS, as in blood vessels, is guanylate cyclase, resulting in an increase of cGMP. Only recently has it been realized that this is the molecular basis of elevations in cGMP observed in cerebellar tissue following application of excitatory amino acids, a biochemical assay of receptor function employed for many years.

Brain NO synthetase has recently been purified and cloned [118] and has been shown to be identical to 'NADPH diaphorase', an enzyme marker utilized for the histochemical localization of some peptide-containing neurons. In the brain, NO synthetase is found in neurons, separate from the cellular elements which contain guanylate cyclase. In the cerebellum, for example, NO synthetase in granule neurons is activated by glutamate released from mossy fibre terminals, although the resultant increase in cGMP occurs primarily in astrocytes and in presynaptic terminals [117]. NO acts thus as a transcellular messenger.

The half-life of NO in biological tissue is about 4 seconds and hence, although freely diffusible through membranes, its sphere of influence is likely to be highly limited. NO is thus an attractive candidate for the postulated trans-synaptic mediator which, after release from postsynaptic cells in response to NMDA receptor activation, may contribute to LTP by increasing the release of glutamate from presynaptic terminals. Indeed, inhibitors of NO synthetase have been shown to block LTP in the hippocampus in vitro [120,121]. Interestingly, however, the converse effect can also be shown, depending on the precise temporal relationship between the presynaptic tetanus and NMDA receptor activation [122], and may reflect direct inhibition of the NMDA receptor by NO [123]. The temporal sequence of pre- and postsynaptic events has long been known to be critical for the induction of LTP, and NO, by exerting differential modulatory actions at pre- and postsynaptic levels, may be a key physiological regulator of this form of synaptic plasticity [122].

NO has additionally been proposed to be a mediator of NMDA-induced neurotoxicity, since it is involved in the generation of highly reactive free radicals which are implicated in lipid peroxidation and ischemic cell damage [119]. One consequence of these findings is that NO synthetase might be seen as an attractive target for the development of novel neuroprotective drugs. Whether, however, NO is produced in the CNS in sufficient concentration to be neurotoxic is unclear. In support of a role for NO in acute neurodegenerative events, NO synthetase inhibitors have been shown to antagonize glutamate-induced neurotoxicity in cortical cell culture [119] and to reduce cerebral damage in a mouse model of stroke [124]. However, contradictory results have been obtained by other groups. For example, in two recent studies, NO synthetase inhibitors were found to have no cerebroprotective effects following focal brain ischaemia in rats ([125]; D. Sauer and P. Allegrini, personal communication). Further studies to define the roles of NO in excitatory amino acid physiology and pathology are required.

Acknowledgements

Thanks to Andy Butler for preparing the figures, to Bev and Vic for typing skills, and to Barb for staying the course.

References

1. Meldrum, B.S., Ed. (1991) Excitatory Amino Acid Antagonists, Blackwell Scientific Publications, Oxford.
2. Lodge, D. and Collingridge, G., Eds. (1990) The Pharmacology of Excitatory Amino Acids: A TIPS Special Report. Elsevier Trends Journals, Cambridge.
3. Fagg, G.E. and Foster, A.C. (1983) Neuroscience 4, 701–719.
4. Mayer, M.L. and Westbrook, G.L. (1987) Prog. Neurobiol. 28, 197–276.
5. Collingridge, G.L. and Lester, R.A. (1989) Pharmacol. Rev. 40, 143–210.
6. Monaghan, D.T., Bridges, R.J. and Cotman, C.W. (1989) Annu. Rev. Pharmacol. Toxicol. 29, 365–402.
7. Wong, E.H.F. and Kemp, J.A. (1991) Annu. Rev. Pharmacol. Toxicol. 31, 401–425.
8. Watkins, J.C. and Evans, R.H. (1981) Annu. Rev. Pharmacol. Toxicol. 21, 165–204.
9. Foster, A.C. and Fagg, G.E. (1984) Brain Res. Rev. 7, 103–164.
10. Watkins, J.C., Krogsgaard-Larsen, P. and Honore, T. (1990) Trends Pharmacol. Sci. 11, 25–33.
11. Schoepp, D., Bockaert, J. and Sladeczek, F. (1990) Trends Pharmacol. Sci. 11, 508–515.
12. MacDonald, J.F., Porietis, A.V. and Wojtowicz, J.M. (1982) Brain Res. 237, 248–253.
13. Nowak, L., Bregestovski, P., Ascher, P., Herbet, A. and Prochiantz, Z. (1984) Nature 307, 462–465.
14. MacDermott, A.B., Mayer, M.L., Westbrook, G.L., Smith, S.J. and Barker, J.L. (1986) Nature 321, 519–522.
15. Dingledine, R., McBain, C.J. and McNamara, J.O. (1990) Trends Pharmacol. Sci. 11, 334–338.
16. Choi, D.W. (1988) Neuron 1, 623–634.
17. Meldrum, B.S. (1990) Cerebrovasc. Brain Metab. Rev. 2, 27–-57.
18. Kelso, S.R., Ganong, A.H. and Brown, T.H. (1986) Proc. Natl Acad. Sci. USA 83, 5326–5330.
19. Garthwaite, G., Hajos, F. and Garthwaite, J. (1986) Neuroscience 18, 437–447.
20. Mueller, W. and Connor, J.A. (1991) Nature 354, 73–75.
21. Bading, H. and Greenberg, M.E. (1991) Science 253, 912–914.
22. Dumuis, A., Sebben, M., Pin, J.P., Haynes, L.W. and Bockaert, J. (1988) Nature 336, 68–-76.
23. Fagg, G.E. and Matus, A. (1984) Proc. Nat. Acad. Sci. USA 81, 6876–6880.
24. Mayer, M.L. and Vycklicky, L. (1989) Proc. Natl Acad. Sci. USA 86, 1411–1415.
25. Iino, M., Ozawa, S. and Tsuzuki, K. (1990) J. Physiol. 424, 151–165.
26. Honore, T., Davies, S.N., Drejer, J., Fletcher, E.J., Jacobsen, P., Lodge, D. and Nielsen, F.E. (1988) Science 241, 701–703.
27. Fagg, G.E. and Massieu, L. (1991) In: Excitatory Amino Acid Antagonists (Meldrum, B.S., Ed.) pp. 39–63, Blackwell Scientific Publications, Oxford.
28. Cotman, C.W., Flatman, J.A., Ganong, A.H. and Perkins, M.N. (1986) J. Physiol. 378, 403–415.
29. Forsythe, I.D. and Clements, J.D. (1990) J. Physiol. 429, 1–16.
30. Trombley, P.Q. and Westbrook, G.L. (1992) J. Neurosci. 12, 2043–2050.
31. Lester, R.A.J. and Jahr, C.E. (1990) Neuron 4, 741–749.
32. Baskys, A. and Malenka, R.C. (1991) Eur. J. Pharmacol. 193, 131–132.
33. Lovinger, D.M. (1991) Neurosci. Lett. 129, 17–21.
34. Miller, R.J. (1991) Trends Pharmacol. Sci. 12, 365–367.

35. Sladeczek, F., Pin, J.P., Recasens, M., Bockaert, J. and Weiss, S. (1985) Nature 317, 717–719.
36. Sugiyama, H., Ito, I. and Hirono, C. (1985) Nature 325, 531–533.
37. Anwyl, R. (1991) Trends Pharmacol. Sci. 12, 324–326.
38. Schoepp, D.D., Johnson, B.G. and Monn, J.A. (1992) J. Neurochem. 58, 1184–1186.
39. Cartmell, J., Kemp, J.A., Alexander, S.P.H., Hill, S.J. and Kendall, D.A. (1992) J. Neurochem. 58, 1964–1966.
40. Charpak, S., Gähwiler, B., Do, K.Q. and Knöpfel, T. (1990) Nature 347, 765–767.
41. Taschenberger, H., Roy, B.L. and Lowe, D.A. (1992) NeuroReport 3, 629–632.
42. Reymann, K.G. and Matthies, H. (1989) Neurosci. Lett. 98, 166–171.
43. Zheng, F. and Gallagher, J.P. (1992) Neuron 9, 163–172.
44. Koh, J.-Y., Palmer, E. and Cotman, C.W. (1991) Proc. Natl Acad. Sci. USA 88, 9431–9435.
45. Opitz, T. and Reymann, K.G. (1991) NeuroReport 2, 455–457.
46. Siliprandi, R., Lipartiti, M., Fadda, E., Sautter, J. and Manev, H. (1992) Eur. J. Pharmacol. 219, 173–174.
47. Chiamulera, C., Albertini, P., Valerio, E. and Reggiani, A. (1992) Eur. J. Pharmacol. 216, 335–336.
48. McDonald, J.W. and Schoepp, D.D. (1992) Eur. J. Pharmacol. 215, 353–354.
49. Fagg, G.E., Olpe, H.R., Pozza, M.F., Baud, J., Steinmann, M., Schmutz, M., Portet, C., Baumann, P., Thedinga, K., Bittiger, H., Allgeier, H., Heckendorn, R., Angst, C., Brundish, D. and Dingwall, J.G. (1990) Br. J. Pharmacol. 99, 791–797.
50. Lowe, D.A., Neijt, H.C. and Aebischer, B. (1990) Neurosci. Lett. 113, 315–321.
51. Benveniste, M. and Mayer, M.L. (1991) Br. J. Pharmacol. 104, 207–221.
52. Fagg, G.E. and Baud, J. (1988) In: Excitatory Amino Acids in Health and Disease (Lodge, D., Ed.), pp. 63–90, Wiley & Sons Ltd., Chichester.
53. Cordi, A.A., Monahan, J.B. and Reisse, J. (1990) J. Receptor Res. 10, 299–315.
54. Johnson, J.W. and Ascher, P. (1987) Nature 325, 529–531.
55. Kleckner, N.W. and Dingledine, R. (1988) Science 238, 355–358.
56. Foster, A.C., Donald, A.E., Grimwood, S., Leeson, P.D. and Williams, B.J. (1991) Br. J. Pharmacol. 102, 64P.
57. Kemp, J.A., Priestley, T., Marshall, G.R., Leeson, P.D. and Williams, B.J. (1991) Br. J. Pharmacol. 102, 65P.
58. Baron, B.M., Harrison, B.L., McDonald, I.A., Meldrum, B.S., Palfreyman, M.G., Salituro, F.G., Siegel, B.W., Slone, A.L., Turner, J.P. and White, H.S. (1992) J. Pharmacol. Exp. Ther. 262, 947–955.
59. Foster, A.C., Kemp, J.A., Leeson, P.D., Grimwood, S., Donald, A.E., Marshall, G.R., Priestley, T., Smith., J.D. and Carling, R.W. (1992) J. Pharmacol. Exp. Ther. 41, 914–922.
60. Saywell, K., Singh, L., Oles, R.J., Vass, C., Leeson, P.D., Williams, B.J. and Tricklebank, M.D. (1991) Br. J. Pharmacol. 102, 66P.
61. Gill, R., Hargreaves, R. and Kemp, J.A. (1991) J. Cereb. Blood Flow Metab. 11 (Suppl.) 2, S304.
62. MacDonald, J.F., Miljkovic, Z. and Pennefather, P. (1987) J. Neurophysiol. 58, 251–266.
63. Singh, L., Wong, E.H.F., Kesingland, A.C. and Tricklebank, M.D. (1990) Br. J. Pharmacol. 99, 145–151.
64. Woods, J.H., Koek, W., France, C.P. and Moerschbaecher, J.M. (1991) In: Excitatory Amino Acid Antagonists (Meldrum, B.S., Ed.), pp. 237–264, Blackwell Scientific Publications, Oxford.
65. Ransom, R.W. and Stec, N.L. (1988) J. Neurochem. 51, 830–836.
66. Williams, K., Dawson, V.L., Romano, C., Dichter, M.A. and Molinoff, P. (1990) Neuron 5, 199–208.
67. Carter, C., Benavides, J., Dana, C., Schoemaker, H., Perrault, G., Sanger, D. and Scatton, B. (1991) In: Excitatory Amino Acid Antagonists (Meldrum, B.S., Ed.), pp. 130–163, Blackwell Scientific Publications, Oxford.
68. Sheardown, M.J., Nielsen, E.O., Hansen, A.J., Jacobsen, P. and Honore, T. (1990) Science 247, 571–574.
69. Smith, S.E., Duermueller, N. and Meldrum, B.S. (1991) Eur. J. Pharmacol. 201, 179–183.

70. Vyklicky, L., Patneau, D.K. and Mayer, M.L. (1991) Neuron 7, 971–984.
71. Smith, S.E. and Meldrum, B.S. (1992) Stroke 23, 861–864.
72. Recasens, M. and Guiramand, J. (1991) In: Excitatory Amino Acid Antagonists (Meldrum, B.S., Ed.), pp. 195–215, Blackwell Scientific Publications, Oxford.
73. Hayashi, Y., Tanabe, Y., Aramori, I., Masu, M., Shimamoto, K., Ohfune, Y. and Nakanishi, S. (1992) Br. J. Pharmacol. 107, 539–543.
74. Sugiyama, H. (1990) Neurochem. Int. 16 (Suppl. I) 17.
75. Chapman, A.G. (1991) In: Excitatory Amino Acid Antagonists (Meldrum, B.S., Ed.), pp. 265–286. Blackwell Scientific Publications, Oxford.
76. Schmutz, M., Portet, C., Jeker, A., Klebs, K., Vassout, A., Allgeier, H., Heckendorn, R., Fagg, G.E., Olpe, H.R. and Van Riezen, H. (1990) Naunyn-Schmiedebergs Arch. Pharmacol. 342, 61–66.
77. McCulloch, J., Bullock, R. and Teasdale, G.M. (1991) In: Excitatory Amino Acid Antagonists (Meldrum, B.S., Ed.), pp. 287–326. Blackwell Scientific Publications, Oxford.
78. Sauer, D. and Fagg, G.E. (1992) In: Excitatory Amino Acid Receptors: Design of Agonists and Antagonists (Krogsgaard-Larsen, P. and Hansen, J.J., Eds.), pp. 13–33. Ellis Horwood Ltd., London.
79. Greenamyre, J.T. and O'Brien, C.F. (1991) Arch. Neurol. 48, 977–981.
80. Greenamyre, J.T. and Young, A.B. (1989) Neurobiol. Aging 10, 593–602.
81. Lipton, S.A. (1992) Trends Neurosci. 15, 75-77.
82. Klockgether, T. and Turski, L. (1990) Ann. Neurol. 28, 539–546.
83. Greenamyre, J.T., Klockgether, T., Turski, L., Zhang, Z., Kurlan, R. and Gash, D.M. (1992). In: Excitatory Amino Acids: Fidia Research Foundation Symposium Series (Simon, R.P., Ed.), Vol. 9, pp. 195–198, Thieme Medical Publishers Inc., New York.
84. Wachtel, H., Kunow, M. and Löschmann, P.-A. (1992) Neurosci. Lett. 142, 179–182.
85. Collingridge, G.L. and Bliss, T.V.P. (1987) Trends Neurosci. 10, 288–293.
86. Monahan, J.B., Handelman, G.E., Hood, W.F. and Cordi, A.A. (1990) Pharmacol. Biochem. Behav. 34, 649–653.
87. Hollmann, M., O'Shea-Greenfield, A., Rogers, S.W. and Heinemann, S. (1989) Nature 342, 643–648.
88. Moriyoshi, K., Masu, M., Ishil, T., Shigemoto, R., Mizuno, N. and Nakanishi, S. (1991) Nature 354, 31–37.
89. Meguro, H., Mori, H., Araki, K., Kushiya, E., Kutsuwada, T., Yamazaki, M., Kumanishi, T., Arakawa, M., Sakimura, K. and Mishina, M. (1992) Nature 357, 70–72.
90. Monyer, H., Sprengel, R., Schoepfer, R., Herb, A., Higuchi, M., Lomeli, H., Burnashev, N., Sakmann, B. and Seeburg, P.H. (1992) Science 256, 1217–1220.
91. Sugihara, H., Moriyoshi, K., Ishii, T., Masu, M. and Nakanishi, S. (1992) Biochem. Biophys. Res. Commun. 185, 826–832.
92. Kutsuwada, T., Kashiwabuchi, N., Mori, H., Sakimura, K., Kushiya, E., Araki, K., Meguro, H., Masaki, H., Kumanishi, T., Arakawa, M. and Mishina, M. (1992) Nature 358, 36–40.
93. Burnashev, N., Schoepfer, R., Monyer, H., Ruppersberg, J.P., Günther, W., Seeburg, P.H. and Sakmann, B. (1992) Science 257, 1415–1418.
94. Mori, H., Masaki, H., Yamakura, T. and Mishina, M. (1992) Nature 358, 673–676.
95. Kumar, K.N., Tilakaratne, N., Johnson, P.S., Allen, A.E. and Michaelis, E.K. (1991) Nature 354, 70–73.
96. Keinanen, K., Wisden, W., Sommer, B., Werner, P., Herb, A., Verdoorn, T.A., Sakmann, B. and Seeburg, P.H. (1990) Science 249, 556–560.
97. Boulter, J., Hollmann, M., O'Shea-Greenfield, A., Hartley, M., Deneris, E., Maron, C. and Heinemann, S. (1990) Science 249, 1033–1037.
98. McNamara, J.O., Eubanks, J.H., McPherson, J.D., Wasmuth, J.J., Evans, G.A. and Heinemann, S. (1992) J. Neurosci. 12, 2555–2562.
99. Wenthold, R.J., Yokoyani, N., Doi, K. and Wada, K. (1992) J. Biol. Chem. 267, 501–507.
100. Lambolez, B., Audinat, E., Bochet, P., Crepel, F. and Rossier, J. (1992) Neuron 9, 247–258.

101. Sommer, B., Keinanen, K., Verdoorn, T.A., Wisden, W., Burnashev, N., Herb, A., Kohler, M., Takagi, T., Sakmann, B. and Seeburg, P.H. (1990) Science 249, 1580–1585.
102. Hollmann, M., Hartley, M. and Heinemann, S. (1991) Science 252, 851–853.
103. Verdoorn, T.A., Burnashev, N., Monyer, H., Seeburg, P.H. and Sakmann, B. (1991) Science 252, 1715–1718.
104. Sommer, B., Kohler, M., Sprengel, R. and Seeburg, P.H. (1992) Cell 57, 11–19.
105. Burnashev, N., Monyer, H., Seeburg, P.H. and Sakmann, B. (1992) Neuron 8, 189–198.
106. Bettler, B., Boulter, J., Hermans-Borgmeyer, I., O'Shea-Greenfield, A., Deneris, E.S., Moll, C., Borgmeyer, U., Hollman, M. and Heinemann, S. (1991) Neuron 5, 583–595.
107. Egebjerg, J., Bettler, B., Hermans-Borgmeyer, I. and Heinemann, S. (1991) Nature 351, 745–748.
108. Werner, P., Voigt, M., Keinanen, K., Wisden, W. and Seeburg, P.H. (1991) Nature 351, 742–744.
109. Bettler, B., Egebjerg, J., Sharma, G., Pecht, G., Hermans-Borgmeyer, I., Moll, C., Stevens, C.F. and Heinemann, S. (1992) Neuron 8, 257–265.
110. Herb, A., Burnashev, N., Werner, P., Sakmann, B., Wisden, W. and Seeburg, P.H. (1992) Neuron 8, 775–785.
111. Wada, K., Dechesne, C.J., Shimasaki, S., King, R.G., Kusano, K., Buonanno, A., Hampson, D.R., Banner, C., Wenthold, R.J. and Nakatani, Y. (1989) Nature 342, 684–689.
112. Gregor, P., Mano, I., Maoz, I., McKeown, M. and Teichberg, V.I. (1989) Nature 342, 689–692.
113. Masu, M., Tanabe, Y., Tsuchida, K., Shigemoto, R. and Nakanishi, S. (1991) Nature 349, 760–765.
114. Tanabe, Y., Masu, M., Ishii, T., Shigemoto, R. and Nakanishi, S. (1992) Neuron 8, 169–179.
114a. Tanabe, T., Nomura, A., Masu, M., Shigemoto, R., Mizuno, N. and Nakanishi, S. (1993) J. Neurosci. 13, 1372–1378.
115. Aramori, I. and Nakanishi, S. (1992) Neuron 8, 757–765.
116. Abe, T., Sugihara, H., Nawa, H., Shigemoto, R., Mizuno, N. and Nakanishi, S. (1992) J. Biol. Chem. 267, 13361–13368.
117. Garthwaite, J. (1991) Trends Neurosci. 14, 60–67.
118. Snyder, S.H. and Bredt, D.S. (1991) Trends Pharmacol. Sci. 12, 125–128.
119. Dawson, V.L., Dawson, T.M., London, E.D., Bredt, D.S. and Snyder, S.H. (1991) Proc. Natl Acad. Sci. USA 88, 6368–6371.
120. Schuman, E.M. and Madison, D.V. (1991) Abstr. Soc. Neurosci. 17, 4.5.
121. Errington, M.L., Li, Y-G., Mattheis, H., Williams, J.H. and Bliss, T.V.P. (1991) Abstr. Soc. Neurosci. 17, 380.17.
122. Izumi, Y., Clifford, D.B. and Zorumski, C.F. (1992) Science 257, 1273–1276.
123. Manzoni, O., Prezeau, L., Marin, P., Deshager, S., Bockaert, J. and Fagni, L. (1992) Neuron 8, 653–662.
124. Nowicki, J.P., Duval, D., Poignet, H. and Scatton, B. (1991) Eur. J. Pharmacol. 204, 339–340.
125. Dawson, D.A., Kusumoto, K., Graham, D.I., McCulloch, J. and Macrae, I.M. (1992) Neurosci. Lett. 142, 151–154.

IV. Neuropeptide Receptors

F. Hucho (Ed.) *Neurotransmitter Receptors*
© 1993 Elsevier Science Publishers B.V. All rights reserved.

CHAPTER 11

Opioid receptors

ERIC A. BARNARD and JOSEPH SIMON

Molecular Neurobiology Unit, Royal Free Hospital School of Medicine (University of London), London NW3 2PF, U.K.

> '..a drowsy numbness pains my sense,
> as though of hemlock I had drunk,
> or emptied some dull opiate to the drains'
>
> John Keats

1. Introduction

The narcotic effects of the extract of the opium poppy were clearly known in classical antiquity. Its use was well-established in ancient Greece; it spread via Arabia later to China and became important in several cultures. The clinical value of opiates as analgesics was understood early in Western medicine:

> 'Among the remedies which it has pleased Almighty God to give man to relieve his sufferings, none is so universal and so efficacious as opium' Sir Thomas Sydenham (1680).

The term 'opiate' was originally in use to designate narcotic drugs derived from opium, i.e. morphine, codeine, and their many semisynthetic derivatives (illustrated in Fig. 1). Later, the word 'opioid' was coined to refer in a generic sense to all drugs, natural and synthetic, which have morphine-related actions, as well as to the endogenous peptides later discovered with such actions. The narcotic analgesics, of which the prototype is morphine, produce a large variety of pharmacological responses by interacting with cell surface receptors specific for endogenous opioids in the brain, spinal cord, and certain peripheral tissues. It became clear that morphine, being an agonist, cross-reacted very poorly with some of the other types of opioids (e.g. dynorphins) and that, as is usual in pharmacology, antagonist activity is a more useful criterion for relationship in these receptors. Hence, sensitivity to the antagonist naloxone (Fig. 1), which competes with all types of opioids, is now the accepted criterion for an opioid receptor.

OPIOID AGONISTS:

Morphine   Etorphine   U-69,593

OPIOID ANTAGONISTS:

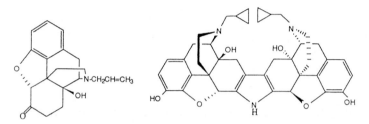

Naloxone   nor-Binaltorphimine

Naltrindrole

BENZOMORPHANS WITH MIXED AGONIST AND ANTAGONIST CHARACTER

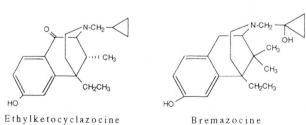

Ethylketocyclazocine   Bremazocine

The celebrated dual effects of opiates in man, namely deadening pain and producing euphoria, indicated that opioid receptors are important in central and spinal pain pathways and in the limbic system. A variety of additional effects of opiates, e.g. on the gastrointestinal tract and on respiratory depression, denotes other locations of opioid receptors, including many sites in the autonomic nervous system. At many of the central and peripheral locations the effect of an opiate agonist is known to be modulatory, acting by inhibiting the release of a classical transmitter. This can occur at many types of nerve terminals, including those releasing acetylcholine, glutamate, noradrenaline, or dopamine [1]. The release of another peptide modulator, e.g. substance P, can also be inhibited [2]. The modulatory effects are also physiologically important in the opioid control of neuroendocrine and corticosteroid release [3,4]. At some sites opioid receptor occupancy leads to the closing of a voltage-dependent Ca^{2+} channel [5,6], while at others it opens a voltage-dependent K^+ channel [7]. At the signalling level these actions of opioids on channels are inhibitory, either by hyperpolarising the target neurone or by shortening the Ca^{2+} component of the presynaptic action potential [6,7]. They can equally occur in postsynaptic actions of opioids. These diverse actions on channels are, as will be discussed later, not direct, but are mediated by some route which involves the action of a guanine-nucleotide-binding protein (G-protein) [6,7].

A direct excitatory action at some locations, e.g. via the closing of a K^+ channel, has also been described [8]. This would, together with the more usual inhibitory actions, form a dual modulatory mechanism which would explain the effects of opioids in both inhibiting transmitter release generally and enhancing it at some sites at higher opioid concentrations, as well as the sometime paradoxical effects of opioids, analgesic or hyperalgesic, and euphoric or aversive, and some aspects of opioid tolerance and addiction [8].

Opioid receptors exert a particular interest within the general study of receptors for several reasons. Firstly, the direct action of a receptor in removing pain or dramatically changing mood provides an unusual link from the molecular level to the level of consciousness. An understanding of the opioid mediation of neural pathways will be particularly illuminating. Secondly, the fact that the complex structure of morphine (Fig. 1) acts as a mimic for the structure of a pentapeptide, enkephalin, at opioid receptors is striking in structure-activity relations of receptors and has opened the field of design of nonpeptide equivalents of natural peptide ligands for receptors. Thirdly, the opiates are involved in a particularly pronounced form of the drug addiction phenomenon, coupled with severe tolerance on continued exposure. These

---

Fig. 1. Structures of some opioid drugs. U-69593 (κ), nor-binaltorphimine (κ), morphine (μ), and naltrindole (δ) are of high or moderately high specificity (Table I), while the others are nonselective. CI-977 (see Table I) has the U-69593 structure with the benzene ring replaced by the bicyclic 4-benzofuranyl group [44].

effects will need to be unravelled at the receptor level. Lastly, it is sobering to realise that morphine is not only the oldest but is still the most effective drug for severe pain, e.g. in hospice use. A major goal of research in opioid receptors will be to find the basis for new types of drug acting there, selective in analgesia and free of the present great drawbacks of resistance, dependence, and tolerance.

2. Multiple opioid receptors and their ligands

2.1. Opioid ligands

The existence of specific opioid-binding sites, inferred originally from the specific behavioural and clinical effects of opiates, was confirmed by their biochemical identification. In 1973, three laboratories independently reported evidence for stereospecific, saturable opioid-binding sites in the mammalian nervous system using radiolabelled agonists [9–11]. These became associated with multiple types of opioid receptors in 1976 on the basis of pharmacological distinctions: the administration of morphine, ketocyclazocine, or N-allylnormetazocine (SKF-10047) to spinal dogs produced very distinct behavioural responses depending on the drug used [12]. Martin et al. [12] used Greek letters to assign putative opioid receptor types: μ- for the receptor that mediated morphine actions and κ- for the receptor that mediated the actions of a benzomorphan compound, ketocyclazocine. Although Martin also proposed another type, called σ- on the basis of the different effects of the analgesic SKF-10047, this receptor is not now considered to be an opioid receptor. The σ-receptors show cross-reactivity with other receptors, including phencyclidine-binding sites, and their effects cannot be antagonized by naloxone.

The landmark discovery in 1975 by Hughes et al. [13] of two endogenous peptides with opioid agonist activity opened up the new field of neuropeptides, of far-reaching significance. These two are leu-enkephalin (Tyr-Gly-Gly-Phe-Leu) and met-enkephalin (Tyr-Gly-Gly-Phe-Met). The 28-residue neuropeptide β-endorphin was soon shown to have a similar type of activity. Another opioid neuropeptide family includes the 17-residue dynorphin A [14] and this was assigned as an endogenous ligand for the κ-receptor. β-Endorphin and dynorphin each contain an enkephalin sequence at the N-terminus. Three opioid precursor polypeptides, each encoded by a separate gene, were identified by DNA cloning. These are processed post-translationally to form respectively the enkephalins, β-endorphin, and the dynorphins, as well as several minor related opioid peptides. These precursors have different distributions in vivo (reviewed by Mansour et al. [15]); this in itself suggests that the different opioid peptides exert different actions.

The pharmacology of the endogenous peptides on isolated organ test preparations led to three types of opioid receptor, μ, κ and δ being needed to account for the responses. The third type, δ, is prominent in the mouse vas deferens [16]. The existence of at least these three (μ, δ, and κ) types of opioid receptor has been confirmed by in-vitro binding studies using radiolabelled opioid agonists and antagonists (reviewed by Corbett et al. [17]). Thus, a number of synthetic or semisynthetic opioid ligands have been devised which show some selectivity for one of the pharmacologically-recognised opioid receptor types. These compounds generally fall into three groups.

1. Ligands based upon the enkephalin structure. These generally show a higher selectivity toward the μ- or the δ-type of opioid receptors (for example DAMGO for μ, and DSLET ([D-Ser^2]-Leu^5-enkephalin), BUBUC, and DPDPE for δ. Some very effective peptides (Table I) are only distantly related to enkephalins (CTOP, DALDA). The selectivity for μ or for δ can in some structures be considerably increased by cyclising a peptide (DPDPE, CTAP). It gives an advantage in potency, also, to include peptidase-resistant features, e.g. D- or non-natural amino acids used in some of the structures.
2. Morphine skeleton-based agonists (e.g. dihydromorphine), their tetracyclic relatives, the oripavines such as etorphine, and their bicyclic relatives, the benzomorphans (e.g. EKC, bremazocine). These are either moderately μ-selective or they distinguish poorly between opioid receptor types (Table I).

TABLE I
Opioid receptor types and their selective ligands

Type	Prototype agonist	Selective agonists[a]	Selective antagonists	Selective radioligands
μ-	Morphine	DAMGO (μ/δ 160) DALDA (μ/δ 1,300)	CTOP (μ/δ 1,800)	[^{3}H]DAMGO
δ-	Leu-enkephalin	DPDPE (δ/μ 560) BUBUC (δ/μ 1,020)	ICI 174864 (μ/δ 94) Naltrindole (μ/δ 100) TIPP (N/S 1410)	[^{3}H]DPDPE [^{3}H]DSBULET
κ-	Ethyl-ketocyclazocine (EKC); Dynorphin A	U-69593 (κ/μ 3670) CI-977 (κ/δ 900)	Nor-binaltorphimine (κ/δ 40)	[^{3}H]U-69593 [^{3}H]CI-977
ε-	β-Endorphin	None known	None known	[^{125}I]-β-endorphin[b]

BUBUC: Tyr-D-Cys(O-tBu)-Gly-Phe-Leu-Thr(O-tBu); CI-977: (5R)-(5α,7α,8β)-N-methyl-N-(7-[1-pyrrolidinyl]-1-oxaspiro[4,5]dec-8-yl)-4benzofuranacetamide; CTOP: D-Phe-Cys-Tyr-D-Trp-Orn-Thr-Pen-Thr-NH$_2$; DAMGO: [D-Ala^2, MePhe^4-Gly-ol^5]enkephalin; DPDPE: [D-Pen^2, D-Pen^5]enkephalin; DALDA, Tyr-D-Arg-Phe-Lys; DSBULET: Tyr-D-Ser(O-tBu)-Gly-Phe-Leu-Thr; ICI 174864: N,N-diallyl-Tyr-Aib-Aib-Phe-Thr; TIPP: H-Tyr-Tic-Phe-Phe-OH; U-69593: 5α,7α,β-(−)-N-methyl-N-[7-pyrrolidinyl)1-oxaspiro94,5)dec-8-yl]benzene acetamide. (Pen = penicillaminyl, linked to the other Pen (or Cys) in the peptide by a disulphide bridge; Aib = amino-iso-butyryl).

[a]In parentheses, the ratio of the binding affinities for the two highest affinity types. Note that varying values for these ratios have been reported, depending on the tissue and assay conditions used. Values here are from Delay-Goyet et al. [42] (BUBUC, DALDA); Portoghese et al. [43] (naltrindole, nor-binaltorphimine), Hunter et al. [44](CI-977, U-69593, DPDPE), Schiller et al. [129] (TIPP) and Corbett et al. [17] (others).
[b]Only in the presence of blocking by agonists for the μ-, δ-, and κ-sites.

3. Other, structurally distinct series of opioid ligands, for example arylacetamides of the type U69593 or CI977 (Table I), which are highly selective for the κ-receptors. Other series, such as anilidopiperidines of the type sufentanil, can be highly selective for the μ-type or certain benzodiazepines (tifluadom) for the κ-type.

Some of the nonpeptide structures are illustrated in Figure 1. Good affinity is not necessarily associated with high binding selectivity. A selection of some of the most selective ligands now available is shown in Table I. An example of the selectivity of appropriate opioid ligands on a single preparation, explored over a wide concentration range in competition-binding experiments, is illustrated in Figure 2A.

Finally, it should be noted that in recent years excitement has been generated by reports of nonpeptide endogenous ligands for opioid receptors, derivatives of morphine actually detected in vertebrate brain (for more references, see [18]). These are present in low amounts and while it has been much discussed whether or not these form another series of endogenous brain opioids, biosynthesis of the morphine skeleton in the mammal has been described [18].

2.2. Types of opioid receptors

The classification of the opioid receptors in situ into δ, μ, and κ (Table I) has become well-established, since it correlates well with the in-vitro (or ex-vivo) binding selectivities of the various ligands and the pharmacological differences seen in different tissue preparations and in behavioural and clinical responses. Thus, the hamster vas deferens responds primarily to δ, and the rabbit vas deferens to κ-agonists [17]. In many other physiological responses, in isolated tissues or in-vivo, as well as in binding profiles, the differential effects of the various opioid agents can only be explained by the heterogeneity of the receptors [16–24].

Further evidence on this heterogeneity comes from brain neuroanatomical distribution patterns. The distribution of the enkephalins and dynorphin differ, as do those of their precursor mRNAs. More selectively, the autoradiographic patterns of the binding of the μ-selective, δ-selective, and κ-selective ligands are in some areas (in the same species) strikingly distinct.

In general, therefore, there is convergence, from multiple types of studies on tissues, on a set of distinct μ-, δ-, and κ-types of opioid receptors. As will be seen below, this can be directly examined at the protein level. Not all of the observations which have been reported on tissues or membranes are consistent, but this is not surprising in view of the anatomical complexity involved, the overlaps of the ligand specificities, and the variations in type usage between animal species (or even strains), changes of affinities with the medium, and other conditions.

The question of the possible existence of further types has been raised. An indication of this comes from the specificity of β-endorphin actions, which does not fit easily into the μ, δ, κ series. A separate, 'epsilon' receptor has been proposed as the site of β-endorphin action, originally on the basis of the high and selective responsiveness of the rat vas deferens to this neuropeptide, differing apparently from the responses of μ- and δ-receptors there [19,20]. Benzomorphans which are κ-agonists elsewhere are antagonists at this putative epsilon site in the rat vas deferens [21]. Autoradiography of [^{3}H]β-endorphin binding in brain also shows a partially different pattern to δ and μ ligand binding [22].

In competition with [^{3}H]EKC, membrane-binding differences have been seen between β-endorphin and some μ-, δ-, and κ-ligands [23]. However, β-endorphin alone does show strong binding (~ 5 nM K_D) activity at μ- and δ-binding sites [24], and the interpretation of the rat vas deferens responses to opioids has been challenged [25]. No selective ligand for the postulated epsilon type has yet been found. In the absence of such, its existence remains an open question.

3. Subtypes of the opioid receptor types

The δ-receptors in the CNS have been proposed to comprise δ_1- and δ_2-subtypes (reviewed in [26], where the original authors and the ligand structures are cited), δ_1-receptors being those where the agonists DPDPE and deltorphin I are much more potent, while δ_2 sites are preferred by DSLET and [D-Ala^2] deltorphin II. Further, cross-tolerance in vivo is lacking between the two types of δ-agonists, suggesting separate receptors. A new antagonist developed by Portoghese and co-workers, 7-benzylidene naltrexone, has likewise been discussed [26] as selective for δ_1-sites, as is also the antagonist [D-Ala^2-Leu^5-Cys^6] enkephalin. However, in the usual peripheral tissue assays these distinctions between δ-receptors are not found and the subtype(s) there may be different.

The μ-receptors have been subdivided by Pasternak and co-workers [27] into μ_1- and μ_2-subtypes; the deduced μ_1-sites bind both morphine or other μ-preferring agonists *and* enkephalins with high affinity (K_d < 1 nM) while the μ_2-sites are the 'classical' μ-receptors and have a significantly lower affinity for enkephalins than for morphine. For example, in calf thalamus μ_2-receptors were deduced to have K_i = 7.2 nM for DADLE and 2.5 nM for morphine [28]. However, these sites differ from the δ-receptor, which has a much lower affinity for morphine and a far higher affinity for DPDPE (reviewed by Pasternak and Wood [27]). The existence of distinct μ_1-sites was also supported by their regional localisation and by their different developmental appearance [26–28]. It must be noted that the μ_1-subtype is apparent at a very low density, even in the richer source of the thalamus [28], and in the rat brain they are only ~ 4% of the total μ-sites, by a careful statistical analysis [30]. This must lower the accuracy of the distinction made. Other authors have reported that the differences in binding that are attributed to μ_1-sites are within the error of the cross-selectivity of the ligands used [31]. Since no ligand highly specific for μ_1-and μ_2-sites has been found so far, it has been pointed out that the subdivision should be regarded as yet as unproven [24]. However, N subtypes of some form are still likely to occur.

Subdivision of the κ-opioid receptors has also been proposed [32–37]. The κ_1-sites differ from the κ_2-sites in their selective binding of the arylacetamide U-69593, whereas the nonselective benzomorphans (EKC, bremazocine) bind to both subtypes in the presence of excess ligands for the μ- and the δ-receptors to suppress the labelling at the latter. Dynorphins also bind to both subtypes with a preference toward the κ_1-subtype. Thus, by its preferential binding of U-69593, the guinea pig cerebellum has mainly κ_1 receptors [34]. By the autoradiography of binding in selective conditions, partially different brain distributions of two subtypes of κ-receptors were found [34,36]. Functional differences have also been used to discriminate κ_1- and κ_2-sites, for example in the opioid-mediated release of corticosterone [37].

The elongated met-enkephalin [Met^5-Arg^6-Phe^7]enkephalin was shown by Castanas et al. [38] to bind apparently at a different site to dynorphin(1-13). Benyhe et al. [39] found the former peptide to be reactive at the abundant κ_2-sites in frog brain.

Another subtype of the κ-receptor (κ_3) has also been proposed, e.g. in calf striatum [40] and in guinea pig spinal cord [41]. This additional subtype, like κ_2, has poor affinity for κ-ligands of the U-69593 type and for the κ-selective antagonist nor-binaltorphimine (norBNI), but the κ_3-subtype retains high affinity for EKC (K_i 1.4 nM), distinguishing it from the κ_2 sites. In this action, however, the κ_3 subtype behaves like an isoform of the μ-receptor.

The identification by competitive binding methods of these various subtypes is not universally accepted. Wood and Traynor [45] showed that the apparent subdivision of κ-sites varied with the competitive binding conditions used and pointed out that such results are equally compatible with a single κ-receptor type which can be changed from a low to a high affinity state. The latter situation has also been deduced to exist for the μ receptor [46]. This interpretation has been specified more fully recently [47]: the nonselective opioid ligand [^{3}H]bremazocine was shown to bind to twice as many sites as *either* other nonspecific opioid ligands such as [^{3}H]EKC *or* the combination of selective μ-, δ- and κ-ligands (Fig. 2). Analysis showed that two affinity states, high and low, for each of those other ligands would explain all of the behaviour seen, if [^{3}H]bremazocine itself binds with about equal affinity to both of the states. The second state was shown to be the G-protein-coupled state of each receptor, since only half of the [^{3}H]bremazocine binding could be inhibited by a guanine nucleotide analogue or by an uncoupling treatment. The same results held for four different tissues which vary greatly in their proportions of the μ-, δ- and κ-types [47]. Bremazocine, therefore, has some properties of an antagonist (binding insensitive to guanine nucleotide) and, indeed, in physiological responses it shows antagonist action at μ sites and agonist action at κ-sites, both peripherally and centrally [48,49].

This explanation would mean that subtypes of the receptors (e.g. κ_1, κ_2) cannot safely be identified by the differences in binding between specific agonist ligands and the non-specific opioid antagonists (naloxone, diprenorphine) or mixed agonist/antagonist ligands (e.g. bremazocine). The ability of the latter classes of ligand, but not the specific agonists, to recognise with high affinity the uncoupled pool of receptors

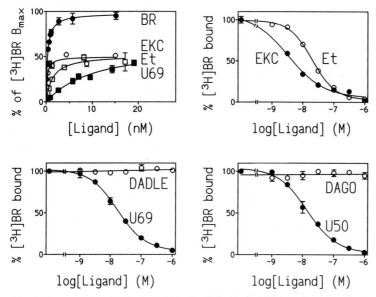

Fig. 2. An example of radioligand-binding behaviour with a single type of opioid site. In this case, in human placental membranes, essentially only the κ-site is present. The results are expressed as a percentage of the [^{3}H]bremazocine control value (in A) or B_{max} value (in B). (Bars or symbol height, ± SEM). (From Richardson et al. [47], by courtesy of the Editors of *Proc. Natl Acad. Sci. USA*). A. Inhibition of [^{3}H]bremazocine binding by the arylacetamides U-50488, U-69593, or PD-117302 (κ-selective agonists), or naloxone (nonspecific antagonist). There is no inhibition by D-Ala^2-D-Leu^4 enkephalin (DADLE) or DAMGO ('DAGO') up to 1000-fold excess concentration, showing that there are no detectable μ- and δ-sites. The [^{3}H]bremazocine concentration was 1.5 nM, but the same results were obtained at 10 nM. B. Saturation curves for [^{3}H]bremazocine (nonspecific mixed agonist-antagonist, ●) and for [^{3}H]ethylketocyclozocine (○) or [^{3}H]etorphine (■) (nonspecific agonists) or [^{3}H]U-69593 (κ-selective agonist). Note how bremazocine occupies nearly twice the number of agonist sites (although all of its binding was shown to be displaceable by naloxone and, at higher ligand concentrations, by these agonists). The two-fold difference is attributed to the G-protein-uncoupled pool of the κ-receptors present [47].

as well as the coupled pool will mimic subtype differentiation by the specific agonists. Evidence for multiple affinity states of opioid receptors due to their G-protein-coupled and -uncoupled pools has been advanced by Werling et al. [50] and for G-protein-linked receptors generally by Jarv et al. [51]. The common experimental paradigm in which two of the types present are blocked by selective ligands, to measure binding at the third, can also produce apparent subtype multiplicity, since the type-selective ligands used are generally agonists. They will, therefore, block only the coupled receptors, leaving the uncoupled receptors of all types (μ, δ, and κ) still available to the universal opioid ligands. Finally, when such paradigms are used in autoradiographic studies some differences in the distributions of apparent subtypes could also arise, since only the most strongly bound ligand fraction survives the tissue-selection autoradiographic procedures and the extent of this will depend upon regional penetrability to ligands. Further, heterogeneity may be mimicked by differences in the local concentrations of endogenous opioid peptides, nucleotides, and cations. This has been investigated in the parallel case of the adenosine A_1 receptor, where it has been shown [52] that in prewashed tissue sections the quantitation of radioligand autoradiography is still compromised by bound endogenous adenosine.

It can be concluded that opioid receptor subtypes cannot be identified with confidence by the methods, reviewed above, by which they have been defined so far. Subtypes are very likely to exist, but they will need to be characterised by molecular cloning: the subtypes to be revealed thus may not be the same as those defined by differential membrane- or tissue-binding studies. A similar situation has been revealed with some other G-protein-linked receptors after their cloning: the cloned subtypes of the muscarinic receptor, m1 to m5, were not those recognised by prior pharmacological distinctions based upon selective binding, nor those of the dopamine receptor, where the 'subtypes' recognised by differential binding in membranes were indeed found to be G-protein-coupled and -uncoupled states of the true isoform [53].

4. Cellular mechanisms of opioid actions

4.1. Interactions with G-proteins

Most nerve cells respond to direct application of opioid drugs by hyperpolarisation, inhibition of cell firing, and, for the presynaptic receptors, an inhibition of neurotransmitter release. As noted above, these effects can be correlated with the observed opening of a K^+ channel or the closing of a Ca^{2+} channel in the neuronal membrane. (The opposite effects on these two channels, seen only at very low opioid concentrations, have also been described in excitatory effects at certain locations by Crain and Shen [8]). The inhibitory effects are produced by each of the μ-, δ- and κ-agonist types on appropriate cells, and although type selectivity for the channel action is often seen (μ and δ for K^+ and κ for Ca^{2+} channels), each opioid receptor type can be linked, in different situations, to either the K^+ channel or the Ca^{2+} channel (which is of the N type) [54–56].

These effects on the membrane channels should be related to a considerable body of evidence that opioid receptor activation leads to the activation of a G-protein. In membrane ligand-binding experiments the binding to the receptor of opioid agonists, but not of antagonists, is greatly weakened by nonhydrolysable analogues of GTP such as Gpp(NH)p or GTPγS. Initially studied with opiates of wide specificity [57,58], this effect was later shown with specific ligands to occur with all three types, μ, δ, and κ [59,60]. This was further specified by means of a specific covalent reaction at G-proteins, using the reagent 5'-p-fluorosulphonylbenzoyl guanosine, which causes a great decrease in the agonist binding by uncoupling the receptor-G-protein complex [61]: this showed that the G-protein linked to the brain μ- and δ-receptors is distinct from that (insensitive to this reaction) linked to another inhibitory neuropeptide receptor, the neurotensin receptor. In a cell line, NG108-15, a neuroblastoma/glyoma hybrid [62] whose membranes contain δ-receptors [62], the pertussis-toxin-catalysed ADP-ribosylation of brain G_i- and G_o-proteins inhibits opioid-stimulated GTPase activity [63]. This toxin sensitivity was also shown directly on μ- and δ-site binding in rat brain membranes [64,65], where the agonist affinity was attenuated and the guanine nucleotide sensitivity was lost. These, with similar results reported at other tissue locations, suggest that a toxin-sensitive G-protein such as a G_i or G_o species is normally coupled to the μ- and δ-receptors.

When the μ receptors purified from rat brain membranes were reconstituted, both G_i (mixed types) and G_o produced partial recovery of agonist binding and agonist-dependent GTPase activity [65]. In intact NG108-15 cells, however, after pertussis-toxin-mediated inactivation, the reconstitution of opioid-activated channel activity primarily required G_i-proteins [56]. GTPase activation was also reconstituted in such cells by G_i plus G_o [66]. It may be that the G-protein specificity is relaxed somewhat when solubilised receptors are tested.

4.2. Coupling to effector systems

There is much evidence that μ- and δ-receptor activation leads to the inhibition of adenylate cyclase (which in other systems is associated with G_i-proteins). This inhibition was first shown for the δ receptors of the NG108-15 cells [62,67] and more recently for μ and δ in rat brain membranes [68,69] and in another neuroblastoma cell line, SH-SY5Y [70], together with pertussis toxin sensitivity of the effect. The μ-opioid-mediated inactivation of the Ca^{2+} current in the latter cells is also pertussis-sensitive and is perpetuated by the presence of GTPγS. However, there is no evidence as yet which shows that the effects of opioid agonists on Ca^{2+} and K^+ channels are a consequence of the decrease in the cAMP level. In fact, the experimental raising of intracellular cAMP does not oppose these channel effects, but, rather, mimics them [55,71]. Hence, it is not clear whether, as in a number of other systems, the Gα-subunit which is activated via the opioid-receptor-agonist complex interacts directly with a channel to gate it or whether the partial inhibition of cAMP formation is secondary to some other second messenger action.

The involvement of κ-receptors in the modulation of adenylate cyclase activity is more controversial. The κ-opioid receptors sometimes do not give evidence for being coupled to adenylate cyclase inhibition, e.g. in embryonic striatal cell cultures, in contrast to the μ- and the δ-receptors there [72], and in general an inhibition of adenylate cyclase by κ-agonists is either present (at ~20–30%) or absent in some tissues or conditions (see also section 7). Attali et al. [73] have reported that κ-opioid agonists (U-50488H, dynorphin) inhibited the depolarization-stimulated *or* cAMP-dependent Ca^{2+} influx in rat spinal cord/dorsal root ganglion co-cultures in a naloxone-reversible, pertussis toxin-sensitive manner. This suggests G_i- or G_o-protein involvement in a transduction pathway in which a κ receptor is coupled negatively to a voltage-dependent Ca^{2+} channel. However, again the raising of intracellular cAMP concentration did not oppose the inhibition of Ca^{2+} current by κ agonists [73]. An associated inhibition of adenylate cyclase (basal *and* forskolin-stimulated) by the κ agonists U-50488 (IC_{50} 100 nM) or MR 2034, both sensitive to the κ antagonist MR 2266, occurred, but reached a maximum of only 20–30% [74]. Likewise, κ agonists close Ca^{2+} channels, of the N-type (in a naloxone-sensitive manner), in mouse dorsal root ganglion neurones, an action mimicked (and not opposed) by cAMP [71]; the same is true in myenteric neurones [5].

Again, therefore, there is no direct linkage that has been shown between the channel effects of κ opioids and the inhibition of adenylate cyclase. Indeed, it has been claimed by Ueda et al. [75] that the κ-selective agonist U-50488H (Fig. 1, Table I) inhibits, very surprisingly, the low K_m GTPase in guinea pig striatum; this effect was antagonised by a κ-opioid antagonist, MR-2266, but not by naloxone, whereas the μ-selective DAGO stimulated it in a naloxone-reversible manner in guinea pig striatum. Activation of κ receptors in guinea pig cerebellar membranes was reported by the same laboratory to produce an inhibition of the GTP-stimulated phospholipase C activity, a parallel unusual effect found also with a very few other receptors, which suggested an alternative signal transduction mechanism for some of the κ receptors [76]. Further discussion of the transduction through κ receptors is given in section 7.

5. *The states of opioid receptors in the membrane*

Fractionation of brain homogenates has shown that the binding sites of the three opioid receptor types are concentrated in the synaptosomal and microsomal centrifugal fractions [77]. The former contains the receptors principally on synaptic and cell body plasma membranes. The 'microsomal' fraction contains intracellular opioid receptors in the endoplasmic reticulum and Golgi zone [78]; it could also contain the receptors from longer dendritic spines, which may become sheared off in the membrane preparation. The intracellular fraction may be in transit to the cell membrane or it may comprise internalised receptors [78].

Binding at opioid receptors shows a monovalent cation inhibition, much more by Na^+ than by K^+, and seen with agonists but not antagonists [10,79]. Na^+ is required for the agonist-mediated adenylate cyclase inhibition. This Na^+ effect is now known, by using selective ligands, to hold for μ- and δ-sites [59,80], but for κ sites, while their univalent cation inhibition remains strong, the selectivity for Na^+ is lost [81]. Na^+ is needed for the guanine nucleotide inhibition of agonist binding, as also is Mg^{2+}, and when buffer effects are excluded the presence of Mg^{2+} stabilises maximal μ- and δ-receptor binding [59]. In these conditions, positively co-operative interactions between μ- and δ- (and between μ and μ) sites can be detected, suggesting that the receptors are assembled into oligomers ([59]; see also section 7). When solubilized in detergent, these oligomers can be separated in complexes with G_o and G_i types of G-proteins, which can be further dissociated to receptor monomers [82,83].

It has been proposed, alternatively, that there is only one basic opioid receptor subunit type, which is interconvertible to different binding isoforms with the addition of ions and nucleotides [84]. However, all of the binding properties and apparent distributions cited in support of this can be explained on the basis of the distinct types with co-operativity in some cases [59]. It cannot be that μ- and δ-types must always be located together, because cell lines are known with only δ receptors (NG108-15) [62] or with a large excess of μ receptors (SH-SY5Y) [70], and likewise with tissues such as the rabbit cerebellum with a very large excess of μ receptors [17]. On the other hand, channels associated with distinct μ- and δ-receptors have been detected on the same neurone [5]. κ-Receptors have a distinct distribution, but in some sites they are co-localised with μ or δ [17]. All of the evidence that we now have supports the idea of separate μ-, δ- and κ-receptors, which in some instances can co-assemble in co-operative hetero-oligomers. Subtypes of these are also likely, but have not been identified clearly yet by binding methods. The new evidence on the separate genes for μ, δ and κ receptors (section 7) finally removes the idea that these types arise by interconvertibility from one protein.

6. Solubilisation and purification of opioid receptors

6.1. Solubilisation

Progress in purification has been considerably slower for the opioid receptors than for many other neurotransmitter and hormone receptors. Several reasons can be suggested for this. Firstly, the opioid receptors have proven very difficult to solubilise, because of the extreme sensitivity of their ligand binding to detergents commonly used to solubilise other cell surface receptors. Secondly, the opioid receptors are generally difficult to characterise in the purified state, their agonist-binding activity being greatly reduced when they become uncoupled from their G-proteins. Reconstitution with both G-proteins and lipids is needed for this to be overcome, and has only recently been achieved. Thirdly, mixtures of the three types, μ, δ, and κ, are commonly present in starting tissues: a few exceptions can be found, e.g. human placenta as a source of only κ receptors [85] and NG108-15 cells for δ receptors [86], but the densities there are low.

The first successful attempt to solubilise active opioid receptors, using Triton X-100, was reported by Bidlack and Abood [87], but in general that detergent has been found to give a low yield of active receptor. Other detergents were tried later and among them the nonionic digitonin and the zwitterionic 3-[(3-cholamidopropyl)-dimethylammonio]-1-propane-sulphonate (CHAPS) are the most effective. Solubilisation of active opioid receptors in digitonin from amphibian, chicken, and mammalian brain has been described by several groups [88–91]. CHAPS was also found to be a useful detergent in solubilisation of the δ-receptors from NG 108-15 cells [86] and subsequently the μ- and κ-receptors from rat brain [91–93]. The opioid receptor activity of the preparation was, however, lower than that of digitonin. In other detergents the yields were lower still. The solubilised receptors in those studies retained stereospecific opioid ligand binding, but usually with decreased agonist affinity relative to the binding of antagonists. The solubilised receptors bind to lectins [89], indicating their glycoprotein nature.

Pretreatment of rat brain membranes with Mg^{2+} prior to the solubilisation and the presence of this cation throughout the digitonin treatment, carried out in TES/KOH buffer in the presence of protease inhibitors, was found to retain both high affinity and capacity of agonist binding to the soluble μ- and δ-receptors [82,90]. The Mg^{2+} ions stabilise the opioid receptor-G-protein complex in the high affinity agonist state, and after solubilisation the μ- and δ-agonist-binding sites retain their positive co-operativity [82] and their full sensitivity to cations, guanine nucleotides, and pertussis toxin, indicating that the opioid receptor in solution still remains functionally coupled to the toxin-sensitive type of G-protein. This interpretation is supported by the demonstration that G_i- and G_o-proteins remain coupled to the μ- and δ-receptors when these are partially purified on a wheat germ agglutinin (WGA) affinity column (giving about 60-fold purification) or by hydrophobic chromatography on phenyl-Sepharose matrix [83].

6.2. Purification

Purification of various opioid receptors has been reported from several laboratories. The common feature of these methods is that all involve affinity chromatography using an opioid ligand immobilised covalently on a support and (usually) eluted with another specific ligand. In some cases this is combined with lectin [89], hydrophobic [83] or hydroxyapatite chromatography to achieve complete purification. By the earlier methods of this type, 500–2000-fold purification was achieved (Table II). In some cases, addition of lipids was made to restore full agonist and stereospecific ligand binding. For a complete purification, if the receptor is a polypeptide of M_r 60,000, and with the usual levels of specific activity in the starting detergent extract, ~ 60,000-fold purification is required.

More recently, purification of the μ-, δ-, and κ-opioid-binding proteins to apparent homogeneity has been achieved (Table II). Usually the high specific activities noted are for the binding of [^{3}H]bremazocine or [^{3}H]naloxone, i.e. not requiring G-protein. Lower affinity binding of specific agonists could also be shown in some cases.

Thus, E.J. Simon's laboratory [98] reported a complete purification of the μ-receptor in digitonin/0.5 M NaCl medium; the product could be covalently cross-linked to ^{125}I-labelled β-endorphin and give 70% displacement of the [^{3}H]bremazocine binding present by 500 nM DAMGO. The μ receptor was also purified by H. Loh's labora-

TABLE II
Affinity purification of opioid-binding proteins from brain tissue

Tissue[a]	Receptor	Coupled ligand	M_r (Kilodaltons, K) of subunits present	Yield (%)	Purification[b] (fold, approx.)	References
Rat	μ	BAM	43 K, 35 K, 23 K	ns	2000	Bidlack et al. [94]
Rat	μ or δ	DALE	62 K	ns	450	Fujioka et al. [95]
Bovine	μ	Hybromet	94 K	0.6	500	Maneckjee et al. [96]
NG108-15	δ	AntiFIT	58 K	3	30,000	Simonds et al. [97]
Bovine	μ	β-naltrexethylene diamine	65 K	5.8	65,000–75,000	Gioannini et al. [98]
Rat	μ	6-Succinylmorphine	58 K	6	68,000	Cho et al. [99]
Rat	μ	DALECK	62 K, 54 K	10	20,000–30,000	Barnard et al. [100]
Rat	μ	GANC	62 K	8	60,000–70,000	Barnard et al. [100], Demoliou-Mason and Barnard [101]
Rat	μ	6-Succinylmorphine	58 K	ns	60,000[c]	Ueda et al. [65]
Frog	κ	DALE	65 K, 58 K	0.07	4300	Simon, J. et al. [102]
Frog	κ	Dynorphin(1-10)	65 K	8	19,600[c]	Simon, J. et al. [103]
Guinea pig	κ	[Ala^{11}]dynorphin(1-11)	62 K	6	15,000–20,000[c]	Simon, J. et al. [104]

AntiFIT: antibody to the fentanyl group; BAM: 14-β-bromo-acetamido morphine; DALE: [D[Ala^2,L-Leu^5]enkephalin; DALECK: DALE-chloromethyl ketone; GANC: 14-β[(-glycyl)-amido(N-cyclopropymethyl)] norcodeinone; Hybromet: 7α-(1R)-hydroxy-1-methyl-3-p[4-(3'bromomercury-2'-methoxy-propoxy)phenyl]-propyl-6,14-endoethenotetrahydro-thebaine; ns, yield not stated.
[a]All are brain membranes, except the NG108-15 cultured cell line.
[b]Fold purification (relative to the starting extract) and yield are obtained after removal of ligand, and in some cases increased purification, by additional steps (e.g. lectin or hydroxyapatite chromatography or gel filtration).
[c]Fold purification is given, when the receptors are reconstituted into lipid micelles, with or without G-proteins (see text).

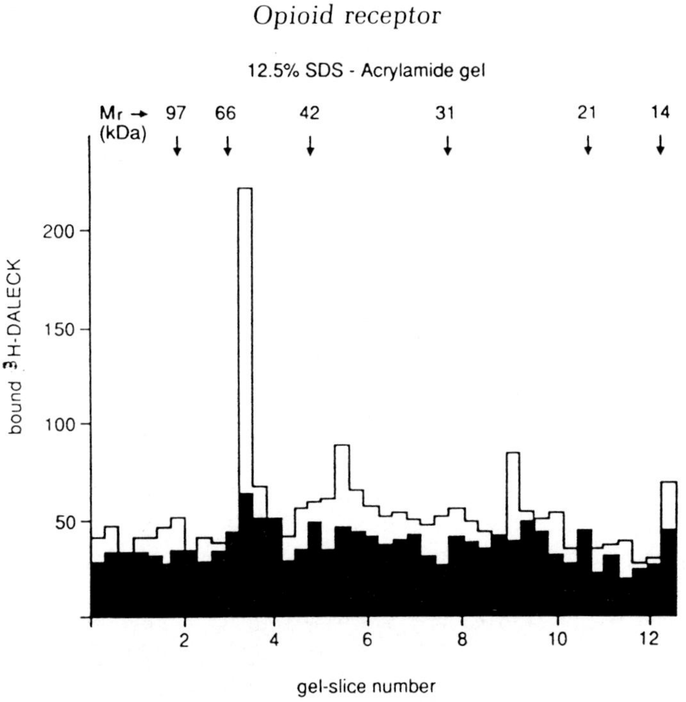

Fig. 3. Identification by affinity labelling of the purified μ-receptor subunit. The receptor was purified from rat brain on the GANC affinity column (see Table II) to yield a single polypeptide of 62,000 M_r, as detectable in Coomassie Blue staining after SDS-gel electrophoresis. A sample (reconstituted in liposomes) reacted (1 hour at 25° C, pH 8.0) with the alkylating μ-specific reagent [111,112] [^{3}H]DALECK (see Section 6.3). Unreacted [^{3}H]DALECK was then inactivated with 1 mM DTT; after denaturation with SDS, electrophoresis was on an SDS/12.5% acrylamide gel. Slices were cut, dissolved, and counted. Nonspecific labelling was measured by a parallel reaction of the protein in the presence of 10^{-6} M unlabelled DALECK and similar processing in a parallel lane. The profiles represent, after total and (black) nonspecific reaction, amounts of [^{3}H] recovered/slice. Arrows indicate the elution profile of parallel molecular weight standards. kDa = kilodaltons. The minor peaks are random noise, since they did not occur at constant positions in repeat experiments, unlike the major sharp peak at 62 kDa. This labelling is prevented by excess DAMGO present during the alkylation but not by excess DSLET or U-50488, showing that it is a μ-receptor subunit of apparent M_r 62 kDa which has been isolated.

tory [99] in Triton X-100 solution, involving removal of the detergent and affinity column elution by NaCl only and requiring the addition of acidic lipids for any binding activity. The receptor was identified by chemical cross-linking of [^{3}H]dihydromorphine or [^{125}I]β-endorphin. Simultaneously, Barnard et al. [100] and Demoliou-Mason and Barnard [101] obtained the complete purification of a μ-receptor in 0.1% digitonin/Mg^{2+} on a column carrying a codeinone derivative [105], followed by WGA lectin-affinity chromatography. Although the purified (62 kD) material exhibited high affinity binding to antagonists, only low affinity μ-agonist binding was present (without reconstitution), attributed to the absence of G-proteins in the pure receptor. The μ-specific peptide alkylating agent DALECK (Section 6.3) reacted with the 62 kD protein that was purified, with competition by μ- (but not δ or κ) ligands (Fig. 3).

Ueda et al. [65] used a combination of a 6-succinylmorphine affinity chromatography [98] (in CHAPS/0.1 mM dithiothreitol] and isoelectric chromatography to obtain a purified μ-receptor preparation. When this preparation was reconstituted in liposomes with purified G-proteins, G_i or G_o, high affinity displacement of [^{3}H]naloxone binding by μ-selective opioid agonists was observed, and was abolished by adding guanine nucleotide. μ-Receptor agonists, but not δ- and κ-agonists, also stimulated the binding of [^{3}H]GppNHp and the low K_m GTPase in the reconstituted preparation, suggesting that the purified μ-receptor protein is functionally coupled to G_i and/or G_o in the reconstituted phospholipid vesicles.

The δ-opioid receptor was purified to apparent homogeneity from NG108-15 cells by Klee's laboratory [97]. First, the δ-receptors on NG108-15 cells were covalently labelled with [^{3}H]methylfenantyl-isothiocyanate ([^{3}H]FIT), a selective covalent label for the δ-sites, which served as a marker to follow the receptors in the subsequent purification steps. The [^{3}H]FIT irreversibly bound to a 58,000-dalton polypeptide. This labelled protein was extracted from the cell membranes with a mixture of Lubrol and CHAPS detergents and was purified in four consecutive steps, using WGA-lectin, immunoaffinity chromatography using an antibody to the labelling group (FIT), followed by preparative SDS-polyacrylamide gel electrophoresis and electroelution. Of course, this approach yields the receptor protein in a permanently inactivated form, but is suitable for obtaining peptide sequences.

For the purification of the κ-receptors, the source is important, since usually κ sites are a minority among μ- and δ-sites. The human placenta, as an exception, has > 95% of its opioid binding in the κ category [85,105] (Fig. 2). However, to date it has not been a source for κ receptor purification, because the density of κ-binding sites in placental tissue is very low (mean B_{max}:< 100 fmol/mg protein) and it varies greatly between individuals [85,105], and because of the high levels of many proteolytic enzymes in this tissue. Other sources with a majority of κ receptors are frog brain membranes (B_{max} for [^{3}H]EKC: 0.7 pmol/mg protein) and membranes from guinea pig cerebellum. In the latter source, competition with [^{3}H]bremazocine binding over a very wide concentration range of competing site-selective ligands has shown that there are in the opioid receptor population 50% κ-sites, 30% δ sites and 20% μ sites [47]. Guinea pig cortex or whole brain, although having only ~ 30% of the opioid sites as κ, can be used successfully for larger scale purification [104].

The purification of the κ-receptor was first reported in 1987 by J. Simon et al. [102] from frog brain membranes. An affinity column consisting of D-Ala^2-Leu-enkephalin (DALE) coupled to Sepharose 6B was used to partially purify opioid receptors from digitonin-solubilised membrane preparations, followed by a further purification by size using gel filtration. This method was improved to obtain a highly purified receptor, using a more κ-selective affinity column prepared by coupling dynorphin(1-

10) to AH-Sepharose 6B. The purified receptor bound [^{3}H]EKC to close to the theoretical value. Gel analysis revealed a single band of apparent M_r 65,000 [103].

The κ-receptors have been purified from mammalian brain by the Barnard laboratory [104]. The κ-receptor protein was purified from the digitonin extract of guinea pig brain membranes on a column containing another immobilised derivative of dynorphin ([Ala^{11}]dynorphin(1-11). The extent of the purification was approx. 20,000-fold, with high affinity binding of [^{3}H]bremazocine but not agonists. When reconstituted with purified G_i- and G_o-proteins into phospholipid vesicles, however, the preparation also bound the κ-agonist ligands dynorphin and U69,593 with high affinity (but not μ- or δ-ligands). It contains a single polypeptide of apparent M_r 62,000.

6.3. Affinity labelling of opioid receptors

By the synthesis of opiates and opioid peptides having an alkylating or photoreactive group attached to their structure, affinity labelling of the receptor at its ligand-binding site can in principle be achieved. This approach with radioactive ligands has been of value biochemically: (a) in identifying the receptor subunit in intact membranes or crude extracts; (b) in confirming the subunit identification in a purified receptor preparation; and (c) with application of an antibody to the attached ligand for immunopurification as described above for FIT [97] (which labels a 58,000-dalton protein in NG108-15 cell membranes, which contain only δ-receptors). Cross-linking of a peptide ligand by an added bifunctional reagent has also been used in this way, but offers more possibilities of additional labelling to other proteins. These methods as applied to opioid receptors are reviewed, with practical details, by Demoliou-Mason and Barnard [101].

The technique of chemically cross-linking opioid peptides to their receptors was first applied by Zukin and Kream [106] using [D-Ala^2-D-Leu^5]-enkephalinamide as the ligand and dimethyl suberimidate (DMS) as a cross-linker. The peptide cross-linked to a protein of 380,000 daltons under nondenaturing conditions and of 35,000 daltons in SDS/gel electrophoresis.

Howard et al. [107] cross-linked [^{125}I]-β-endorphin to opioid receptors of rat brain membranes. The specific cross-linking of this ligand revealed four bands (65,000, 53,000 (major), 38,000, and 25,000 daltons). On the basis of protection by selective ligands, they deduced that the 65,000-dalton band contains the μ- and the 53,000-dalton band the δ-receptor protein. Only the 53,000-dalton band was observed in cross-linked membranes from NG 108-15 cells.

A photo-reactive derivative of the δ-selective ligand DTLET, Tyr-D-Thr-Gly-Phe(pN$_3$)-Leu-Thr was used in radioactive form [108,109]. In rat brain membranes, gel analysis showed the covalently bound radioactivity to be specifically incorporated into polypeptides of apparent M_r 40,000–50,000.

In 1981 Venn and Barnard [110] synthesized an alkylating enkephalin analogue,

Tyr-D-Ala-Gly-Phe-LeuCH$_2$Cl (DALECK). This ligand was shown capable of binding both reversibly and (at pH 8 or above) irreversibly to block opioid receptors in membranes. This reagent was subsequently applied by Newman and Barnard [111] in tritiated form. It was demonstrated thus that the irreversible binding at pH 8 is selective for the μ-receptors. This is to be expected since removal of the terminal carboxyl from enkephalin in general makes them much more selective for μ-sites. SDS/gel electrophoresis performed under reducing conditions revealed that the alkylating reagent has become incorporated into a single protein species of apparent M_r 58,000 [111]. In the affinity labelling of this polypeptide (M_r 58,000) from rat brain membranes [112], the μ specificity was confirmed by showing that with 2 nM [^{3}H]reagent, DAMGO protected with a K_i of 10 nM, whereas δ- and κ-selective ligands yielded very high K_i values (1200–1400 nM). [^{3}H]DALECK reacted likewise with a single subunit of similar size in the completely purified μ receptor from rat brain. This is shown in Figure 3, where this type of analysis is illustrated.

Irreversible, specific affinity reagents can also be used to occlude, in vitro and in vivo, specific populations of the opioid receptors, both for probing drug specificity and for spare receptor studies. An example is β-funaltrexamine, an alkylating naloxone derivative, which blocks μ receptors [113], but with incomplete selectivity. DALECK can also produce a μ-receptor occlusion in vivo [110]. Using a test system of four isolated tissue assays, it was shown that DALECK exerts a very strong μ-agonist potency in vivo [114]. Thus, in the guinea pig ileum, against naloxone the Ke value for DALECK was 2.0 ± 0.37 nM and the agonist potency was 1.7 times that of DAMGO. In the mouse vas deferens the potency was 70 times that of DTLET, whereas the δ-antagonist ICI 174804 showed a Ke of 305 nM against DALECK.

The range of subunit sizes found for μ- and δ-receptors by affinity labelling is likely to be a reflection of experimental problems in such studies, rather than true biological variation. Proteolysis is the most probable cause of the lower molecular weights seen in some studies. Cross-linking of subunits to dimers or higher polymers can also occur as an artefact in gel analysis. The consistent labelling of a polypeptide of M_r 58,000–62,000, under conditions where special precautions are taken to prevent those artefacts, is noteworthy. This agrees also with the sizes of 58,000–65,000 daltons found in the highest purifications (Table II). The variation within that range is understandable, since this varies with the extent of disulphide reduction before gel analysis. High concentrations of, for example, dithiothreitol are needed for the full effect and the apparent M_r here varies with its concentration, as has been observed by Gioannini et al. [98] and confirmed in other preparations [100,104,112]. This latter anomaly is characteristic of hydrophobic polypeptides with strong internal disulphide bonds.

In summary, the evidence suggests that subunits of ~60,000 daltons occur in the μ-, δ-, and κ-receptors. This will not be the true protein molecular weight, due to the considerable weight of attached carbohydrate and to the anomalies of migration of strongly hydrophobic polypeptides in SDS/polyacrylamide gel electrophoresis. This observed size is, in fact, common to many of the G-protein-linked receptors.

7. Molecular biology of opioid receptors

The evidence reviewed above (Sections 4 and 5) shows that opioid receptors can be expected to be members of the G-protein-linked class of membrane receptors. More than a hundred types of these are known, including those for the other neuropeptides where investigated. Molecular cloning has revealed them to comprise each one type of subunit, containing seven hydrophobic, presumably membrane-spanning α-helices. While the great majority are in one superfamily, with the homology between them varying from 25% to 60%, at present two other superfamilies, each with a few of these receptors, are known, with no sequence homology shared between them (reviewed in [115]). Nevertheless, they all have the seven-transmembrane-domain hydrophobicity pattern.

It is, therefore, difficult to predict which known G-protein-linked receptors will share sufficient sequence identity with an opioid receptor to permit cloning by crosshybridisation from the cDNA sequence of the former. Even if only receptors linked to G_i- or G_o-types, or neuropeptide receptors, are taken, attempts to perform such cross-hybridisation in several laboratories have been unsuccessful so far.

The most direct route to the opioid receptor cDNA cloning is to obtain some peptide sequences (to construct hybridisation screening probes) from the proteins purified as noted in Table II. However, the yields are small and the proteins are very hydrophobic, and progress by this route has been difficult. One protein purified as an opioid receptor, that of Cho et al. [99] (Table II), led to positive clones by this route. This led to the cloning by Schofield et al. [116] of a cDNA encoding an extracellularly located glycoprotein of 345 amino acids. However, this protein did not possess any membrane-spanning domain, but contained a carboxy-terminal sequence characteristic of membrane attachment through phosphatidylinositol linkage. This protein shows no homologies to any known G-protein-coupled receptors. It displays homology to several members of the immunoglobulin superfamily, and especially to the neural cell adhesion molecule (NCAM). The authors concluded, based on these sequence homologies, that this protein (opioid-binding protein-cell adhesion molecule: OBCAM) could play a role in either cell recognition and adhesion, peptidergic ligand binding, or both. They were not able to express this protein, neither in mammalian cell lines nor in oocytes.

Xie et al. more recently obtained a clone by an expression cloning route aimed at finding a κ receptor cDNA [117]. They used a cultured mammalian cell line transfected with cDNAs derived from human placental mRNA and screened by a 'panning' ligand-binding procedure. The cloned cDNA encodes a 440-residue protein of the seven-transmembrane-domain receptor family. However, this has 93% identity to a known receptor, the human neurokinin B receptor and does not show in expression studies κ selectivity in the opioid binding.

Another version of the transfection/ligand-binding expression cloning approach has recently been successful in two laboratories. Evans et al. [118] and Kieffer et al.

[119] simultaneously described a δ receptor clone, each starting from the same source, the neuroblastoma-glioma cell line NG108-15. The latter is known to express only the δ type in its native opioid receptors. Both groups employed radiolabelled enkephalin analogues for screening of COS cells transfected with pools of cDNAs derived from a library. The two groups obtained the same protein sequence of 372 residues, with 7 hydrophobic segments at positions consistent with those in the rest of the superfamily (Fig. 4). The ligand affinities of the protein expressed in COS cells are essentially the same as those for their binding to the δ receptor on the NG108-15 cells, while μ and κ selective ligands had low activity.

A cDNA encoding a κ receptor has been cloned from mouse brain by Yasuda et al. [120]. They had cloned, using PCR primers based on the somatostatin receptor sequence, two related 'orphan receptor' cDNAs, which were recognised from the afore-

```
mouse δ Opioid Receptor    MELVPSARAELQSS PL                         VNLSDAFPS       25
rat    μ Opioid Receptor   MDSSTGPGNTSDCSD-LA QASCS PAPGSWLNLSHVDGNQSDPC             43
mouse  κ Opioid Receptor   MESPIQIFRGDPC  -TCSPSACLLPWSSSWFP                         31

                                                    TM 1
DOR    APPSAGANASGSPGARSAS SLAL  AIAITALYSAVCAVGLLGNVLVMFGIVRYTKLK                   81
MOR    GLNRTGLGGNDSLCPQTGSPSMVT  AIT-M-L--I-CV---F--F---YV-V----M-                  100
KOR    NWAESDENGSVGSEDQQLESAHISPAIPVI-T-V--V-FV---V--S----FV-I----M-                 91

           TM 2                                TM 3
DOR    TATNIYIFNLALADALATSTLPFQSAKYLMETWPFGELLCKAVLSIDYYNMFTSIFTLTM                  141
MOR    -----------------A-S-L----VN---GT----TI---I-I--------------CT                 160
KOR    -----------------V-T-M----AV---NS----DV---I-I--------------TM                 151

                                         TM 4
DOR    MSVDRYIAVCHPVKALDFRTPAKAKLINICIWVLASGVGVPIMVMAVTQPRDGA    VVCM                199
MOR    --------------------RN--IV-V-N-I-S-AI-LPVMFMAT-KY-QGS     ID-T                218
KOR    -------------------LK--II-I-I-L-A-SV-ISAIVLGG-KV-EDVDVIE-S                    211

                                    TM 5
DOR    LQFPSP  SWYWDTVTKICVFLFAFVVPILIITVCYGLMLLRLRSVRLLSGSKEKDRSLR                  257
MOR    -T-SHP  T-Y-ENLL-----I---IM-I---T---G--I---K---M----K----N--                  276
KOR    -Q-PDDEYS- -DLFM-----V---VI-V---I---T---I---K---L----R----N--                 270

                 TM 6                            TM 7
DOR    RITRMVLVVVGAFVVCWAPIHIFVIVWTLVDINRRDPLVVAALHLCIALGYANSSLNPVL                  317
MOR    ---RM-----AV-IV--T----YVIIKA-ITI PETTFQTVSWHF------T--C-----                  335
KOR    ---KL-----AV-II--T----FILVEA-CSTSHSTA ALSSYYF------T--S-----                  329

DOR    YAFLDENFKRCFRQLCRTPCGRQEPGSLRRPRQATTRERVTACTPSDGPGGGAAA                       368
MOR    -------------EF-IPTSSTI-QQNST-V-Q -NREHPSTANTVDRTNHQLENLEAET                  394
KOR    -------------DF-FPIKMRM-RQSTN-V-N -VQDPASMRDVGGMNKPV                          377

MOR    APLP                                                                          398
```

Fig. 4. A comparison of the protein sequences of the mouse δ, rat μ and mouse κ opioid receptors. The one-letter protein code is used. The alignments were made to maximise homology. The seven hydrophobic domains are overlined.

mentioned δ receptor sequence to encode opioid receptors. One of these was the same δ receptor and the other (380 residues) was found (in transfected COS cells) to bind κ ligands and not μ or δ ligands. The affinities are very strong, with e.g. IC_{50} values (using 1 nM [^{3}H]U-69593) of 1 nM for U-50488 or norBNI and 0.4 nM for dynorphin A.

A μ opioid receptor sequence (from rat brain) has also been obtained by Chen et al. [121], by cross-hybridisation based on the mouse δ sequence. On expression in COS cells it showed the expected selectivity of μ-ligand binding. The rodent δ, κ and μ receptors share 60–64% amino acid identity. This is concentrated principally in the seven transmembrane (TM) domains (Fig. 4) (with TM4 less conserved).

With the usually-assumed topology of the G-protein-coupled receptors (with the N-terminus extracellular), the second and third cytoplasmic loops are also very strongly conserved, suggesting that these regions are involved in processes of intracellular signalling and regulation common to the different types of opioid receptors.

It is interesting to find that the opioid receptors show no particular homology to receptors for other neuropeptides, except for somatostatin (where the identity is ~ 35%). The next nearest receptor sequence is that for angiotensin (31%) and the chemotactic N-formyl peptide and interleukin-8 (21%). The sequences fall into the main superfamily of the three G-protein-linked receptor superfamilies [115], but are in a separate branch of it, containing those just mentioned. A dendrogram showing the relationship to some other branches is given by Kieffer et al. [119]. The sequences of the two proteins obtained in previous attempts at opioid receptor DNA cloning, as discussed above [116,117], are very far removed from those of these true opioid receptors. OBCAM [116] can be considered as an unrelated protein which was present in a partly-purified opioid receptor preparation. The identity of the neurokinin-receptor-like protein of Xie et al. [117] is at present unknown.

The opioid receptor sequences are close to the shortest known among the G-protein-linked receptors, and their hydrophobic character is particularly high. Several cysteines occur which either, by homology, are likely to form an internal disulphide bond [118] or may form inter-subunit bonds. These features can explain the difficulties experienced (see sections 6.2 and 6.3) in purifying the opioid receptors and also in reducing them, and likewise in obtaining peptides from them amenable to screening-probe design. The sequences of the δ, μ and κ types are such that they cannot arise from alternative splicing of a common precursor mRNA and must be from independent genes. The evidence does not exclude the possibility [59] of the co-assembly of different subunit types at certain of the receptor locations: this was discussed in section 5, and also has been proposed for a μ-δ assembly associated with striatal dopamine receptors [122]. That complex is discriminated by the much lower effect on it of the selective antagonists (Table I) CTOP and naltrindole. Cross-modulation of μ receptors by certain agonists or antagonists highly selective for δ receptors at various locations has been found, to support this idea of 'δ-complexed μ receptors', as reviewed by Traynor and Elliot [26]. Such co-assemblies might involve cross-linking through certain of the cysteines present.

The molecular weights of the three polypeptides so far known are all about 42,000 daltons. There are several putative extracellular glycosylation sites and the sizes found from protein purification and affinity labelling (section 6) when due precautions are taken, of 53,000–62,000 daltons, are as expected for a usual level of glycosylation of such polypeptides. Large size differences specific for the three types are not predicted, and the values below 40,000 in the literature from gel analyses must arise from proteolysis.

It is of particular interest, in view of the open questions on the transduction routes of opioid receptors (section 4.2), to learn the transduction properties seen with the subtypes of recombinant opioid receptors. When the δ receptor was transiently expressed by transfection into COS cells, a δ-specific agonist produced a strong, naloxone-sensitive inhibition of adenylate cyclase (which had been stimulated by forskolin) [118,120]. The same effect, but to a lesser degree (18% inhibition, at 100 nM DAMGO) was found with the similarly-expressed μ receptor [121]. Interestingly, on the κ receptor in the same expression system [120], the selective agonist U-50448 also gave inhibition (50% at 1 μM). The results with the μ and δ receptors agree with those found on those in native systems (section 4.2).

With the κ receptor, the adenylate cyclase inhibition now seen parallels that found in some biochemical studies (section 4.2), e.g. on spinal cord/dorsal root ganglion cells [74]. In the latter, it has recently been shown [123] that the κ-agonist-induced decrease in cAMP leads to a block of the phosphorylation of the synaptic vesicle protein synapsin I, which could underlie the opioid inhibition of transmitter release, i.e. at presynaptic κ receptors. Adenylate cyclase inhibition due to κ receptor stimulation has also been found at some sites in the brain, as in the κ-receptor-rich [17] guinea pig cerebellum, in some conditions [124] but not in others [76,125]. However, there is also evidence for κ receptors coupling to Ca^{2+} channels via a pathway independent of cAMP decrease (section 4.2; see also [126]). Interestingly, in embryonic rat brain cells cultured for 21 days, a specific κ-agonist stimulation of phosphoinositide formation was found, but at 7 days in culture this produced a strong inhibition thereof [127], suggesting that native κ receptor subtypes differ in their transductions. In slices from some regions of the adult rat brain, a stimulation of phosphoinositide turnover by κ agonists was found [128]. It cannot be assumed that the transduction seen in the artificial system of transfected kidney cells, which will depend upon the effectors available therein, is necessarily the same as that for native κ receptors in all locations. Differences in both local effectors and in receptor subtypes may give diverse transduction behaviour at different locations.

8. Subtypes of the opioid receptor types at the molecular level

Certain of the ligand binding differences which have been used previously to define subtypes of μ, δ and κ receptors may be truly recognising different subunit sequences, but this has to be carefully distinguished from the apparent differences due to the pools of G-protein-coupled and -uncoupled receptors [47], as discussed in section 3. The availability of the three types of clone will now permit the true definition of the subtypes. There is already evidence that some cross-hybridise to related sequences. As these are cloned and expressed, and their *in situ* distributions are compared, we can look forward to the recognition of a set of defined subtypes for the μ, δ and κ types, and perhaps others capable of being revealed only by the high resolving power of molecular biology.

References

1. Werling, L.L., Frattali, A., Porthogese, P.S., Takemori, A.E. and Cox, B.M. (1988) J. Pharmacol. Exp. Ther. 246, 282–286.
2. Leeman, S.E. (1980) Nature 286, 155–157.
3. Howlett, T.A. and Rees, L.H. (1986) Annu. Rev. Physiol. 48, 527–537.
4. Milanés, M.V., Gonzalvez, M.L., Fuente, T. and Vargas, M.L. (1991) Neuropeptides 20, 95–102.
5. Cherubini, E. and North, R.A. (1985) Proc. Natl Acad. Sci., USA 82, 1860–1863.
6. Gross, R.A. and McDonald, R.L. (1987) Proc. Natl Acad. Sci. USA 84, 5469–5473.
7. North, A.R., Williams, J.T., Surprenant, A. and Christie, M.J. (1987) Proc. Natl Acad. Sci. USA 84, 5487–5491.
8. Crain, S.M. and Shen, K.F. (1990) Trends Pharmacol. Sci. 11, 77–81.
9. Pert, C.B. and Snyder, S.H. (1973) Science 179, 1011–1014.
10. Simon, E.J., Hiller, J.M. and Edelman, I. (1973) Proc. Natl Acad. Sci. USA. 70, 1947–1949.
11. Terenius, L. (1973) Acta Pharmacol. Toxicol. 32, 317–320.
12. Martin, W.R., Eades, C.G., Thompson, J.A., Huper, R.E. and Gilbert, P.E. (1976) J. Pharmacol. Exp. Ther. 197, 517–532.
13. Hughes, J., Smith, T.W., Kosterlitz, H.W., Fothergill, L.A., Morgan, B.A. and Morris, H.R. (1975) Nature 258, 555–579.
14. Chavkin, C., James, I.F. and Goldstein, A. (1982) Science 215, 413–415.
15. Mansour, A., Khachaturian, H., Lewis, M.E., Akil, H. and Watson, S.J. (1988) Trends Neurosci. 11, 308–314.
16. Lord, J.A.H., Waterfield, A.A., Hughes, J. and Kosterlitz, H.W. (1977) Nature 267, 495–499.
17. Corbett, A.D., Patterson, S.J. and Kosterlitz, H.W. (1991) In: Handbook of Experimental Pharmacology (Herz, A., Akil, H. and Simon, E.J., Eds), Vol. Opioids, Springer-Verlag, Heidelberg.
18. Weitz, C.J., Faull, K.F. and Goldstein, A. (1987) Nature 330, 674–677.
19. Garzen, J., Schulz, R. and Herz, A. (1985) Mol. Pharmacol. 28, 1–9.
20. Schulz, R., Faase, E., Wüster, M. and Herz, A. (1979) Life Sci. 24, 843–850.
21. Gillan, M.G.C., Kosterlitz, H.W. and Magnan, J. (1981) Br. J. Pharmacol. 72, 13–15.
22. Goodman, R.R., Houghton, R.A. and Pasternak, G.W. (1983) Brain Res. 288, 334–337.
23. Nock, B., Giordano, A.L., Cicero, T.J. and O'Connor, L.H. (1990) J. Pharmacol. Exp. Ther. 254, 412–419.
24. Goldstein, A. and Naidu, A. (1989) Mol. Pharmacol. 36, 265.
25. Sheehan, M.J., Hayes, A.G. and Tyers, M.B. (1988) Eur. J. Pharmacol. 154, 237–245.
26. Traynor, J.R. and Elliott, J. (1993) Trends Pharmacol. Sci. 14, 84–86.
27. Pasternak, G.W. and Wood, O. (1986) Life Sci. 38, 1889–1898.
28. Clark, J.A., Haighten, R. and Pasternak, G.W. (1988) Mol. Pharmacol. 34, 308–317.
29. Zhang, A.Z. and Pasternak, G.W. (1981) Eur. J. Pharmacol. 73, 29–40.

30. Lutz, R.A., Cruciani, R.A., Munson, P.J. and Rodbard, D. (1986) Life Sci. 36, 2233–2238.
31. Sarne, Y. and Kenner, A. (1987) Life Sci. 41, 555–562.
32. Pfeiffer, A., Pasi, A., Mehraein, P. and Herz, A. (1981) Neuropeptides 2, 89–97.
33. Attali, B., Guarderes, C., Mazurguil, H., Audigier, Y. and Cros, J. (1982) Neuropeptides 3, 53–64.
34. Zukin, R.S., Eghbali, M., Olive, D., Unterwald, E.M. and Tempel, A. (1988) Proc. Natl Acad. Sci. USA 85, 4061–4065.
35. Tiberi, M. and Magnan, J. (1990) Eur. J. Pharmacol. 188, 379–389.
36. Nock, B., Rajpara, A., O'Connor, L.H. and Cicero, T.J. (1988) Eur. J. Pharmacol. 154, 27–34.
37. Iyengar, S., Kim, H.S. and Wood, P.L. (1986) Life Sci. 39, 637–644.
38. Castanas, E., Giraud, P., Bourhuim, N., Cantau, P. and Olivier, C. (1984) Neuropeptides 5, 133–136.
39. Benyhe, S., Varga, E., Hepp, J., Magyar, A., Borsodi, A. and Wollemann, M. (1990) Neurochem. Res. 15, 894–904.
40. Clark, J.A., Lin, L., Price, M., Hersh, B., Edelson, M. and Pasternak, G.W. (1989) J. Pharmacol. Exp. Ther. 251, 461–468.
41. Tiberi, M. and Magnan, J. (1990) Mol. Pharmacol. 37, 694–703.
42. Delay-Goyet, P., Seguin, C., Gacel, G. and Roques, B.P. (1988) J. Biol. Chem. 263, 4124–4130.
43. Portoghese, P.S., Sultana, M. and Takemori, A.E. (1988) Eur. J. Pharmacol. 146, 185–186.
44. Hunter, J.C., Leighton, G.E., Meecham, K.G., Boyle, S., Horwell, D.C., Rees, D.C. and Hughes, J. (1990) Br. J. Pharmacol. 101, 183–189.
45. Wood, M.S. and Traynor, J.R. (1989) J. Neurochem. 53, 173–178.
46. Carroll, J.A., Shaw, J.W. and Wickenden, A.D. (1988) Br. J. Pharmacol. 94, 625–631.
47. Richardson, A., Demoliou-Mason, C. and Barnard, E.A. (1992) Proc. Natl Acad. Sci. USA 89, 10198–10202.
48. Corbett, A.D. and Kosterlitz, H.W. (1986) Br. J. Pharmacol. 89, 245–249.
49. Morris, B.J. and Herz, A. (1989) Neuroscience 29, 433–442.
50. Werling, L.L., Puttfarcken, P.S. and Cox, B.M. (1988) Mol. Pharmacol. 33, 423–431.
51. Jarv, J., Hedlund, B. and Bartfai, T. (1979) J. Biol. Chem. 254, 5595–5598.
52. Parkinson, F.E. and Fredholm, B.B. (1992) J. Neurochem. 58, 941–949.
53. Leff, S.E. and Creese, I. (1985) Mol. Pharmacol. 27, 184–192.
54. Surprenant, A., Shen, K.-Z., North, R.A. and Tatsumi, H. (1990) J. Physiol. 431, 585–608.
55. Seward, E., Hammond, C. and Henderson, G. (1991) Proc. R. Soc. B. 244, 129–135.
56. Hescheler, J., Rosenthal, W., Trautwein, W. and Schultz, G. (1987) Nature 325, 445–447.
57. Blume, A.J. (1978) Proc. Natl. Acad. Sci. USA 75, 1713–1717.
58. Childers, S.R. and Snyder, S.H. (1980) Mol. Pharmacol. 16, 69–76.
59. Demoliou-Mason, C.D. and Barnard, E.A. (1986) J. Neurochem. 46, 1118–1128.
60. Francés, B., Moisand, C. and Meunier, J-C. (1985) Eur. J. Pharmacol. 117, 223–232.
61. Wong, Y.H., Demoliou-Mason, C.D., Hanley, M.R. and Barnard, E.A. (1990) J. Neurochem. 54, 39–45.
62. Klee, W.A. and Nirenberg, M. (1974) Proc. Natl Acad. Sci. USA 71, 3474–3477.
63. Burns, D.L., Hewlett, E.A., Moss, J. and Vaughan, M. (1983) J. Biol. Chem. 258, 1435–1438.
64. Wong, Y.H., Demoliou-Mason, C.D. and Barnard, E.A. (1988) J. Neurochem. 51, 114–121.
65. Ueda, H., Harada, H., Nozaki, H., Kadata, T., Ui, M., Satoh, M. and Tagaki, H. (1988) Proc. Natl Acad. Sci. USA 83, 7013–7017.
66. Costa, T., Lang, J., Gless, C. and Herz, A. (1990) Mol. Pharmacol. 37, 383–394.
67. Koski, G. and Klee, W.A. (1981) Proc. Natl Acad. Sci. USA., 78, 4309–4313.
68. Cooper, D.M.F., Londos, C., Gill, D.I., and Rodbell, M. (1982) J. Neurochem. 38, 1164–1167.
69. Childers, S.R. (1988) J. Neurochem. 50, 543–553.
70. Yu, V. and Sadeé, W. (1988) J. Pharmacol. Exp. Ther. 245, 350–355.
71. Gross, R.A. and Macdonald, R.L. (1989) J. Neurophysiol. 61, 97–105.
72. Chneiweiss, M., Glowinski, J. and Premont, J. (1988) J. Neurosci. 8, 3376–3382.

73. Attali, B., Saya, D., Yevl Nah, S. and Vogel, Z. (1989) J. Biol. Chem. 264, 347–353.
74. Attali, B., Saya, D. and Vogel, Z. (1989) J. Neurochem. 52, 360–369.
75. Ueda, H., Misawa, H., Fukushima, N. and Takagi, H. (1987) Eur. J. Pharmacol. 138, 129–132.
76. Misawa, H., Ueda, H. and Satoh, H. (1990) Neurosci. Lett. 112, 324–327.
77. Glasel, J.A., Venn, R.F. and Barnard, E.A. (1980) Biochem. Biophys. Res. Commun. 95, 263–268.
78. Roth, H.L. and Coscia, C.J. (1984) J. Neurochem. 42, 1677–1684.
79. Pert, C.B., Pasternak, G.W. and Snyder, S.H. (1974) Science 182, 1359–1361.
80. Zajac, J-M. and Roques, B.P. (1985) J. Neurochem. 44, 1605–1614.
81. Patterson, S.J., Robson, L.E. and Kosterlitz, H.W. (1986) Proc. Natl Acad. Sci. USA 83, 6216–6220.
82. Demoliou-Mason, C.D. and Barnard, E.A. (1986) J. Neurochem. 46, 1129–1136.
83. Wong, Y.H., Demoliou-Mason, C.D. and Barnard, E.A. (1989) J. Neurochem. 52, 999–1009.
84. Bowen, W.D. and Pert, C.B. (1982) Cell. Mol. Neurobiol. 2, 115–128.
85. Richardson, A., Brugger, F., Demoliou-Mason, C.D. and Barnard, E.A. (1989) Adv. Biosci. 75, 13–17.
86. Simonds, W.F., Koski, G., Streaty, R.A., Hjelmeland, L.M. and Klee, W.A. (1980) Proc. Natl Acad. Sci. USA, 77, 4623–4627.
87. Bidlack, J.M. and Abood, L.G. (1980) Life Sci. 27, 331–340.
88. Ruegg, U.T., Hiller, J.M. and Simon, E.J. (1980) Eur. J. Pharmacol. 64, 367–368.
89. Howells, R.D., Gioannini, T.L., Hiller, J.M. and Simon, E.J. (1982) J. Pharmacol. Exp. Ther. 222, 629–634.
90. Demoliou-Mason, C.D. and Barnard, E.A. (1984) FEBS Lett. 170, 378–382.
91. Simon, J., Benyhe, S., Abutidze, K., Borsodi, A., Szücs, M., Tóth, G. and Wollemann, M. (1986) J. Neurochem., 46, 695–701.
92. Chow, T. and Zukin, R.S. (1983) Mol. Pharmacol. 24, 203–212.
93. Quirion, R., Bowen, W.D., Herkenham, M. and Pert, C.B. (1982) Cell. Mol. Neurobiol. 2, 333–346.
94. Bidlack, J.M., Abood, L.G., Osei-Gymah, P. and Archer, S. (1981) Proc. Natl Acad. Sci. USA 78, 636–639.
95. Fujioka, T., Inoue, F. and Kiyama, M. (1985) Biochem. Biophys. Res. Commun. 131, 640–646.
96. Maneckjee, R., Archer, S. and Zukin, R.S. (1988) J. Neuroimmunol. 17, 199–208.
97. Simonds, W.F., Burke, T.R., Rice, K.C., Jacobson, A.E. and Klee, W.A. (1985) Proc. Natl Acad. Sci. USA 82, 4974–4978.
98. Gioannini, T.L., Howard, A.D., Hiller, J.M. and Simon, E.J. (1985) J. Biol. Chem. 260, 15117–15121.
99. Cho, T.M., Hasegawa, J.I., Ge, B.L. and Loh, H.H. (1986) Proc. Natl Acad. Sci. USA, 83, 4138–4142.
100. Barnard, E.A., Demoliou-Mason, C.D. and Wong, Y.H. (1986) NIDA Research Series 75, 612.
101. Demoliou-Mason, C.D. and Barnard, E.A. (1990) In: Receptor Biochemistry, A Practical Approach (Hulme, E.C., Ed.), pp. 99-124, IRL Press, Oxford.
102. Simon, J., Benyhe, S., Hepp, J., Khan, A., Borsodi, A., Szücs, M., Medzihradszky, K. and Wollemann, M. (1987) Neuropeptides 10, 19–128.
103. Simon, J., Benyhe, S., Hepp, J., Varga, E., Medzihradszky, K., Borsodi, A. and Wollemann, M. (1990) J. Neurosci. Res. 25, 549–555.
104. Simon, J., Buchet-Braunstein, M-J., Demoliou-Mason, C.D., Hepp, J. and Barnard, E.A. (1992) Unpublished observations.
105. Ahmed, M.S. (1983) Membr. Biochem. 5, 35–47.
106. Zukin, R.S. and Kream, R.M. (1979) Proc. Natl Acad. Sci. USA 76, 1595–1597.
107. Howard, A.D., De La Baume, S., Gioannini, T.L., Hiller, J.M. and Simon, E.J. (1985) J. Biol. Chem. 260, 10833–10839.
108. Zajac, J-M., Rosténe, W. and Roques, B.P. (1987) Neuropeptides 9, 295–307.
109. Bochet, P., Icard, C.L., Pasquini, E., Garbay-Jaurequiberry, C., Baudet, A., Roques, B.P. and Rossier, J. (1988) Mol. Pharmacol. 34, 436–443.

110. Venn, R.F. and Barnard, E.A. (1981) J. Biol. Chem. 256, 1529–1532.
111. Newman, E.L. and Barnard, E.A. (1984) Biochemistry 23, 5385–5389.
112. Newman, E.L., Borsodi, A., Toth, G., Hepp, F. and Barnard, E.A. (1986) Neuropeptides 8, 305–315.
113. Ward, S.J., Portoghese, P.S. and Takemori, A.E. (1982) J. Pharmacol. Exp. Ther. 220, 494–502.
114. Peers, E.M., Rance, M.J., Barnard, E.A., Haynes, A.S. and Smith, C.F. (1983) Life Sci., Suppl. 1, 33, 439–444.
115. Barnard, E.A. (1992) Trends Biochem. Sci. 17, 368–373.
116. Schofield, P.R., McFarland, K.C., Hayflick, J.S., Wilcox, J.N., Cho, T.M., Roy, S., Nee, N.M., Loh, H.H. and Seeburg, P.H. (1989) Eur. Mol. Biol. Organ. (EMBO) J. 8, 489–495.
117. Xie, Guo-Xi, Miyajima, A. and Goldstein, A. (1992) Proc. Natl Acad. Sci. USA 89, 4124–4128.
118. Evans, C.J., Keith, D.E. Jr., Morrison, H., Magendzo, K. and Edwards, R.H. (1992) Science 258, 1952–1955.
119. Kieffer, B.L., Befort, K., Gaveriaux-Ruff, C. and Hirth, C.G. (1992) Proc. Natl Acad. Sci. USA 89, 12048–12052.
120. Yasuda, K., Raynor, K., Kong, H., Breder, C.D., Takeda, J., Reisine, T. and Bell, G.I. (1993) Proc. Natl Acad. Sci. USA 90, 6736–6740.
121. Chen, Y., Mestek, A., Liu, J., Hurley, J.A. and Yu, L. (1993) Mol. Pharmacol. 44, 8–12.
122. Schoffelmeer, A.N.M., De Vries, T.J., Hogenboom, F., Hruby, V.J., Portoghese, P.S. and Mulder, A.H. (1992) J. Pharmacol Exp. Ther. 263, 20–27.
123. Nah, S.Y., Saya, D., Borg, J. and Vogel, Z. (1993) Proc. Natl Acad. Sci. USA 90, 4052–4056.
124. Konkoy, C.S. and Childers, S.R. (1989) Mol. Pharmacol. 36, 627–634.
125. Polastron, J., Boyer, M.J., Querternont, Y., Thouvenot, J.B., Meunier, J.C. and Jauzac, P. (1990) J. Neurochem., 54, 562–570.
126. Gross, R.A., Moises, H.C., Uhler, M.D. and Macdonald R.L. (1990) Proc. Natl Acad. Sci. USA 87, 7025–7029.
127. Barg, J., Belcheva, M.M., Rowinski, J. and Coscia, C.J. (1993) J. Neurochem. 60, 1505–1511.
128. Periyasamy, S. and Hoss, W. (1990) Life Sci. 47, 219–224.
129. Schiller, P.W., Nyguyen, T.M.D., Weltrawska, G., Wilkes, D.C., Marsden, B.J., Lemieux, C. and Chung, N.N. (1992) Proc. Natl. Acad. Sci. USA 89, 11871–11875.

CHAPTER 12

Guanylyl cyclases as effectors of hormone and neurotransmitter receptors

DORIS KOESLING, EYCKE BÖHME and GÜNTER SCHULTZ

*Institut für Pharmakologie, Freie Universität Berlin, Thielallee 67-73,
D-14195 Berlin, Germany*

1. Introduction

Guanylyl cyclases, which exist in membrane-bound and cytosolic forms, are involved in signal transduction across membranes and within cells in the central nervous system and the periphery. Their enzymatic product, cGMP, acts as an intracellular messenger regulating cGMP-sensitive protein kinases, ion channels and phosphodiesterases [1–3]. Although the role of cGMP is in general not well-understood, cGMP is an established mediator in the control of smooth muscle tone, platelet aggregation, and retinal phototransduction [4]. The membrane-bound guanylyl cyclases are directly activated by different peptides and, therefore, belong to the group of receptor-linked enzymes [5]. The peptides stimulating membrane-bound guanylyl cyclases serve as hormones, as they are secreted in different tissues and transported in the circulation to their target tissues in the periphery and act as neurotransmitters in the brain. In contrast to the membrane-bound guanylyl cyclases, the soluble form of the enzyme is indirectly modulated by some neurotransmitters and hormones. As soluble guanylyl cyclase is activated by nitric oxide (NO), stimulation of receptors that lead to a Ca^{2+}-induced stimulation of NO synthase will result in activation of soluble guanylyl cyclase [6]. NO itself has recently been proposed to represent a new type of neurotransmitter, which being membrane-permeant can activate soluble guanylyl cyclase in the same and in adjacent cells [7].

This Chapter gives an overview of guanylyl cyclases, emphasizing structural features, as the recent progress on primary structures has helped us to understand the mechanisms of physiological regulation of the various membrane-bound and soluble guanylyl cyclases (Table I).

TABLE I
Guanylyl cyclase: structures and regulations

Subcellular localization		Activators	Source	Structure	Molecular mass cDNA-deduced	Cyclase domains	Transmembrane domains
Plasma membrane	GC-A:	ANP	Mammalian tissues	Monomer	115	1	1
	GC-B:	CNP	Mammalian tissues	Monomer	114	1	1
	GC-C:	STa + guanylin	Mammalian intestine	Monomer	121	1	1
		Resact	Sea urchin	Monomer	123	1	1
		Ca^{2+} ($\downarrow$)	Mammalian retina	Monomer			
Cytosol	$\alpha_1\beta_1$	NO, NO-containing compounds	Mammalian tissues	Heterodimer	77.5 + 70.5	2	–
	α_2 ?	NO, NO-containing compounds	Mammalian tissues	Heterodimer	81.7 + ?	2	–
	? β_2	?	Mammalian tissues	?	? + 76.3	?	

ANP, atrial natriuretic peptide; CNP, C-type natriuretic peptide; STa, heat-stable enterotoxin of *E. coli*; $\downarrow$ decreased activity by Ca^{2+}; NO, nitric oxide.

2. Membrane-bound guanylyl cyclases

2.1. Regulation

Research on membrane-bound guanylyl cyclase was greatly stimulated by the findings that some peptide hormones (Table II) activate membrane-bound forms of guanylyl cyclase, but not the cytosolic ones. The first described member of a family of polypeptide hormones regulating salt and water balance and blood pressure in mammals, atrial natriuretic peptide (A-type natriuretic peptide: ANP), was shown to interact with two main classes of natriuretic peptide receptors. These two receptors were initially distinguished by chemical crosslinking and receptor purification [8], and only the less abundant receptor subtype was shown to activate a membrane-bound guanylyl cyclase [9]. During isolation, the membrane-bound guanylyl cyclase copurified with an ANP receptor, as shown by the ability of the purified guanylyl cyclase to bind ANP [10,11]. The final evidence that one protein serves as a receptor for ANP and contains guanylyl cyclase activity and that catalytic activity is increased

TABLE II
Primary structures of peptides that activate membrane-bound guanylyl cyclases

ANP	S L R R S S C F G G R M D R I G A Q S G L G C N S F R Y
BNP	S P K T M R D S G C F G R R L D R I G S L S G L G C N V L R R Y
CNP	G L S K G C F G L K L D R I G S M S G L G C
STa	N T F Y C C E L C C N P A C A G C Y
Guanylin	P N T C E I C A Y A A C T G C
Resact	C V T G A P G C V G G G R L -NH$_2$

ANP, atrial natriuretic peptide; BNP, brain natriuretic peptide; CNP, C-type natriuretic peptide.
Given are the porcine sequences of ANP, BNP, and CNP; STa is heat-stable enterotoxin from *Escherichia coli*; guanylin was purified from a human cell line; resact is synthesized by the sea urchin, *A. punctulata*.

by hormone binding came from the first cloned membrane-bound guanylyl cyclase expressed in eukaryotic cells [12,13]. Subsequent to the identification of ANP, two further members of the family of natriuretic peptides, brain natriuretic peptide (B-type natriuretic peptide: BNP) and C-type natriuretic peptide (CNP), were found [14,15]. Although BNP was first isolated from brain, BNP and ANP are primarily cardiac hormones, which are released from the right atrium of the heart, while the expression of the newly discovered hormone CNP appears to be limited to the nervous system.

To date three different natriuretic peptide receptors have been identified by molecular cloning. Two of them, the atrial natriuretic receptor type A (ANPR-A) and type B (ANPR-B) contain membrane-bound guanylyl cyclases, which are also referred to as GC-A and GC-B (Fig. 1). ANPR-A responds to stimulation by ANP but is not a target for CNP stimulation. ANPR-B has the highest affinity for CNP, which apparently represents the physiological ligand [16]; in accordance with the exclusive expression of CNP in brain, detectable levels of mRNA coding for ANPR-B were also found in brain, although a direct linkage between ANPR-B and CNP has not been established [17]. The third receptor, termed atrial natriuretic peptide receptor type C (ANPR-C) imposes the fewest structural constrains of all the receptors on ligands it recognizes in a binding assay; all three known natriuretic peptides bind to ANPR-C with affinities in the low nanomolar range. ANPR-C is homologous to ANPR-A and ANPR-B throughout the extracellular domain, but in contrast to the guanylyl cyclase-containing receptors, it possesses only a very short cytoplasmic domain (37 amino acids) and, therefore, does not signal by formation of cGMP. A role of ANPR-C in clearing natriuretic peptide from the circulation has been postulated, but there is also evidence that the receptor might function through G-proteins to inhibit adenylyl cyclase and to activate the phosphoinositol pathway [18,19].

Besides the ANP-induced stimulation of membrane-bound guanylyl cyclase, ATP

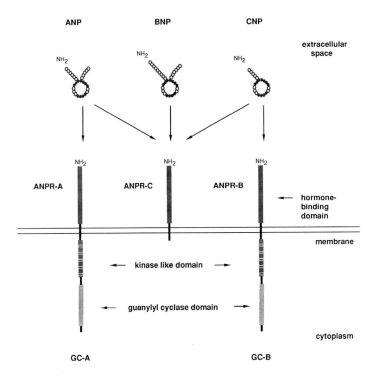

Fig. 1. Hormone specificity of natriuretic peptide receptors. Circles indicate the amino acids of the natriuretic peptides; identical residues among the three peptides are shaded. The ring structure in each peptide is formed by intramolecular disulfide linkage between two cysteine residues. ANP, natriuretic peptide type A; BNP, natriuretic peptide type B; CNP, natriuretic peptide type C; ANPR-A, atrial natriuretic receptor type A; ANPR-B, atrial natriuretic receptor type B; ANPR-C, atrial natriuretic receptor type C; GC-A, membrane-bound guanylyl cyclase type A; GC-B, membrane-bound guanylyl cyclase type B.

is well-known to markedly potentiate cGMP formation [20]. The effectiveness of poorly hydrolyzable analogues of ATP (ATPγS, AMP-PNP) suggests that this activation does not include a transfer of a phosphoryl residue but rather binding of the nucleotide. Studies of expressed ANPR-A revealed that binding of an adenine nucleotide is an absolute requirement for ANP activation of ANPR-A [21]. The adenine nucleotide-sensitive modulation of membrane-bound guanylyl cyclase activity appears to be similar to the guanine nucleotide-dependent activation of the adenylyl cyclases, but represents a novel mechanism, as in the membrane-bound guanylyl cyclases the domains responsible for the nucleotide-sensitivity and the enzymatic activity are located in the same molecule.

A guanylyl cyclase-containing peptide receptor was also found in sea urchins. The chemotactic peptide resact, which is synthesized by sea urchin eggs and causes changes in sperm motility, was shown to stimulate membrane-bound guanylyl cyclase of the sea urchin sperm [22], cells which have a very highly active membrane-bound guanylyl cyclase and do not contain a soluble form of the enzyme. This finding greatly stimulated research on guanylyl cyclases since the sea urchin sperms were used as the source for the first purification of the membrane-bound guanylyl cyclase in quantities sufficient to obtain partial amino acid sequence information.

A membrane-bound guanylyl cyclase in the intestinal mucosa has been known to be stimulated by heat-stable enterotoxins, which are small diarrhea-causing peptides produced by *Escherichia coli* [23]. The resulting elevations of cGMP are followed by effects on transport of chloride and other ions. As attempts to purify the enterotoxin-activated guanylyl cyclase failed, it was not until the expression of the cDNA coding for this guanylyl cyclase that this enzyme was shown to be a member of this peptide-regulated guanylyl cyclase family [24]. A physiological activator of the enzyme has apparently been identified. A peptide was purified from jejunum and termed guanylin, whose amino acid sequence reveals a high degree of homology with heat-stable enterotoxins of *E. coli* and which is able to stimulate the enterotoxin-sensitive guanylyl cyclase [25]. The occurrence of this guanylyl cyclase type may not be limited to intestinal mucosa, since in the kidneys of opossum and kangaroo, exposure to heat-stable enterotoxin also leads to elevation of cyclic GMP levels [26].

One membrane-bound guanylyl cyclase, probably differing from the other membrane-bound guanylyl cyclases described so far, occurs in retinal rod outer segments [27]. Although the purified enzyme exhibits properties similar to the other membrane-bound guanylyl cyclases [28], it is not known if the retinal membrane-bound enzyme belongs to the group of receptor-linked enzymes. In contrast to a missing effect on other membrane-bound guanylyl cyclases, Ca^{2+} has an inhibitory effect on this enzyme, which is mediated by the Ca^{2+}-sensitive regulatory protein, recoverin [29,30]. Visual excitation in retinal rod cells is mediated by a cascade of events which lead to increased hydrolysis of cyclic GMP and the consequent closure of cyclic GMP-activated cation channels, whereby the influx of Na^+ and Ca^{2+} is blocked [31]. The lowering of cytosolic Ca^{2+} stimulates the retinal guanylyl cyclase, and through increased resynthesis of cyclic GMP the dark state is recovered.

In contrast to its effect on the enzyme in retina, Ca^{2+} leads to stimulation of membrane-bound guanylyl cyclase in protozoans. The Ca^{2+}-sensitive protein mediating the response to the cation appears to be calmodulin [32]. Ca^{2+}/calmodulin-activated guanylyl cyclases have so far not been detected in vertebrates.

2.2. Structure

As already stated before, molecular cloning of membrane-bound guanylyl cyclases started with the help of partial amino acid sequences obtained from sea urchin guanylyl cyclase [33,34]. The deduced amino acid sequence predicted a protein with a single transmembrane domain, an extracellular ligand-binding domain, and an intracellular catalytic domain. A comparison of the amino acid sequence exhibited a region homologous to the protein-tyrosine kinase domain of the growth factor receptors [33]. Although the cDNA clone coding for the sea urchin guanylyl cyclase has never been functionally expressed, it was used as a probe to screen for mammalian membrane-bound guanylyl cyclases. Low stringency screening with this probe yielded GC-A [12,13] and a little later GC-B [35,36]. By this time, the family of guanylyl cyclases and adenylyl cyclases was well-established, and conserved amino acid sequences from the putative catalytic domain shared between all guanylyl and adenylyl cyclases were used in the polymerase chain reaction to find a cDNA clone coding for the membrane-bound guanylyl cyclase stimulated by the enterotoxin of E.coli [24]. All three cloned mammalian membrane-bound guanylyl cyclases were successfully expressed in eukaryotic cells and were shown to be stimulated by the respective peptide hormone. As discussed previously, GC-A serves as a receptor for ANP, and cyclic GMP-forming activity was increased by hormone binding [12,13]; the enterotoxin of E. coli was shown to bind to GC-C and to increase catalytic activity [24]; GC-B is specifically stimulated by the recently discovered CNP [16].

To compare the sequences of the membrane-bound guanylyl cyclases, the polypeptide chains are divided into three domains: the N-terminal ligand-binding domain, the intracellular kinase-like domain, and the putative catalytic domain, which is also found in the subunits of soluble guanylyl cyclase and in two hydrophilic domains of the adenylyl cyclases [37]. The putative catalytic or cyclase domains reveal the highest degree of similarities, with 91% identical amino acids shared by GC-A with GC-B and 55% with GC-C. 72% of the identical amino acids are found in the protein-tyrosine kinase-like domains of GC-A and GC-B, whereas GC-C reveals 39% and 35% identical amino acids to GC-A and GC-B, respectively. The homologies in the extracellular ligand-binding domain are clearly lower, 43% between GC-A and GC-B and only 10% between these forms and GC-C. The membrane-bound guanylyl cyclase of the sea urchin shows homologies to the mammalian enzymes in the intracellular part of the molecules, whereas the extracellular ligand-binding domain is apparently unrelated. These numbers reflect the close functional relationship between GC-A and GC-B and suggest that GC-C has branched off early during evolution and has developed separately. The differences in the extracellular ligand-binding domain of the membrane-bound guanylyl cyclases reflect their affinity for different peptide hormones.

3. Soluble guanylyl cyclase

3.1. Regulation

Since the early 1970s, regulations of cellular cyclic GMP concentrations by various hormones and neurotransmitters have been demonstrated (for refs. see [1]). The continued failure of hormonal stimulation of the enzyme in broken cell preparations led to the hypothesis that soluble guanylyl cyclase is regulated by hormonal factors through indirect mechanisms, for example by intermediates of the Ca^{2+}- and O_2-dependent biosynthesis of eicosanoids [38]. In the late 1970s, NO-containing compounds were demonstrated to increase cGMP levels and to be potent activators of soluble guanylyl cyclase [39–41]. As the soluble enzyme form was characterized as a heme-containing protein, it was proposed that NO-containing compounds or NO interact with the prosthetic heme group, thereby stimulating the enzyme [42]. The physiological significance of NO-induced activation of soluble guanylyl cyclase was not clear until the endothelium-derived relaxing factor (EDRF), which causes vasodilation *via* stimulation of soluble guanylyl cyclase and subsequent elevation of the cyclic GMP concentration, was identified as NO [43]. Subsequently, the formation of NO from arginine has been demonstrated in a variety of tissues [6] and up to six isoforms of NO-forming enzymes have been postulated [44]. The major isoform of the NO synthases, designated type I, was purified from cerebellum [45–47] and was shown to catalyze the conversion of L-arginine into NO and citrulline in a $Ca^{2+}/$calmodulin- and NADPH-dependent manner with the consumption of molecular oxygen and with tetrahydrobiopterine and flavins as additional cofactors [48]. Hormones which bind to G-protein-coupled receptors stimulating phospholipase C and signal through breakdown of phospholipids lead to an inositol 1,4,5-triphosphate-mediated rise in the cytoplasmic Ca^{2+} concentration. In NO synthase-containing tissues, the elevation of Ca^{2+} stimulates the Ca^{2+}/calmodulin-dependent enzyme, which catalyzes the conversion of arginine into NO and citrulline. The enzymatically produced NO finally activates soluble guanylyl cyclase, probably by interaction with its heme prosthetic group as mentioned earlier (Fig. 2). Thus, soluble guanylyl cyclase appears to be an effector molecule for endogenous NO. In addition to NO's role in the mediation of endothelial relaxation and in immune response of macrophages (for refs, see [6]), NO probably is a central and peripheral neuronal messenger ([49] and refs. therein). In the peripheral nervous system, there is evidence that the previously unidentified neurotransmitter released by nonadrenergic and non-cholinergic nerves (NANC) is in fact NO (see Table II in [7]). In the central nervous system NO is probably involved in classical anterograde neuronal signalling and might also be the retrograde messenger in long-term potentiation (LTP) [50]. The role of NO in LTP is an interesting example of the unorthodox neurotransmitter function of NO. LTP is a term used to describe the sustained increase in synaptic transmission that occurs in synaptic circuits of the hippocampus in response to brief increases in activity and has been proposed as a mechanism of memory. Although the initiation of LTP is proposed to be caused by repetitive activation of postsynaptic N-methyl-D-aspartate (NMDA) receptors and subsequent influx of Ca^{2+} through NMDA receptor-operated channels, the maintenance of LTP appears to be dependent on the release of a retrograde messenger by the postsynaptic cells, acting on presynaptic terminals to increase the amount of transmitter being released. This retrograde messenger is likely to be membrane-permeant and able to reach the presynaptic neuron by diffusion. There is indirect evidence that NO may be this retrograde mes-

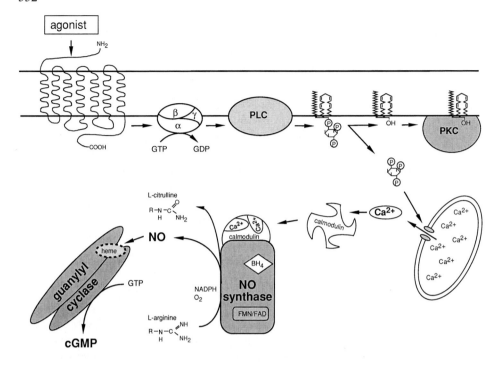

Fig. 2. Signal transduction cascade involved in the agonist-induced activation of phospholipase (PI response) and eventually of soluble guanylyl cyclase. $\alpha\beta\gamma$, subunits of a G-protein; PLC, phospholipase C; PKC, protein kinase C. For further explanations see text.

senger, as inhibitors of NO synthase and hemoglobin, which traps NO, block LTP in brain slices and exogenous NO increases presynaptic release of transmitter in cultured cells. Although the identification of NO as the retrograde messenger seems tempting, it must be considered tentative until it is demonstrated that NO can be synthesized in the appropriate postsynaptic neuron and that exogenously added NO can completely simulate the natural induction of LTP.

3.2. Structure

In contrast to membrane-bound guanylyl cyclases, soluble guanylyl cyclase consists of two different subunits designated α and β. With the help of peptide sequences obtained from the purified $\alpha_1\beta_1$ heterodimer [51,52], both subunits have been cloned and sequenced [37,53–55]. A comparison of the deduced amino acid sequences revealed similarities between the subunits with about 34% identical amino acids over the whole length of the polypeptide chains. The homology is relatively low in the N-terminal parts and pronounced in the central and C-terminal regions (Fig. 3); the latter also show homologies with the putative catalytic domains of the membrane-

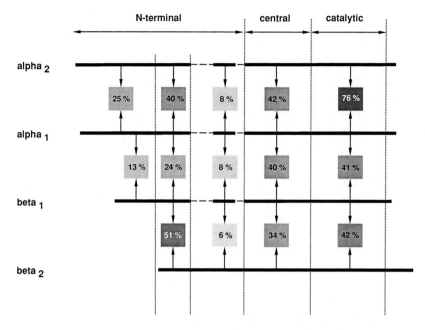

Fig. 3. Comparison of the subunits of soluble guanylyl cyclase. The polypetide chains are shown as a bold line; gaps are indicated by a dotted line. The numbers give the amount of identical amino acids shared between the respective subunits in the regions indicated by vertical lines. The C-termini of the subunits of soluble guanylyl cyclase are unrelated. The regions chosen for comparison are: amino acids 1–161 of the α_2-, 1–126 of the α_1-, and 1–56 of the β_1-subunit; 162–268 of the α_2-, 127–234 of the α_1-, 57–164 of the β_1-, and 1–101 of the β_2-subunit; 269–321 of the α_2, 235–282 of the α_1-, 165–212 of the β_1-, and 102–208 of the β_2-subunit; 322–483 of the α_2-, 283–441 of the α_1-, 213–383 of the β_1-, and 205–368 of the β_2-subunit; 484–698 of the α_2-, 442–660 of the α_1-, 384–605 of the β_1-, and 369–581 of the β_2-subunit.

bound guanylyl cyclases. Expression experiments showed that both subunits are necessary to yield a functionally active enzyme [56,57]. The homology between the subunits suggests that both are involved in the catalytic function of the enzyme.

In addition to the two subunits of the purified enzyme, two other subunits of soluble guanylyl cyclase have been identified using degenerate oligonucleotides deduced from conserved amino acid sequences to prime in the polymerase chain reaction. One of these subunits revealed more homology towards the β_1- than the α_1-subunit and was therefore designated β_2 [58]. In comparison with the other subunits of soluble guanylyl cyclase, the β_2-subunit shows some structural differences since it lacks the first 62 amino acids of the β_1-subunit but extends 86 amino acids beyond the C-terminus of β_1 and contains a consensus sequence for isoprenylation/carboxymethylation. The other new subunit of soluble guanylyl cyclase shows more similarity towards the α_1- than the β_1-subunit and was therefore designated α_2 [59]. Noticeable in the comparison of the amino acid sequences of the subunits is a stretch of about 100 amino acids in the N-terminal region which reveals a significantly higher degree of identical amino acids between α_1 and α_2 and between β_1 and β_2 than between α_1 and β_1 (see Fig. 3). One can speculate that this region is responsible for the respective properties of a α- or a β-subunit.

In expression experiments the α_2-subunit was shown to represent an isoform of the α_1-subunit as coexpression with the β_1- but not the α_1-subunit resulted in the formation of a catalytically active enzyme [59]. Although the $\alpha_2\beta_1$ heterodimer exhibited 3- to 6-times lower basal activity than the $\alpha_1\beta_1$ heterodimer, their regulation is similar, as the $\alpha_2\beta_1$ heterodimer was stimulated by sodium nitroprusside to a degree comparable with that of the $\alpha_1\beta_1$ heterodimer. Coexpression of the β_2-subunit with either one of the other subunits, i.e. α_1, β_1 and α_2, did not yield a functionally active enzyme (unpublished results). Thus, either the β_2-subunit is able to form a heterodimer only with another, yet unidentified α-subunit or the β_2-subunit is post-translationally modified and these modifications are not performed in the expression system used. The possibility that the β_2-subunit is a 'non active subunit' and inhibits the dimerization between active subunits cannot be excluded so far. The reasons for the existence of several subunits of soluble guanylyl cyclase are not clear. Immunological studies of the physiologically occurring heterodimers are necessary to identify the corresponding subunit in the respective heterodimer, and functional characterisation of the heterodimers should elucidate possible differences in regulation.

4. Related proteins

Several proteins exhibit sequence homologies with guanylyl cyclases encircling possible functionally related domains, while other proteins share analogies in the structural arrangements of the molecules. As already mentioned before, there are homologies in the ligand-binding domains of GC-A and GC-B with the N-terminal extracellular region of the ANP receptor ANPR-C, which is not very surprising as all three receptors are stimulated by natriuretic peptides [12] (see Fig. 1).

The putative catalytic domain of the guanylyl cyclases is also found in both cytosolic regions of all nonbacterial adenylyl cyclases sequenced so far [60–63]. The primary structures of the mammalian adenylyl cyclases predict proteins with 12 membrane-spanning regions and two large hydrophilic domains, which are similar to each other and share homologies with the C-terminal part of all guanylyl cyclases. As the regulation of all these cyclases is very different (coupling to G-proteins, direct hormone binding and NO-heme interaction for adenylyl cyclase, plasma membrane-bound and cytosolic guanylyl cyclases, respectively), but the reactions catalyzed by the enzymes are quite similar, we presume that the homologous region common to all cyclases represents a domain involved in the formation of cyclic nucleotides. Interestingly, the heterodimeric soluble forms of the guanylyl cyclase and the adenylyl cyclases possess two of these putative catalytic domains suggesting the necessity of two of those homologous domains for catalytic function. This would imply dimerization of membrane-bound guanylyl cyclases, and indeed there are recent results indicating that the active form of the membrane-bound guanylyl cyclase is a homodimer [64]. All membrane-bound guanylyl cyclases show a weak homology to the cytosolic tyrosine kinase domain of the growth factor receptors [33]. Although tyrosine kinase activity of membrane-bound guanylyl cyclases has not been demonstrated, the region appears to play a role in regulating enzyme activity since deletion of the region homologous to the tyrosine kinases results in a permanent ANP- and ATP-independent activation of membrane-bound guanylyl cyclase [65]. Recently, a GxGxxxG motif has been identified to be responsible for the binding of the adenosine nucleotide [66] (see also under regulation of membrane-bound guanylyl cyclases).

In addition to the homologies to the membrane-bound guanylyl cyclases, the protein tyrosine kinase-linked growth factor receptors [67] exhibit structural characteristics similar to the membrane-bound guanylyl cyclases (Fig. 4). Both enzyme families consist of various membrane-bound forms, in which different extracellular receptor domains are linked to a common catalytic domain. Besides these membrane-bound receptor-linked forms, tyrosine kinases and guanylyl cyclases occur in cytosolic forms, which exhibit a catalytic domain homologous to that of the membrane-bound

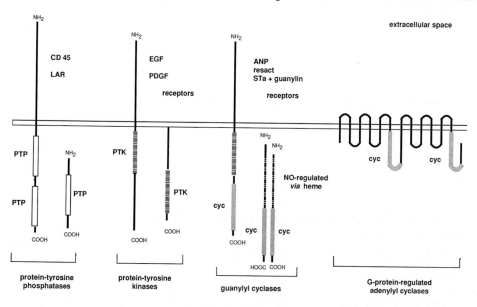

Fig. 4. Structurally related receptor-linked enzyme families. Shown are the membrane-bound and cytosolic forms of protein-tyrosine phosphatases, protein-tyrosine kinases, guanylyl cyclases, and the G-protein-regulated membrane-bound adenylyl cyclase. Labelled boxes indicate sequence homologies. PTP, putative catalytic domain of protein-tyrosine phosphatases; PTK, putative catalytic domain of protein-tyrosine kinases; cyc, putative catalytic domain of guanylyl and adenylyl cyclases; CD 45, leukocyte common antigen; LAR, leukocyte common antigen-related molecule; EGF, epidermal growth factor; PDGF, platelet-derived growth factor; ANP, atrial natriuretic peptide; ST, heat-stable enterotoxin of *Escherichia coli*. For further explanations see text.

forms. The third group in the family of receptor-linked enzymes are protein tyrosine phosphatases, which display similar structural characteristics as the protein tyrosine kinases and the guanylyl cyclases, although the mechanism of activation of these enzymes is not understood yet; soluble protein tyrosine phosphatases occur here as well [68,69].

5. Summary

So far three mammalian membrane-bound guanylyl cyclases, which differ in their peptide-binding domains, but share a common intracellular catalytic domain, have been identified by molecular cloning, and the existence of further members of the enzyme family is certain. The multiple forms appear to represent different receptor families. Therefore, different extracellular signals, such as the natriuretic peptides, the enterotoxin of *E. coli*, guanylin, and probably other not yet identified peptides, use the same intracellular signal-generating system to elicit cell type-specific responses. In contrast, the importance of the existence of multiple forms of soluble guanylyl cyclase subunits is not clear, as various heterodimers may exhibit different sensitivities to NO and may vary in their tissue distribution and subcellular localisation.

References

1. Walter, U. (1989) Rev. Physiol. Biochem. Pharmacol. 113, 42–88.
2. Tremblay, J., Gerzer, H. and Hamet, P. (1988) Adv. Second Messenger Phosphoprotein Res. 22, 319–383.
3. Kaupp, U.B. (1991) Trends Neurosci. 14, 150–157.
4. Goy, M.F. (1991) Trends Neurosci. 14, 293–299.
5. Chinkers, M. and Garbers, D.L. (1991) Annu. Rev. Biochem. 60, 553–575.
6. Moncada, S., Palmer, M.J. and Higgs, E.A. (1991) Pharmacol. Rev. 43, 109–134.
7. Garthwaite, J. (1991) Trends Neurosci. 14, 60–67.
8. Takayanagi, R., Snajdar, R.M., Imada, T., Tamura, M., Pandey, K.N., Misono, K.S. and Inagami, T. (1987) Biochim. Biophys. Acta 144, 244–250.
9. Waldman, S.A., Rapoport, R.M. and Murad, F. (1984) J. Biol. Chem. 259, 14332–14334.
10. Kuno, T., Andresen, J.W., Kamisaki, Y., Waldman, S.A., Chang., L.Y., Saheki, S., Leitman, D.C., Nakane, M. and Murad, F. (1986) J. Biol. Chem. 261, 5817–5823.
11. Paul, A.K., Marala, R.B., Jaiswal, R.K. and Sharma, R.K. (1987) Science 235, 1224–1226.
12. Chinkers, M., Garbers, D.L., Chang, M.-S., Lowe, D.G., Chin, H., Goeddel, D.V. and Schulz, S. (1989) Nature 338, 78–83.
13. Lowe, D.G., Chang, M.-S., Hellmiss, R., Chen, E., Singh, S., Garbers, D.L. and Goeddel, D.V. (1989) Eur. Mol. Biol. Organ. (EMBO) J. 8, 1377–1384.
14. Sudoh, T., Kangawa, K., Minamino, N. and Matsuo, H. (1988) Nature 332, 78.
15. Sudoh, T., Minamino, N., Kangawa, K. and Matsuo, H. (1990) Biochem. Biophys. Res. Commun. 168, 863–870.

16. Koller, K.J., Lowe, D.G., Bennett, G.L., Minamino, N., Kangawa, K., Matsuo, H. and Goeddel, D.V. (1991) Science 252, 120–123.
17. Wilcox, J.W., Augustine, A., Goeddel, D.V. and Lowe, D.G. (1991) Mol. Cell. Biol. 11, 3454–3462.
18. Anand-Srivastava, M.B., Srivastava, A.K. and Cartin, M. (1987) J. Biol. Chem. 262, 4931–4934.
19. Hirata, M., Chang, C.H. and Murad, F. (1989) Biochim. Biophys. Acta 1010, 346–351.
20. Chang, C.-H., Kohse, K.P., Chang, B., Hirata, M., Jiang, B., Douglas, J.E. and Murad F. (1990) Biochim. Biophys. Acta 1052, 159–165.
21. Chinkers, M., Singh, S. and Garbers, D.L. (1991) J. Biol. Chem. 266, 4088–4093.
22. Shimomura, H., Dangott, L.J. and Garbers, D.L. (1986) J. Biol. Chem. 261, 15778–15782.
23. Hughes, J.M., Murad, F., Chang, B. and Guerrant, R.L. (1978) Nature 271, 755–756.
24. Schulz, S., Green, C.K., Yuen, P.S.T. and Garbers, D.L. (1990) Cell 63, 941–948.
25. Currie, M.G., Fok, K.F., Kato, J., Moore, R.J., Hamra, F.K. Duffin, K.L. and Smith, C.E (1992) Proc. Natl Acad. Sci. USA 89, 947–951.
26. Forte, L.R., Krause, W.J. and Freeman, R.H. (1988) Am. J. Physiol. 255, F1040–F1046.
27. Fleischman, D., Denisevich, M., Raveed, D. and Pannbacker R.G. (1980) Biochim. Biophys. Acta 630, 176–186.
28. Koch, K.-W. (1991) J. Biol. Chem. 266, 8634–8637.
29. Lambrecht, H.G. and Koch, K.-W. (1991) Eur. Mol. Biol. Organ. (EMBO) J. 10, 793–798.
30. Dizhoor, A.M., Ray, S., Kumar, S., Niemi, G., Spencer, M., Brolley, D., Waslsh, K.A., Philipov, P.P., Hurley, J.B. and Stryer L. (1991) Science 251, 915–918.
31. Fesenko, E.E., Kolesnikov, S.S. and Lyubarsky, A.L. (1985) Nature 313, 310–313.
32. Schultz, J.E. and Klumpp, S. (1984) Adv. Cylic Nucleotide Protein Phosphorylation Res. 17, 275–283.
33. Singh, S., Lowe, D.G., Thorpe, D.S., Rodrigues, H., Kuang, W.-J., Dangott, L.J., Chinkers, M., Goeddel, D.W. and Garbers D.L. (1988) Nature 334, 708–712.
34. Thorpe, D.S. and Garbers, D.L. (1989) J. Biol. Chem. 264, 6545–6549.
35. Schulz, S., Singh, S., Bellet, R.A., Singh, G., Tubb, D.J., Chin, H. and Garbers, D.L. (1989) Cell 58, 1155–1162.
36. Chang, M.-S., Lowe, D.G., Lewis, M., Hellmiss, R., Chen, E. and Goeddel, D.V. (1989) Nature 341, 68–72.
37. Koesling, D., Harteneck, Ch., Humbert, P., Bosserhoff, A., Frank, R., Schultz, G. and Böhme, E. (1990) FEBS Lett. 266, 128–132.
38. Spies, C., Schultz, K.-D. and Schultz, G. (1980) Naunyn-Schmiedebergs Arch. Pharmacol. 311, 71–77.
39. Schultz, K.-D., Schultz, J. and Schultz, G. (1977) Nature 265, 750–751.
40. Murad, F., Mittal, C.K., Arnold W.P., Katsuki, S. and Kimura, H. (1978) Adv. Cyclic Nucleotide Res. 9, 175–204.
41. Böhme, E., Graf, H. and Schultz, G. (1978) Adv. Cyclic Nucleotide Res. 9, 131–143.
42. Gerzer, R., Böhme, E., Hofmann, F. and Schultz, G. (1981) FEBS Lett. 132, 71–74.
43. Palmer, R.M.J., Ferrige, A.G. and Moncada, S. (1987) Nature 327, 524–526.
44. Förstermann, U., Schmidt, H.W., Pollock, J.S., Sheng. H., Mitchell, J.A., Warner, T.D., Nakane, M. and Murad, F. (1991) Biochem. Pharmacol. 42, 1849–1857.
45. Bredt, D.S. and Snyder, S.H. (1990) Proc. Natl Acad. Sci. USA 87, 682–685.
46. Mayer, B., John, M., and Böhme, E. (1990) FEBS Lett. 277, 215–219.
47. Schmidt, H., Pollock, J.S., Nakane, M., Gorsky, L.D., Förstermann, U. and Murad, F. (1991) Proc. Natl Acad. Sci. USA 88, 365–369.
48. Mayer, B., John, M., Heinzel, B., Werner, E.R., Wachter, H., Schultz, G. and Böhme, E. (1991) FEBS Lett. 288, 187–191.
49. Vincent, S.R. and Hope, B.T. (1992) Trends Neuro. Sci. 15, 108–113.
50. Garthwaite, J., Charles, S.L. and Chess-Williams, R. (1988) Nature 336, 385–388.
51. Kamisaki, Y., Saheki, S., Nakane, M., Palmieri, J., Kuno, T., Chang, B., Waldman, S.A. and Murad, F. (1986) J. Biol. Chem. 261, 7236–7241.

52. Humbert, P., Niroomand, F., Fischer, G., Mayer, B., Koesling, D., Hinsch, K.-H., Gausepohl, H., Frank, R., Schultz, G. and Böhme, E. (1990) Eur. J. Biochem. 190, 273–278.
53. Koesling, D., Herz, J., Gausepohl, H., Niroomand, F., Hinsch, K.-D., Mülsch, H., Böhme, E., Schultz, G. and Frank, R. (1988) FEBS Lett. 239, 29–34.
54. Nakane, M., Saheki, S., Kuno, T., Ishii, K. and Murad, F. (1988) Biochem. Biophys. Res. Commun. 157, 1139–1147.
55. Nakane, M., Arai, K., Saheki, S., Kuno, T., Buechler, W. and Murad, F. (1990) J. Biol. Chem. 265, 16841–16845.
56. Harteneck, Ch., Koesling, D., Söling, A., Schultz, G. and Böhme, E. (1990) FEBS Lett. 272, 221–223.
57. Buechler, W.A., Nakane, M. and Murad, F. (1991) Biochem. Biophys. Res. Commun. 174, 351–357.
58. Yuen, P.S.T., Potter, L.R. and Garbers, D.L. (1990) Biochemistry 29, 10872–10878.
59. Harteneck, Ch., Wedel, B., Koesling, D., Malkewitz, J., Böhme, E. and Schultz, G. (1991) FEBS Lett. 292, 217–222.
60. Krupinski, J., Coussen, F., Bakalyar, H.A., Tang, W.-J., Feinstein, P.G., Orth, K., Slaughter, C., Reed, R.R. and Gilman, A.G. (1989) Science 244, 1558–1564.
61. Bakalyar, H.A., and Reed, R.R. (1990) Science 250, 1403–1406.
62. Feinstein, A.G., Schrader, K.A., Bakalyar, H.A., Tang, W.-J., Krupinski, J., Gilman, A.G. and Reed, R.R. (1991) Proc. Natl Acad. Sci. USA 88, 10173–10177.
63. Gao, B. and Gilman, A.G. (1991) Proc. Natl Acad. Sci. USA 88, 10178–10182.
64. Thorpe, D.S., Niu, S. and Morkin, E. (1991) Biochem. Biophys. Res. Commun. 180, 538–544.
65. Chinkers, M. and Garbers, D.L. (1989) Science 245, 1392–1394.
66. Goraczniak, R.M., Duda, T. and Sharma, K. (1992) Biochem. J. 282, 533–537.
67. Ullrich, A. and Schlessinger, J. (1990) Cell 61, 203–212.
68. Fischer, E.H., Tonks, N.K., Charbonneau, H., Cirirelli, M.F., Cool, D.E., Diltz, C.D., Krebs, E.G. and Walsh, K.A. (1990) Adv. Second Messenger Phosphoprotein Res. 24, 273–279.
69. Kaplan, R., Morse, B., Huebner, K., Croce, C., Howk, R., Ravera, M., Ricca, G., Jaye, M. and Schlessinger, J. (1990) Proc. Natl Acad. Sci. USA 87, 7000–7004.

F. Hucho (Ed.) *Neurotransmitter Receptors*
© 1993 Elsevier Science Publishers B.V. All rights reserved.

CHAPTER 13

The elucidation of neuropeptide receptors and their subtypes through the application of molecular biology

WOLFGANG MEYERHOF, MARK G. DARLISON and DIETMAR RICHTER

Institut für Zellbiochemie und klinische Neurobiologie, UKE, Universität Hamburg, Martinistrasse 52, D-20251 Hamburg, Germany

1. Introduction

Neuropeptides are found widely distributed throughout the animal kingdom where they play a key role, in both the central nervous system and peripheral endocrine systems, in the regulation of physiology and behavior. One of the best characterized peptides is oxytocin. This was discovered at the beginning of this century when it was noted that pituitary extracts stimulated uterus contraction and milk ejection in the experimental animal [1,2], two of the classical endocrine responses induced by oxytocin. Later it was found that hypothalamic neurons possess all of the properties of a secretory cell [3,4]; this observation both enabled and stimulated research into the biosynthesis of oxytocin and many other hormones present in peptidergic neurons of the central nervous system.

The existence of specific receptors for neuropeptides was initially inferred from ligand-binding studies both biochemically on membrane preparations and autoradiographically on brain sections [5,6]. For example, high-affinity binding sites for the hypothalamic-neurohypophyseal peptide oxytocin in synaptic membranes from various brain regions have been described [7]. In addition, quantitative light-microscopic autoradiography and ligand-binding assays have demonstrated specific binding sites for this neuropeptide in such diverse brain regions as the hypothalamus, parts of the forebrain, the olfactory system, and the brainstem [8,9]. Further support for the existence of specific receptors comes from studies on the cellular actions of neuropeptides. For instance, extracellular recordings, from slices of rat brainstem, have indicated that oxytocin is able to excite most neurons of the dorsal motor nucleus of the vagus nerve and of the bed nucleus of the stria terminalis [10,11]. In addition, this peptide increases the firing rate of nonpyramidal neurons in the CA1 field of the rat and guinea pig hippocampus [12].

Despite a considerable body of circumstantial evidence, the existence of specific neuropeptide receptors has, until recently, been somewhat controversial. It has been argued, for example, that since neuropeptides co-localize with classical neurotransmitters in synaptic vesicles [13], they might bind to the receptors for these neurotransmitters. It has also been suggested that neuropeptides could activate intracellular signal transduction pathways in a receptor-independent manner; in fact, it has recently been proposed that this is the physiological mechanism of action of substance P on rat peritoneal mast cells (reviewed in [14]).

Unambiguous evidence for the existence of neuropeptide receptors was obtained only in 1987 when the predicted primary and secondary structures of the first neuropeptide receptor, namely that for substance K, was reported by the group of Shigetada Nakanishi [15] who isolated the corresponding complementary DNA (cDNA) by expression cloning using *Xenopus* oocytes. This study revealed that the substance K receptor had a structure, exhibiting seven putative membrane-spanning domains, similar to that proposed for rhodopsin [16], the β_2-adrenoceptor [17], and the M_1 and M_2 muscarinic acetylcholine receptors [18,19], the sequences of which had been deduced by cDNA cloning studies less than a year earlier. Since the effects of these five receptors are mediated by their interaction with guanine nucleotide-binding proteins (G-proteins), and since they also exhibit amino acid sequence similarity to one another, they are collectively known as G-protein-coupled receptors. The subsequent cloning of cDNAs for a variety of other neuropeptide receptors has revealed that, with the notable exception of the receptor [20] for atrial natriuretic peptide (ANP), these are all members of the G-protein-coupled receptor family.

It is beyond the scope of this Chapter to review, in detail, the biochemical and pharmacological properties of all of the receptors that are activated by peptide ligands (for example, neuropeptide receptors, growth factor receptors, cytokine receptors, etc.). Rather, we have concentrated on those neuropeptide receptors that are predicted to contain seven membrane-spanning domains and whose sequences have been elucidated by the application of recombinant DNA techniques. A list of such receptors for which a cDNA has been isolated, and their mechanism of signal transduction, is given in Table I. In this Chapter, the salient features of neuropeptide receptors are discussed and specific examples are used to illustrate particular points. In this way, we hope to provide a broad perspective of this important class of receptors whose members may be found throughout the central nervous system and in peripheral tissues where they are involved in a variety of fundamental biological processes.

2. Receptor identification and purification

Despite the initial identification of neuropeptide receptors by radioligand-binding studies about fifteen years ago, their subsequent purification has proved extremely

TABLE I
Neuropeptides and their receptors

Neuropeptide	Receptor subtype	Source of cDNA clone	Coupling mechanism in expression systems	References
Angiotensin II	AT_1	Bovine, rat	PLC activation	42, 43
Bombesin	GRP-preferring	Swiss 3T3 cells	PLC activation	25
Bombesin	NMB-preferring	Rat	PLC activation	35
Bradykinin	B_2	Rat	PLC activation	38
Calcitonin	–	Porcine	Activation of adenylate cyclase	76
Cholecystokinin	CCK_A	Rat	PLC activation	26
Neuromedin K	–	Rat	PLC activation	77
Neurotensin	–	Rat	PLC activation	24
Oxytocin	–	Human	PLC activation	39
Parathyroid hormone and parathyroid hormone-related peptide[1]	–	Opossum, rat	PLC activation *and* activation of adenylate cyclase	78, 79
Somatostatin	SSTR1 or SSR-14	Human, murine, rat	N.D.	30, 31
Somatostatin	SSTR2 or SSR-14	Human, murine, rat	N.D.	30, 31
Somatostatin	SSR-28	Rat	N.D.	33
Substance K	–	Bovine	PLC activation	15
Substance P	–	Human, rat	PLC activation	36, 80
TRH	–	Murine	PLC activation	37
Vasopressin	V_{1a}	Rat	PLC activation	41
Vasopressin	V_2	Human, rat	Activation of adenylate cyclase	81, 82

Only those neuropeptide receptors for which a cDNA has been isolated, and for which the activating ligand is known, are listed here. [1]Note: a cDNA has been isolated [78] that encodes a receptor that can be activated by either parathyroid hormone or parathyroid hormone-related peptide.
GRP, gastrin-releasing peptide; N.D., not determined; NMB, neuromedin B; PLC, phospholipase C; TRH, thyrotropin-releasing hormone.

difficult and their biochemical characterization has, until the application of molecular biological methods, been very slow. The three main reasons for this are the very low amounts of receptor protein in tissues, the paucity of specific antagonists that can be used as affinity reagents with which to purify the receptors, and their strong susceptibility to proteolysis during isolation. One of the few neuropeptide receptors to be purified to apparent homogeneity is that for the tridecapeptide neurotensin. Initially, covalent labelling of neurotensin to rat brain membranes [21] resulted in the identification of a polypeptide of 49,000–51,000 daltons. Subsequently, two groups who reported on the purification of neurotensin receptors, identified two different proteins. On sodium dodecyl sulfate/polyacrylamide gels, one receptor from bovine brain had a molecular weight of 72,000 daltons [22], while the other from mouse brain had a molecular weight of 100,000 daltons [23]. The affinities of the two putative receptors were also different, with the bovine preparation having an approximately twenty-fold lower affinity (K_d = 5.5 nM) for neurotensin than the protein purified from the mouse (K_d = 0.26 nM). Eventually, expression cloning of a rat cDNA in *Xenopus* oocytes [24] revealed that the functional neurotensin receptor has a predicted molecular weight of 47,000 daltons, which does not include an allowance for sugar residues that are found attached to all membrane-associated receptors. When the cDNA was expressed in mammalian cells, the receptor displayed a K_D for neurotensin of 0.19 nM, which is very similar to that of the protein purified from mouse brain. However, for the mouse protein to be the counterpart of the receptor whose cDNA was cloned from the rat, it would have to contain up to 50,000 daltons of sugar residues and/or migrate anomalously in sodium dodecyl sulfate/polyacrylamide gel electrophoresis. The relationship of the neurotensin receptor deduced from cDNA cloning and the putative receptor purified from bovine brain remains unclear.

To date, we are only aware of two different cDNA clones that have been obtained directly as a result of the purification of a neuropeptide receptor; these encode the gastrin-releasing peptide-preferring bombesin receptor [25] and the cholecystokinin (CCK) type A (CCK_A) receptor [26]. In these two cases, the proteins were purified to near homogeneity from Swiss 3T3 cells and rat pancreas, respectively, thus permitting microsequencing of receptor peptides and the subsequent synthesis of oligonucleotide probes, based on the deduced chemical sequence, for cDNA library screening.

3. Receptor pharmacology

The paucity of specific neuropeptide agonists and antagonists has not only prevented the affinity purification of many neuropeptide receptors, but has also restricted studies on the possible existence of receptor subtypes. Where specific receptor probes are available, binding studies performed with them should be treated with great caution.

While differences in the affinities for receptor ligands, in either different brain regions or in different tissues, can be taken as a strong indication of the existence of receptor subtypes, a lack of heterogeneity in the binding experiments is not proof of the absence of subtypes. The best illustration of this is the case of the muscarinic acetylcholine receptors (reviewed in this Volume) which were originally distinguished as the M_1 and M_2 types on the basis of their affinities for pirenzepine [27]. Subsequently, DNA cloning experiments have identified a minimum of five receptor subtypes (see Chapter 7, this Volume).

The occurrence of subtypes of the biogenic amine receptors (for example, adrenergic and muscarinic receptors) is now well-established (see Chapters 5, 7, 8 and 9, this Volume); however, this question has not been addressed for many neuropeptide receptors. One case where receptor heterogeneity has been demonstrated is that of the vasopressin receptor. Using specific ligands, the existence of three subtypes, namely V_{1a}, V_{1b} and V_2, which have been demonstrated to be expressed either in brain or in peripheral tissues (for a review, see [28]), has been deduced. For CCK it is known that two receptor subtypes, a peripheral type (the CCK_A receptor) and a central nervous system type (the CCK_B receptor), occur [29]. For somatostatin, three receptor cDNAs, which presumably represent subtypes, have been cloned from mouse, rat, and man [30–33]. When these cDNAs are expressed individually in cultured mammalian cells, one of the receptor types that is inserted into the membrane binds somatostatin-14 with a higher affinity than that for somatostatin-28, whereas the second receptor type prefers somatostatin-28 over somatostatin-14 and the third type displays no selectivity. The isolation and functional expression of cDNA clones have also revealed the existence of two bombesin receptors, one of which is preferentially activated by gastrin-releasing peptide [25,34] and the other by neuromedin-B [35].

One notable neuropeptide receptor for which there is ample biochemical and pharmacological evidence in support of subtypes is the opioid receptor (see Chapter 11, this Volume); for this, at least three subtypes, namely μ, κ and δ, have been described. However, to date, and despite intensive efforts in several laboratories, their existence has not been confirmed by the isolation of the corresponding cDNA clones. With the rapid advances being made through the application of recombinant DNA methods, the full complement of receptors for each neuropeptide (including the opioids) should soon become apparent.

4. Receptor molecular biology

As of April 1992, the sequences of nearly twenty neuropeptide receptors, whose ligands are known, have been elucidated. These have all been deduced from cloning studies that involve one of three basic strategies. Perhaps the most elegant (but technically demanding) approach involves the expression of RNA, obtained by the in-vitro transcription of pools of cDNA clones, in *Xenopus* oocytes followed by the electrophysiological testing for responses that are elicited by specific peptides. As mentioned at the beginning of this Chapter, the first neuropeptide receptor cDNA, namely that for the substance K receptor [15], was isolated using this strategy. Subsequently, cDNAs for the substance P [36], thyrotropin-releasing hormone [37], neurotensin [24], B_2 bradykinin [38], and oxytocin [39] receptors have been isolated similarly. We consider this to be the method of choice for the isolation of neuropeptide receptor cDNAs since, in principle, the only requirement is that responses can be detected in oocytes after the injection of poly(A)$^+$ RNA from a brain region or tissue that contains the receptor. Since the electrophysiological assay is based on the activation of endogenous ion channels by calcium ions that are mobilized as a consequence of inositol phospholipid turnover, it is ideally suited to the study of those neuropeptide receptors which are coupled to this second messenger pathway [40]. A variation of this method has recently been used to isolate a cDNA for the rat V_{1a} vasopressin receptor [41]. Instead of an electrophysiological assay, aequorin luminescence was employed to detect changes in calcium levels in oocytes that had been injected with pools of in-vitro-transcribed RNA.

A second strategy that has been used to isolate neuropeptide receptor cDNAs is based upon the transfection of mammalian cells with a library of clones followed by the selection ('panning') of those that can now bind a specific radioactive ligand. An example of this approach is the isolation of angiotensin II AT_1 receptor clones from libraries made from either bovine adrenal zona glomerulosa cells [42] or rat aortic vascular smooth muscle cells [43]. As with the *Xenopus* oocyte expression system, the transfection strategy does not require any knowledge of the receptor of interest. However, it does necessitate a radiolabelled (preferably iodinated) receptor ligand and a knowledge of the tissue location of the receptor.

The third strategy involves the polymerase chain reaction (PCR) amplification of partial cDNAs and is based on the assumption that the neuropeptide receptor of interest is a member of the family of G-protein-coupled receptors. Essentially, pools of degenerate oligonucleotide primers, that correspond to sequences encoding conserved polypeptide sequences, are used with *Thermus aquaticus (Taq)* DNA polymerase to amplify short segments of DNA from first-strand cDNA. Normally, the primer sequences are based on DNA sequences that encode membrane-spanning domains which are usually the most conserved portions of G-protein-coupled receptors. Using this approach, a variety of partial cDNAs that encode receptors of unknown function are isolated [44]. To search for novel neuropeptide receptor cDNAs, it is usual to determine the sites of expression of the genes corresponding to the partial cDNAs by Northern blotting and/or in-situ hybridization on brain or tissue sections. A prerequisite, therefore, is knowledge of the tissue distribution of the receptor of interest. To formally identify the exact nature of the receptor cDNA clone, it is necessary to express it in *Xenopus* oocytes or in mammalian cells and show either a response to the applied neuropeptide or binding of a specific ligand. Complementary DNAs for somatostatin receptors [30–33] have been cloned in this way. A related approach to this is to use known G-protein-coupled receptor cDNA probes to screen cDNA libraries at reduced stringency to isolate clones that encode receptors that have related amino acid sequences.

In general, neuropeptide receptors comprise between 350 and 450 amino acid residues and have a short amino-terminal extracellular domain that has a small number of sites for the addition of N-linked sugars, seven membrane-spanning segments, and a carboxy-terminus of variable length that is intracellular (see Fig. 1). However, despite their similar predicted topologies, neuropeptide receptors, perhaps surprisingly, share very little sequence identity. The greatest similarity between any two receptors is seen in the transmembrane domains and here, typically, only 25–30% identity is found (Fig. 2).

While it has been shown that the ligands for the adrenergic and muscarinic acetylcholine receptors make contact with amino acid residues within membrane-spanning domains (reviewed in [45]), little is known about the locations of the binding

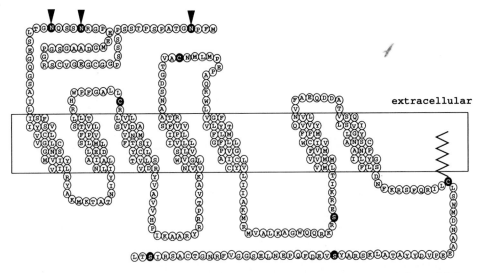

Fig. 1. Topology of a typical neuropeptide receptor in the cell membrane. The sequence and predicted seven membrane-spanning domain (TMI to TMVII) structure of the rat somatostatin type 1 (SSR-14) receptor [30] is shown. Amino acid residues, in single letter code, are represented by circles. Highlighted in black are three asparagine residues (N4, N44, and N48) in the amino-terminal extracellular domain that are sites for the addition of N-linked sugars (shown as black triangles), two cysteine residues (C130 and C208) in the extracellular loops between TMII and TMIII, and TMIV and TMV, that probably form a disulfide bridge, three serine residues (one in the intracellular loop between TMV and TMVI and two in the intracellular carboxy-terminal domain) that are sites for phosphorylation by either calmodulin-dependent protein kinase (S364), protein kinase A (S389), or protein kinase C (S265), and a cysteine residue (C339) that is located 12 amino acids after TMVII and which is presumed to be palmitoylated (shown as a zig-zag line). The palmitoylation is thought to result in the association of this part of the polypeptide chain with the membrane and, hence, the formation of an additional intracellular loop.

```
                         MAAVTYPSSVPTTLDPGNASSAWPLDTSLGNASAGTSLAGLAVSGILISLVY
                                   MALNSSAEDGIKRIQDDCPKAGRHSYIFVMIPTLY
             MHLNSSVPQGTPGEP..11..MEATFLALSLSNGSGNTSESDTAGPNSDLDVNTDIYSKVLVTAIY
                              MDNVLPMDSDLFPNISTNTSESNQFVQPTWQIVLWAAAY
                         MPPRSLPNLSLPTEASESELEPEVWENDFLPDSDGTTAELVIRCVIPSSLY
                                    MENDTVSEMNQTELQPQAAVALEYQVVT..ILLVVII
                                      MFNITTQALGSAHNGTSFEVNCPDTEWWSWLNAIQAPFLW
                MSHSPARQHLVESSRMDVVDSLLMNGSNITPPCELGLENETLFCLDQPQPSKEWQSALQILLY
                         MEGALAANWSAEAANASAAPPGAEGNRTAGPPRRNEALARVEVAVL
           TM I                                TM II
     LVVCVVGLLGNSLVIYVVLRHT---SSPSVTSVYILNLALADELFMLGLPFLAAQNALSY--WPFG
     SIIFVVGIFGNSLVVIVIYFYM---KLKTVASVFLLNLALADLCFLLTLPLWAVYTAMEY-RWPFG
     LALFVVGTVGNSVTAFTLARKKSLQSLQSTVHYHLGSLALSDLLILLLAMPVELYNF..6..WAFG
     TVIVVTSVVGNVVVIWIILAHK---RMRTVTNYFLVNLAFAEACMAAFNTVVNFTYAVHN-VWYYG
     LIIISVGLLGNIMLVKIFLTNS---TMRSVPNIFISNLAAGDLLLLLTCVPVDASRYFFD-EWVFG
     C---GLGIVGNIMVVLVVMRTK---HMRTPTNCYLVSLAVADLMVLVAAGLPNITDSIYG-SWVYG
     -VLFLLAALENIFVLSVFCLHT---KNCTVAEIYLGNLASADLILACGLPFWAITIANNF-DWLFG
     SIIFLLSVLGNTLVITVLIRNK---RMRTVTNIFLLSLAVSDLMLCLFCMPFNLIPNLLK-DFIFG
     CLILLLALSGNACVLLALRTTR---QKHSRLFFFMKHLSIADLVVAVFQVLPQLLWDITF--REYG
                    TM III                              TM IV
     S-LMCRLVMAVDGINQFTSIFCLTVMSVDRYLAVVHPTRSARWRTAPVARMVSAAVWVASAVVVLP
     N-HLCKIASASVSFNLYASVFLLTCLSIDRYLAIVHPMKSRLRRTMLVAKVTCIIIWLMAGLASLP
     D-AGCRGYYFLRDACTYATALNVASLSVERYLAICHPFKAKTLMSRSRTKKFISAIWLASALLAIP
     L-FYCKFHNFFPIAALFASIYSMTAVAFDRYMAIIHPL--QPRLSATATKVVFVIWVLALLLASP
     K-LGCKLIPAIQLTSVGVSVPTLTALSADRYAIV-NPMDM-QTSGVVLWTSVAVGIWVVSVLLAVP
     Y-VGCLCITYLQYLGINASSCSITAFTIERYIAICHPIKAQFLCTFSRAKKIIIFVWAFTSIYCML
     E-VLCRVVNTMIYMNLYSSICFLMLVSIDRYLALVKTMSMGRMRGVRWAKLYSLVIWSCTLLLSSP
     S-AVCKTTTYFMGTSVSVSTFNLVAISLERYGAICRPLQSRVWQTKSHALKVIAATWCLSFTIMTP
     PDLLCRLVKYLQVVGMFASTYLLLLMSLDRCLAICQPL---RSLRRRTDRLAVLATWLGCLVASAP
                                         TM V
     VVVFSGVPRGM--------STCHMQW--PEAAAWRTAFIIYTAALGFFGPLLVICLCYLLIVVKVR
     AVIHRNVYFIENTNI----TVCAFHY-ESRNSTLPIGLGLTKNILGFLFPFLIILTSYTLIWKALK
     MLFTMGLQNRSGDGTHPGGLVCTPIV-DTA---TVKVVIQVNTFMSFLFPMLVISILNTVIANKLT
     QGYYSTTETMPSR------VVCMIEWPE-HPNRTEKAYHICVTVLIYFLPLLVIGYAYTVVGITIL
     EAVFSEVARIGSSDN--SSFTACIPYPQTDE--LHPKIHSVLIFLVYFLIPLVIISISYYYHIAKTLI
     WFFLLDLNISTYK--NAVVVSCGYKISRN----YYSPIYLMDFGVFYVVPMILATVLYGFIARILF
     MLVFRTMKDYR---EEGHNVTCVIVYPSRS---WEVFTNMLLNLVGFLLPLSIITFCTVRIMQVLR
     YPIYSNLVPFTK-NNNQTANMCRFLLPSDA---MQQSWQTFLLLILFLLPGIVMVVAYGLISLELY
     QVHIFSLREVAD-----GVFDCWAVFIQPW--GPKAYITWITLAVYIVPVIVLATCYGLISFKIW
                       TM VI
     ST..20..RRSERRVTRMVVAVVALFVLCWMPFYLLNIVNVVCPLPEE------PAFFGLYFLVVA
     KAYEIQKNKPRNDDIFRIIMAIVLFFFFSWVPHQIFTFLDVLIQLGVIHDCKISDIVDTAMPITIC
     VM..32..VQALRHGVLVLRAVVIAFVVCWLPYHVRRIMFCYISDEQWTTFLF-DFYHYFYMLTNA
     WA..14..VSAKRKVVKMMIVVVCTFAICWLPFHVFFLLPYINPDLYK-----KFIQQVYLASMW
     RS..15..METRKRLAKIVLVFVGCFVFCWFPNHILYLYRSFNYKEIDP----SLGHMIVTLVARV
     LN..27..ASSRKQVTKMLAVVVILFALLWMPYRTLVVVNSFLSSPF-------QENWKLLKCRI
     NN...7..VQTEKKATVLVLAVLGLFVLCWFPFQISTFLDTLLRIGVLSGCWNERAVDIVTQISSY
     QG..66..LIAKKRVIRMLIVIVVLFFLCWMPIFSANAWRAYDTVSAEK-----HLSGTPISFILL
     QN..35..SKAKIRTVKMTFIIVLAFIVCWTPFFFVQMWSVWDANAP-------KEASAFIIVML
     TM VII
     LPYANSCANPILYGFLSYRFKQGFRRILLRPS..73..EATAGDKASTLSHL      RAT-SSR-28
     IAYFNNCLNPLFYGFLGKKFKKYFLQLLKYIP..24..SSSAKKPASCFEVE       RAT-VAT1
     LFYVSSAINPILYNLVSANFRQVFLSTLACLC..22..NHAFSTSATRETLY        RAT-NT
     LAMSSTMYNPIIYCCLNDRFRLGFKHAFRCCP..69..TMTESSSFYSNMLA       RAT-SUBP
     LSFSNSCVNPFALYLLSESFRKHFSNQLCCGQ..33..VLLNGHSTKQEIAL       RAT-NMB
     CIYLNSAINPVIYNLMSQKRFAAFRKLCNCKQ..40..TKVSFDDTCLASEN      MOUSE-TRH
     VAYSNSCLNPLVYVIVGKRFRKKSREVYQAIC..26..RQIHKLQDWAGNKQ       RAT-BK2
     LSYTSSCVNPIIYCFMNKRFRLGPMATFPCCP..25..RYSYSHMSTSAPPP       RAT-CCK
     LASLNSCCNPWIYMLFTGHLFHELVQRFLCCS..27..RSSSQRSCSEPSTA       HUMAN-OT
```

sites for the various neuropeptides on their receptors. Since certain neuropeptides are similar in size to the biogenic amines such as serotonin, and since it is well-established that the biological activity of many neuropeptides resides in specific portions of these molecules, it is plausible that they will bind in a similar manner to their receptors, i.e. they will bond with the side chains of amino acids that are located within membrane-spanning domains. A variation on the G-protein-coupled receptor model is that seen with the receptors for thyrotropin and luteinizing hormone-human choriogonadotrophic hormone, that are much larger in size than other neuropeptide receptors and contain 700–750 amino acids. The extra residues are found at the amino-termini of these receptors where they have been implicated in the binding of their respective large glycoprotein hormones [46].

As with other G-protein-coupled receptors, the neuropeptide receptors discussed here contain two invariant cysteine residues, one between the second and third membrane-spanning domains and the other between the fourth and fifth membrane-spanning domains. Both of these are predicted to be extracellular and may, thus, form a disulfide bond in a manner analogous to that found in a similar position in the β-adrenoceptors [47].

The intracellular loop between the fifth and sixth membrane-spanning domains of the different neuropeptide receptors, whose primary structures have been deduced by cDNA cloning, is variable in length and contains consensus sequences for phosphorylation by a variety of kinases. For example, the thyrotropin-releasing hormone (TRH) receptor has three sites for the possible addition of phosphate moieties by protein kinase C [37]. However, we know of no evidence in support of either a modulation of neuropeptide receptor activity in response to phosphorylation or the exis-

---

Fig. 2. Alignment of a selection of G-protein-coupled neuropeptide receptor sequences. The sequences, in single letter code, of the rat somatostatin type 3 receptor (RAT-SSR-28: [30]), the rat vascular type 1 angiotensin II receptor (RAT-VAT1: [43]), the rat neurotensin receptor (RAT-NT: [24]), the rat substance P receptor (RAT-SUBP: [36]), the rat neuromedin B-preferring bombesin receptor (RAT-NMB: [35]), the mouse thyrotropin-releasing hormone receptor (MOUSE-TRH: [37]), the rat bradykinin B_2 receptor (RAT-BK2: [38]), the rat cholecystokinin type A receptor (RAT-CCK: [26]), and the human oxytocin receptor (HUMAN-OT: [39]) are shown. Dashes denote gaps that have been introduced into the sequences in order to maximize the alignment. The seven putative membrane-spanning domains (TMI to TMVII) are marked by lines above the sequences, and consensus sequences (N-X-S and N-X-T, where X is any amino acid) in the proposed amino-terminal extracellular domains for N-linked glycosylation are underlined. Numbers within the sequences (before TMI, between TMV and TMVI, and after TMVII) refer to the number of additional residues that are present in each polypeptide chain but which are not shown; in these three regions there is no similarity between the different sequences. Residues that occur in the same relative position in all nine sequences are boxed. Three conserved cysteine residues are shown in bold. Two of these, which reside between TMII and TMIII and between TMIV and TMV are predicted to be extracellular and probably together form a disulfide bond; the third, located approximately ten residues after TMVII, which is found in seven of the nine sequences is presumed to be a target for palmitoylation (see text). Note that the conserved amino acids referred to here are not unique to neuropeptide receptors, but are present in almost all G-protein-coupled receptors (see other chapters, this Volume).

tence of specific neuropeptide receptor kinases. In the case of the β-adrenoceptors, phosphorylation of residues on the cytoplasmic side of the membrane is thought to play a role in the modulation of receptor function. This phosphorylation is achieved by a receptor-specific kinase named βARK, an acronym for β-adrenergic receptor kinase, and leads to the binding of a β-arrestin and the resultant uncoupling of the receptor from its transduction pathway (reviewed in [48]).

Studies on the coupling of a particular neuropeptide receptor, that is located within the vertebrate central nervous system, to one or more specific G-proteins, is complicated because of the cellular complexity of the brain and the difficulty in identifying a homogeneous population of neurons there. The majority of investigations on receptor coupling have, therefore, been performed on clonal cell lines. Those derived from the anterior pituitary (for example, GH3 and AtT20 cells) are of particular interest since they secrete peptide hormones, such as growth hormone, prolactin, and adrenocorticotrophic hormone (ACTH), possess receptors for neuropeptides, and are electrically active.

The binding of a neuropeptide to its receptor results, indirectly, in either the stimulation of inositol phospholipid turnover, or an activation or inhibition of the activity of adenylate cyclase. While the exact nature of the G-protein that is involved in the former process is uncertain, the general consensus view, derived from studies using pertussis toxin and cholera toxin (which ribosylate G_i and G_o, and G_s proteins, respectively) is that it is neither G_i, G_o, or G_s; in fact, recent evidence [49,59] points to the involvement of G_q. In contrast, the activation or inhibition of adenylate cyclase occurs through the interaction of the neuropeptide receptor with one of several stimulatory (G_s) or inhibitory (G_i) G-proteins. In addition, the opening and closing of potassium and calcium channels can be modulated by the action of a neuropeptide at its receptor through an interaction with G_i and G_o proteins, respectively (see Fig. 3).

Depending upon the complement of G-proteins, at least some neuropeptide receptors can mediate more than one response within a cell. For example, the TRH receptor has been shown, at least in GH cells (which derive from the rat anterior pituitary), to be able to cause the closure of potassium channels ([51], and references cited therein) and stimulate inositol trisphosphate formation [52]. The identification of some of the G-proteins that mediate receptor/effector coupling when a neuropeptide binds to a receptor has been determined biochemically. For example, it has been shown by the injection of an anti-sense oligonucleotide specific for $G_{\alpha o2}$ [53], which blocks the ability of the receptor to close calcium channels, that the somatostatin receptor present on GH3 cells couples to this G-protein. In addition, it has been shown using membrane patches and purified G_{i3}-protein that somatostatin induces the opening of a 55 picosiemens potassium channel [54].

It is likely that all neuropeptide receptors interact with G-proteins, which intracellularly transduce the signal of peptide binding, and that this site of interaction is probably within the loop between the fifth and sixth membrane-spanning domains and/or within the intracellular carboxy-terminal portion of the receptor molecule.

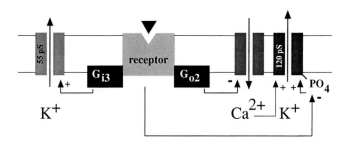

Fig. 3. Schematic representation of a neuropeptide receptor in the cell membrane showing some of the effector systems with which it may couple. In clonal anterior pituitary GH cells, the physiological role of somatostatin receptors is the inhibition of prolactin release. This is achieved through somatostatin binding to receptors on these cells and causing an inhibition of action potentials apparently via an increase in the potassium conductance and the closure of voltage-gated calcium channels. The consequent reduction in intracellular calcium levels results in inhibition of the exocytosis of hormone-containing granules. The figure depicts the possible effects of somatostatin binding to receptors on GH cells. Thus, receptor activation can lead to the opening of a 55 picosiemens potassium channel, through an association of the receptor with a G_{i3}-protein. Interaction of the receptor with a G_{o2}-protein can result in the closure of voltage-gated calcium channels. In addition, somatostatin can stimulate the opening of voltage- and calcium-activated large conductance (120 picosiemens) potassium channels via dephosphorylation of a cAMP-dependent protein kinase phosphorylation site (represented by PO_4) that is presumed to be either on the ion channel itself (as shown) or on a closely-associated regulatory molecule [75].

While studies on chimeric $\alpha_2\beta_2$-adrenergic [55] and M_1 and M_2 muscarinic receptors [56] strongly indicate that the long cytoplasmic loop contains determinants for the specificity of G-protein binding, and hence the selection of the transduction pathway, no direct evidence is yet available for neuropeptide receptors.

Studies on some of the more well-characterized G-protein-coupled receptors, such as rhodopsin and the β-adrenoceptors, have shown that a cysteine residue, located approximately ten amino acid residues after the last membrane-spanning domain, is palmitoylated. Inspection of the sequences of the family of neuropeptide receptors (Fig. 2) reveals that almost all contain a cysteine residue in this region which may be similarly post-translationally modified. A few notable exceptions are the angiotensin II AT_1 receptor [42,43] and the products of the *mas* oncogene [57], the RTA (rat thoracic aorta) gene [58], and the *mrg* (*mas*-related) gene [59] which have both significant sequence and structural similarity to G-protein-coupled receptors. Mutational analysis of the β_2-adrenoceptor [60] and of the 5-HT_2 receptor [61] results in either a markedly reduced or almost complete loss of receptor function when the cysteine residue is replaced by either a glycine or a serine, respectively. It is presumed that palmitoylation, which appears to be an absolute requirement for proper receptor function, results in an association of the carboxy-terminal portion of these proteins with the cell membrane to form an additional cytoplasmic loop that may play a role in G-protein coupling.

5. Physiological roles of neuropeptide receptors

While the isolation of cDNA clones that encode neuropeptide receptors has validated prior autoradiographic and ligand-binding studies that suggested the presence of specific receptors for various neuropeptides, there is a paucity of information on the relationship of 'cloned' receptors, which can be studied in isolation, with the presence and function of those receptor molecules on specific neurons within the central nervous system. Indeed, the specific locations and roles of neuropeptide receptors within brain are poorly understood; for example, it is generally assumed that these receptors occur on neurons, yet recent reports [62,63] have indicated that somatostatin receptors may also be found on astrocytes (see Fig. 4). Furthermore, we are unaware of any evidence in support of the location of neuropeptide receptors at synaptic junctions. It is, thus, unclear whether they play a direct role in neurotransmission. However, it is evident that certain peptides can play a role in the modulation of neurotransmission. For example, there is a growing body of evidence that indicates that oxytocin may play such a role in the central nervous system, where it is involved in drug addiction, ethanol tolerance, learning and memory, sexual and maternal behavior (reviewed in [64]), and in the inhibition of food intake [65].

The identification of an oncogene termed *mas* [57] that encodes a protein that has seven putative membrane-spanning domains and sequence similarity to G-protein-coupled receptors, and which has been suggested to be an angiotensin receptor [66], raises the possibility that dysfunction of a neuropeptide receptor, or its gene, could result in tumorigenesis. However, while the *mas* gene was initially identified as a weakly transforming oncogene in a mammalian transfection assay, it appears to be expressed in many areas of the brain [67] without associated tumor induction. Thus, the oncogenic potential of *mas* is perhaps only revealed when it is expressed in the wrong cellular environment, when it is overexpressed, when the encoded protein is uncoupled from its normal intracellular pathway(s), or when the putative receptor is able to constitutively activate a particular second messenger system. Evidence for the latter is provided by the demonstration that transfection of the cDNA that encodes the 5-HT_{1c} receptor (which is coupled to inositol phosphate and calcium mobilization into NIH3T3 cells results in a transformed phenotype [68].

It has also been observed that transfection of cDNAs encoding either the thyrotropin receptor or the β_2-adrenoceptor, both of which result in the stimulation of adenylate cyclase, into mouse Y1 cells and the stimulation of the expressed receptors by an appropriate ligand causes the appearance of foci [69]. This effect can be reproduced by forskolin, which directly activates adenylate cyclase, and the application of ACTH, thereby activating endogenous receptors that are coupled to the stimulation of adenylate cyclase. Transfection of the α_{1B}-adrenoceptor cDNA into fibroblasts also results in agonist-dependent focus formation. Furthermore, mutation of this receptor in the intracellular loop region between the fifth and sixth membrane-spanning domains enhances the ability of agonists to induce foci [70]. In addition, it has

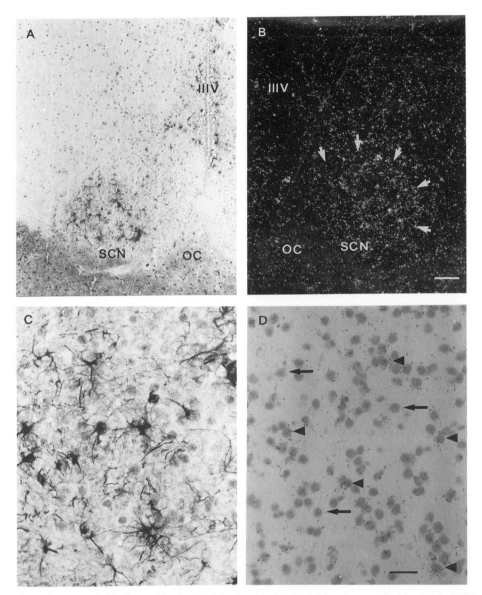

Fig. 4. Somatostatin binding and in-situ hybridization localization of the somatostatin-14 receptor (SST1) mRNA in rat hypothalamus. Coronal sections of the hypothalamus were either labelled with a somatostatin-14-gold conjugate (A and C) as described by Mentlein et al. [62] and Krisch et al. [63] or hybridized with a [^{35}S]-labelled anti-sense complementary RNA specific for the rat SST1 receptor (B, dark-field; D, light-field). The results demonstrate firstly that somatostatin-14 binding sites and SST1 receptor mRNA are co-localized within the SCN (delineated by arrows in B). Secondly, from the higher magnification (C and D), it can be concluded that not all of the cells contain receptor mRNA and that the ligand binding occurs on astrocytes. In D, arrowheads and arrows indicate cells that contain or lack SST1 receptor mRNA, respectively. For A and B, bar = 65 μm; for C and D, bar = 20 μm. IIIV, third ventricle; OC, optic chiasma; SCN, suprachiasmatic nucleus.

been observed in a number of human pituitary tumors that the genes encoding the $G_{\alpha s}$-proteins contain mutations that modify the ability of the protein products to hydrolyze GTP [71]. The consequence of these mutations is that the signalling pathways to which neuropeptide receptors couple are continuously activated; this represents an additional mechanism by which interference of the signal transduction pathway may lead to oncogenesis.

6. *Ontogeny of somatostatin receptors in the rat brain*

One striking feature of genes that are expressed within the central nervous system is that they are frequently found to be differentially regulated during ontogeny, and it has been argued that molecules encoded by such genes may be important for the development of the brain. For instance, in the case of the neuropeptide somatostatin, immunoreactive material first appears in the rat cerebellum on gestational day 16, prior to synaptogenesis. The number of somatostatin-containing structures is greatest between birth and postnatal day 7 and declines thereafter such that immunoreactivity is barely detectable in adult rats [72]. Thus, the peptide is maximally present when the cerebellum begins to function around postnatal day 3, a time during which synaptogenesis is taking place (for a review see [73]). When synaptic transmission is established between postnatal weeks 2 and 3, somatostatin immunoreactivity is hardly detectable. The transient presence of the neuropeptide is paralleled by the transient expression of its binding sites. Receptors for somatostatin are initially detected on neonatal day 15, their number then increases dramatically between postnatal days 4 and 13, and declines from postnatal days 13 to 23 [74]. While the peptide is found in the ascending fibers, its binding sites are present in the transient external granule cell layer where intense cell proliferation occurs. The disappearance of somatostatin receptors parallels the involution of the external granule cell layer which starts at postnatal day 10. The transient appearance of somatostatin-binding sites could reflect changes in either the half-life of the receptor or of the corresponding mRNA, in gene transcription rates, or the birth and death of receptor-expressing cells. Northern blot analysis of cerebellar mRNA extracted from rats at different postnatal stages of development, using specific radiolabelled probes derived from cDNAs that encode a somatostatin-14-preferring receptor (SSR-14) and a somatostatin-28-preferring receptor (SSR-28) [30,33], shows that the situation is even more interesting (Fig. 5). While SSR-14 mRNA levels are maximal on postnatal day 1 and gradually decline thereafter, being barely detectable in adults, SSR-28 mRNA is first detected on postnatal day 7 and is present at a constant level from day 14 onwards. The period during which SSR-14 mRNA is abundant corresponds to the time during which neuronal circuits are being formed and cell proliferation and migration takes place. The SSR-28 mRNA levels, however, reach a plateau only when these processes are essentially complete. Although the SSR-14 mRNA levels in the cerebellum de-

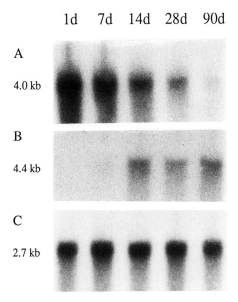

Fig. 5. Detection of somatostatin-14 (SSR-14) and somatostatin-28 (SSR-28) receptor mRNAs in the cerebellum during rat postnatal development. Poly(A)$^+$ RNA was extracted from cerebelli obtained from rats of different postnatal ages (indicated in days), fractionated by agarose gel electrophoresis, and transferred to a nylon membrane. The filter was then sequentially hybridized with cDNA probes specific for the SSR-14 mRNA (A) and the SSR-28 mRNA (B). The filter was also probed with a cDNA for an immunoglobulin heavy-chain binding protein (C) to confirm that each lane contained a comparable amount of RNA. The sizes of the transcripts that are detected with the different probes are indicated.

crease dramatically with age, significant amounts of this transcript can still be detected in the adult rat, by in-situ hybridization, in the medial cerebellar nucleus; in contrast, SSR-28 mRNA can only be detected in the granule and Purkinje cell layers. These observations point to the existence of a finely tuned balance of the somatostatin peptide-receptor system, and suggest that somatostatin has both a neurotrophic and a neuromodulatory role in the process of synaptogenesis. The receptor DNA probes, recently provided by molecular biological studies, are important tools that can now be used to address developmental questions in far greater detail than has been possible to date.

7. Concluding remarks

Usually the motivating force behind the isolation of a cDNA clone that encodes a particular receptor is a desire to study the biochemical and pharmacological properties of the encoded protein in a defined system (for example, in transfected mammalian cells and/or in *Xenopus* oocytes). However, although receptors are frequently considered as independently functioning entities, they actually interact with a variety of cellular molecules, such as G-proteins, kinases (and presumably phosphatases), and cytoskeletal proteins. Many of these may eventually be found to be receptor- or cell-specific. Thus, it is important to study receptors that are expressed from cloned cDNAs in an environment that approximates as closely as possible the milieu in which it occurs in vivo. Clearly, therefore, the most appropriate cell for expression is that in which the receptor is normally found. Unfortunately, these cells already contain the receptor whose cDNA has been cloned, and it is not currently possible to distinguish the effects due to the presence of the expressed receptor from those caused by the receptor that is naturally found there.

Investigations on signal transduction pathways are difficult to perform in complex tissues such as brain. However, neuropeptide receptor DNAs are now powerful tools for the study of receptor/effector coupling in a defined environment, for example by their transfection into appropriate mammalian cells. In addition, the clones that have been, and are being, isolated can be used for the generation of transformed cell lines that express a particular receptor. These may prove of immense value to the pharmaceutical industry for the development of specific agonists and/or antagonists that have therapeutic potential in man. While certain neuropeptides are used clinically (for example, analogues of vasopressin are used to treat patients suffering from *diabetes insipidus*), the exact molecular basis for the therapeutic benefit of many of these is unknown. A greater understanding of the number and pharmacological properties of particular neuropeptide receptor subtypes, and their independent expression in a suitable system, may permit the development of highly specific drugs.

Acknowledgements

We thank Drs S. Fehr (Hamburg) and B. Krisch (Anatomisches Institut der Universität Kiel) for providing us with the photomicrographs shown in Figure 4 and I. Wulfsen (Hamburg) for Figure 5. The work carried out in the authors' laboratory, and referred to in this Chapter, was supported by the Deutsche Forschungsgemeinschaft (SFB 232/B4 to W.M. and D.R.).

References

1. Dale, H.H. (1906) J. Physiol. 34, 163–206.
2. Ott, I. and Scott, J.C. (1910) Proc. Soc. Exp. Biol. Med. 8, 48–49.
3. Scharrer, E. and Scharrer, B. (1940) Res. Publ. Assoc. Res. Nerv. Ment. Dis. 20, 170–194.
4. Bargmann, W. (1966) Int. Rev. Cytol. 19, 183–201.
5. Lord, J., Waterfield, A., Hughes, J. and Kosterlitz, H. (1977) Nature 267, 495–499.
6. Young, I.W. and Kuhar, M. (1979) Brain Res. 179, 255–270.
7. Audigier, S. and Barberis, C. (1985) Eur. Mol. Biol. Organ. (EMBO) J. 4, 1407–1412.

8. Tribollet, E., Barberis, C., Jard, S., Dubois-Dauphin, M. and Dreifuss, J.J. (1988) Brain Res. 442, 105–118.
9. Di Scala-Guenot, D., Strosser, M.T., Freund-Dercier, M.J. and Richard, P. (1990) Brain Res. 524, 10–16.
10. Charpak, S., Armstrong, W.E., Mühlethaler, M. and Dreifuss, J.J. (1984) Brain Res. 300, 83–89.
11. Ingram, C.D., Cutler, K.L. and Wakerley, J.B. (1990) Brain Res. 527, 167–170.
12. Raggenbass, M., Tribollet, E., Dubois-Dauphin, M. and Dreifuss, J.J. (1989) Proc. Natl Acad. Sci. USA 86, 750–754.
13. Hökfelt, T., Fuxe, K. and Pernow, B. (Eds) (1986) Coexistence of Neuronal Messengers: a New Principle in Chemical Transmission. Elsevier, Amsterdam.
14. Mousli, M., Bueb, J.-L., Bronner, C., Rouot, B. and Landry, Y. (1990) Trends Pharmacol. Sci. 11, 358–362.
15. Masu, Y., Nakayama, K., Tamaki, H., Harada, Y., Kuno, M. and Nakanishi, S. (1987) Nature 329, 836–838.
16. Nathans, J. and Hogness, D.S. (1983) Cell 34, 807–814.
17. Dixon, R.A.F., Kobilka, B.K., Strader, D.J., Benovic, J.L., Dohlman, H.G., Frielle, T., Bolanowski, M.D., Bennett, C.D., Rands, E., Diehl, R.E., Mumford, R.A., Slater, E.E., Sigal, I.S., Caron, M.G., Lefkowitz, R.J. and Strader, C.D. (1986) Nature 321, 75–79.
18. Kubo, T., Fukuda, K., Mikamii, A., Maeda, A., Takahashi, H., Mishina, M., Haga, T., Haga, K., Ichyama, A., Kangawa, T., Kojima, M., Matsuo, H., Hirose, T. and Numa, S. (1986) Nature 323, 411–416.
19. Peralta, E.G., Winslow, J.W., Peterson, G.L., Smith, D.H., Ashenazi, A., Ramachandran, J., Schimerlik, M.I. and Capon, D.J. (1987) Science 236, 600–605.
20. Chinkers, M., Garbers, D.L., Chang, M.-S., Lowe, D.G., Chin, H., Goeddel, D.V. and Schulz, S. (1989) Nature 338, 78–83.
21. Mazella, J., Kitabgi, P. and Vincent, J.-P. (1985) J. Biol. Chem. 260, 508–514.
22. Mills, A., Demoliou-Mason, C.D. and Barnard, E.A. (1988) J. Biol. Chem. 263, 13–16.
23. Mazella, J., Chabry, J., Zsurger, N. and Vincent, J.-P. (1989) J. Biol. Chem. 264, 5559–5563.
24. Tanaka, K., Masu, M. and Nakanishi, S. (1990) Neuron 4, 847–854.
25. Battey, J.F., Way, J.M., Corjay, M.H., Shapira, H., Kusano, K., Harkins, R., Wu, J.M., Slattery, T., Mann, E. and Feldmann, R.I. (1991) Proc. Natl Acad. Sci. USA 88, 395–399.
26. Wank, S.A., Harkins, R., Jensen, R.T., Shapira, H., de Weerth, A. and Slattery, T. (1992) Proc. Natl Acad. Sci. USA 89, 3125–3129.
27. Hammer, R., Berrie, C.P., Birdsall, N.J.M., Burgen, A.S.V. and Hulme, E.C. (1980) Nature 283, 90–92.
28. Jard, S., Barberis, C., Audigier, S. and Tribollet, E. (1987) Prog. Brain Res. 72, 173–186.
29. Moran, T.H. and McHugh, P.R. (1990) In: A. Björklund, T. Hökfelt and M.J. Kuhar (Eds), Handbook of Chemical Neuroanatomy Vol. 9, pp. 455–476, Elsevier, Amsterdam.
30. Meyerhof, W., Paust, H.-J., Schönrock, C. and Richter, D. (1991) DNA Cell Biol. 10, 689–694.
31. Yamada, Y., Post, S.R., Wang, K., Tager, H.S., Bell, G.I. and Seino, S. (1992) Proc. Natl Acad. Sci. USA 89, 251–255.
32. Kluxen, F.-W., Bruns, C. and Lübbert, H. (1992) Proc. Natl Acad. Sci. USA 89, 4618–4622.
33. Meyerhof, W., Wulfsen, I., Schönrock, C., Fehr, S. and Richter, D. (1992) Proc. Natl Acad. Sci. USA 89, 10267–10271.
34. Spindel, E.R., Giladi, E., Brehm, T.P., Goodman, R.H. and Segerson, T.P. (1990) Mol. Endocrinol. 43, 1950–1963.
35. Wada, E., Way, J., Shapira, H., Kusano, K., Lebacq-Verheyden, A.M., Coy, D., Jensen, R. and Battey, J. (1991) Neuron 6, 421–430.
36. Yokota, Y., Sasai, Y., Tanaka, K., Fujiwara, T., Tsuchida, K., Shigemoto, R., Kakizuka, A., Ohkubo, H. and Nakanishi, S. (1989) J. Biol. Chem. 264, 17649–17652.

37. Straub, R.E., Frech, G.C., Joho, R.H. and Gershengorn, M.C. (1990) Proc. Natl Acad. Sci. USA 87, 9514–9518.
38. McEachern, A.E., Shelton, E.R., Bhakta, S., Obernolte, R., Bach, C., Zuppan, P., Fujisaki, J., Aldrich, R.W. and Jarnagin, K. (1991) Proc. Natl Acad. Sci. USA 88, 7724–7728.
39. Kimura, T., Tanizawa, O., Mori, K., Brownstein, M.J. and Okayama, H. (1992) Nature 356, 526–529.
40. Meyerhof, W., Morley, S., Schwarz, J. and Richter, D. (1988) Proc. Natl Acad. Sci. USA 85, 714–717.
41. Morel, A., O'Carroll, A.-M., Brownstein, M.J. and Lolait, S.J. (1992) Nature 356, 523–526.
42. Sasaki, K., Yamano, Y., Bardhan, S., Iwai, N., Murray, J.J., Hasegawa, M., Matsuda, Y. and Inagami, T. (1991) Nature 351, 230–233.
43. Murphy, T.J., Alexander, R.W., Griendling, K.K., Runge, M.S. and Bernstein, K.E. (1991) Nature 351, 233–236.
44. Libert, F., Parmentier, M., Lefort, A., Dinsart, C., van Sande, J., Maenhaut, C., Simons, M.-J., Dumont, J.E. and Vassart, G. (1989) Science 244, 569–572.
45. Probst, W.C., Snyder, L.A., Schuster, D.I., Brosius, J. and Sealfon, S.C. (1992) DNA Cell Biol. 11, 1–20.
46. Braun, T., Schofield, P.R. and Sprengel, R. (1991) Eur. Mol. Biol. Organ. (EMBO) J. 10, 1885–1890.
47. Dixon, R.A., Sigal, I.S., Candelore, M.R., Register, R.B., Blake, A.D. and Strader, C.D. (1987) Eur. Mol. Biol. Organ. (EMBO) J. 6, 3269–3275.
48. Dohlman, H.G., Thorner, J., Caron, M.G. and Lefkowitz, R.J. (1991) Annu. Rev. Biochem. 60, 653–688.
49. Smrcka, A.V., Hepler, J.R., Brown, K.O. and Sternweis, P.C. (1991) Science 251, 804–807.
50. Taylor, S.J., Chae, H.Z., Rhee, S.G. and Exton, J.H. (1991) Nature 350, 516–518.
51. Bauer, C.K., Meyerhof, W. and Schwarz, J.R. (1990) J. Physiol. 429, 169–189.
52. Gershengorn, M.C. (1989) In: Secretion and its Control (Oxford, G.S. and Armstrong, C.M., Eds), pp. 1–15, Rockefeller University Press, New York.
53. Kleuss, C., Hescheler, J., Ewel, C., Rosenthal, W., Schultz, G. and Wittig, B. (1991) Nature 353, 43–48.
54. Birnbaumer, L., Abramowitz, J. and Brown, A.M. (1990) Biochim. Biophys. Acta 1031, 163–224.
55. Kobilka, B.K., Kobilka, T.S., Daniel, K., Regan, J.W. and Caron, M.G. (1988) Science 240, 1310–1316.
56. Kubo, T., Bujo, H., Akiba, I., Nakai, J., Mishina, M. and Numa, S. (1988) FEBS Lett. 241, 119–125.
57. Young, D., Waitches, G., Birchmeier, C., Fasano, O. and Wigler, M. (1986) Cell 45, 711–719.
58. Ross, P.C., Figler, R.A., Corjay, M.H., Barber, C.M., Adam, N., Harcus, D.R. and Lynch, K.R. (1990) Proc. Natl Acad. Sci. USA 87, 3052–3056.
59. Monnot, C., Weber, V., Stinnakre, J., Bihoreau, C., Teutsch, B., Corvol, P. and Clauser, E. (1991) Mol. Endocrinol. 5, 1477–1487.
60. O'Dowd, B.F., Hnatowich, M., Caron, M.G., Lefkowitz, R.J. and Bouvier, M. (1989) J. Biol. Chem. 264, 7564–7569.
61. Buck, F., Meyerhof, W., Werr, H. and Richter, D. (1991) Biochem. Biophys. Res. Commun. 178, 1421–1428.
62. Mentlein, R., Buchholz, C. and Krisch, B. (1990) Cell Tissue Res. 262, 431–443.
63. Krisch, B., Buchholz, C. and Mentlein, R. (1991) Cell Tissue Res. 263, 253–263.
64. Kovacs, G.L. (1986) In: Neurobiology of Oxytocin (Ganten, D. and Pfaff, D., Eds), pp. 91–128, Springer-Verlag, Berlin-Heidelberg.
65. Olson, B.R., Drutarosky, M.D., Stricker, E.M. and Verbalis, J.G. (1991) Endocrinology 129, 785–791.
66. Jackson, T.R., Blair, L.A.C., Marshall, J., Goedert, M. and Hanley, M.R. (1988) Nature 335, 437–440.
67. Bunnemann, B., Fuxe, K., Metzger, R., Mullins, J., Jackson, T.R., Hanley, M.R. and Ganten, D. (1990) Neurosci. Lett. 114, 147–153.
68. Julius, D., Livelli, T.J., Jessell, T.M. and Axel, R. (1989) Science 244, 1057–1062.

69. Maenhaut, C. and Libert, F. (1990) Exp. Cell Res. 187, 104–110.
70. Allen, L.F., Lefkowitz, R.J., Caron, M.G. and Cotecchia, S. (1991) Proc. Natl Acad. Sci. USA 88, 11354–11358.
71. Landis, C.A., Masters, S.B., Spada, A., Pace, A.M., Bourne, H.R. and Vallar, L. (1989) Nature 340, 692–696.
72. Inagaki, S., Shiosaka, S., Takatsuki, K., Iida, H., Sakanama, M., Senba, E., Hara, Y., Matsuzaki, T., Kawai, Y. and Tohyama, M. (1982) Dev. Brain Res. 3, 509–527.
73. Altman, J. (1982) Exp. Brain Res. Suppl. 6, 8–49.
74. Gonzalez, B.J., Leroux, P., Laquerriere, A., Coy, D.H., Bodenant, C. and Vaudry, H. (1988) Neuroscience 29, 629–644.
75. White, R.E., Schonbrunn, A. and Armstrong, D.L. (1991) Nature 351, 570–573.
76. Lin, H.Y., Harris, T.L., Flannery, M.S., Aruffo, A., Kaji, E.H., Gorn, A., Kolakowski, L.F., Lodish, H.F. and Goldring, S.R. (1991) Science 254, 1022–1024.
77. Shigemoto, R., Yokota, Y., Tsuchida, K. and Nakanishi, S. (1990) J. Biol. Chem. 265, 623–628.
78. Jüppner, H., Abou-Samra, A.-B., Freeman, M., Kong, X.F., Schipani, E., Richards, J., Kolakowski L.F., Jr, Hock, J., Potts J.T., Jr, Kronenberg, H.M. and Segre, G.V. (1991) Science 254, 1024–1026.
79. Abou-Samra, A.-B., Jüppner, H., Force, T., Freeman, M.W., Kong, X.-F., Schipani, E., Urena, P., Richards, J., Bonventre, J.V., Potts J.T., Jr, Kronenberg, H.M. and Segre, G.V. (1992) Proc. Natl Acad. Sci. USA 89, 2732–2736.
80. Takeda, Y., Chou, K.B., Takeda, J., Sachais, B.S. and Krause, J.E. (1991) Biochem. Biophys. Res. Commun. 179, 1232–1240.
81. Birnbaumer, M., Seibold, A., Gilbert, S., Ishido, M., Barberis, C., Antaramian, A., Brabet, P. and Rosenthal, W. (1992) Nature 357, 333–335.
82. Lolait, S., O'Carroll, A.-M., McBride, O.W., Konig, M., Morel, A. and Brownstein, M.J. (1992) Nature 357, 336–339.

Subject Index

A71623, 26
A-channel (I_A), 52
Abecarnil, 29
Acetylcholine, 107, 130
Acetylcholinesterase, 114, 118
AcGRP$_{20-26}$ ethyl ester, 23
AChR, 11, 113, 114
  active state, 114
  agonist, 115
  allosteric properties, 114, 125
  amino acid sequences, 120, 121
  antagonists, competitive, 115, 125
  antagonists, noncompetitive, 115, 126
  brain, 131
  α-bungarotoxin, 42, 115, 130, 132
  α-bungarotoxin binding proteins, 130
  α-bungarotoxin binding site, 127
  κ-bungarotoxin, 42, 130, 132
  carbohydrate, 118
  desensitized state, 114
  development, 128
  Drosophila, 132
  electron microscopy, 118, 119
  end-plate, 114
  evolutionary aspects, 121
  extrasynaptic, 126, 128, 129
  FTIR spectroscopy, 124
  functional topography, 125
  glycosylation, 123, 124
  in-situ hybridization, 131
  insects, 132
  ion channel, 114, 115, 121, 125, 126
  localization, 131
  locusts, 42, 132
  M_r, 118
  muscle-like, 123, 128
  neuronal, 114, 123, 130
  non-α subunits, 124
  peripheral (or muscle-like), 114, 123, 128
  pharmacology, 115
  phosphorylation, 124
  phosphorylation sites, 128
  primary structure, 121
  quaternary structure, 118
  resting state, 114
  secondary structure, 124
  snake, 132
  subsynaptic, 128
  subunits, 120, 121
  α-subunits, 124, 125, 130
  toxicology, 115
  transmembrane folding, 124
Acidic amino acid receptors, 283
Acidic amino acids, 267
Adenosine, 57
Adenosine receptors, 17
Adenylate, 223
Adenylyl cyclase, 8, 100, 105, 163, 253, 259
Adipocytes
  β-adrenoceptors, 20
ADP, 46
Adrenaline, 18–20
β-Adrenoceptor kinase (βARK), 103, 104, 139, 153, 158, 162
β$_1$-Adrenoceptor-G$_s$, 149
α-Adrenoceptors, 13, 18, 138, 147
α$_1$-Adrenoceptors, 18, 57
α$_2$-Adrenoceptors, 13, 19, 57, 147
β-Adrenoceptors, 10, 13, 20, 57, 100–105, 137, 138
  cAMP-dependent protein kinase, 139
  phosphorylation, 153
  sequestration, 155
  structures, 146
  turnover, 157
AFDX116, 39
ω-Agatoxin, 51
Agonists, 5, 69, 113
  inverse, 5, 6
  partial, 6
Agrin, 128
AH6809, 45
Alprenolol, 20
Alternative splicing, 8
Alzheimer's disease, 282
9-Aminoacridine, 53
D-α-Aminoadipate, 268
3-Aminopropylphosphinic acid, 29
4-Aminopyridine, 55
[L-α-Aminosuberic acid 7,23']-βANP$_{7-28}$, 22

AMP, 46, 100, 105, 106, 139, 243
8-bromo-cAMP, 229
AMP-dependent protein kinase, see PKA, 103
AMP formation, 288
AMPA, 31, 269
AMPA receptor, 268, 271, 285
   Ca^{2+} permeability, 287
   distribution, 271
   Na^+ and K^+ permeability, 271
   pharmacology, 279
AMPA receptor antagonists, 281, 282
Amygdala, 240
Amylin, 25
Analgesics, 297
Anandamide, 25
ANAPP$_3$, arylazidoaminopropionyl ATP, 46
ANF, 9
Angelman's syndrome, 196
Angiotensin I, 21, 57
Angiotensin II (AII), 21
Angiotensin II AT$_1$ receptor, 341
Angiotensin III (AIII), 21
Angiotensin receptors, 21
Anilidopiperidines, 302
Aniracetam, 279
ANP, 22
Antagonists, 5, 69, 113
   cholinergic, 113
   competitive, 5
   noncompetitive, 5
4AP, 52–54
L-AP4, 269
L-AP4 receptor, 268, 272
   presynaptic autoreceptor, 273
D-AP5, 31
Apamin, 53
Aplysia, 55
APNEA, 17
Arachidonic acid release, mGluR1, 288
ARC239, 19
ARIA, 130
βARK, see β-Adrenoceptor kinase
β-Arrestin, 153
Arylacetamides, 43, 302
Aspartate transcarbamylase, 7
Association rate constant, 63, 64, 85, 87
Atenolol, 20
ATP, 46
α-Atrial natriuretic peptide (ANP), 22, 326
Atrial natriuretic peptide receptors, 22
Atropine, 3, 201

Avermectins, 184
AVP, 49

Ba^{2+}, 52–54
Baclofen, 183
L-Baclofen, 29
Barbiturates, 184
BB-823, 44
Benzilate, 201
7-Benzilidine-7-dehydronaltrexone, 43
Benzodiazepines, 7, 9, 29, 184
Benzomorphans, 43, 298
Benzothiazepines, 51
Betaxolol, 20
BH-eledoisin, 48
BH-NMB, 23
BHT920, 19
Bicuculline, 29, 183
BIMU8, 36
Binding assay, 62, 65, 96
   experimental strategies, 72–92
   experimental techniques, 92–97
   pharmacological effect, 66
   separation techniques, 73, 92–95
   solubilized receptors, 93, 95
Binding data, 76
   Eadie–Hofstee plot, 91
   graphical evaluation, 76, 91
   numerical analysis, 77, 91, 92
   Scatchard plot, 76
Binding equilibrium, see Equilibrium, 63, 72
Binding experiments, 72, 88
Binding sites, 62, 72
   heterogeneity, 80
   interacting, 83
Biochemical responses, 200
Bisoprolol, 20
BK channel, 53
Bombesin, 23
[D-Phe^6,Cpa^{14},ψ13-14]Bombesin$_{6-14}$, 23
[Tyr^4]Bombesin, 23
Bombesin receptors, 23, 338
   gastrin-releasing peptide, 343
Bradykinin (BK), 24, 57
[Hyp^3,Tyr(Me)^8]Bradykinin, 24
[Phe^8,ψ(CH$_2$–NH)Arg^9]Bradykinin, 24
Sar[D-Phe^8]Bradykinin$_{1-8}$, 24
Bradykinin receptors, 24
Brain natriuretic peptide (BNP), 22, 327
Bremazocine, 298
BRL37344, 20

Bromocriptine, 27
BUBUC, 301
α-Bungarotoxin, 42, 115, 130, 132
  AChR, 115
  binding proteins, 130
  binding sites, 127
κ-Bungarotoxin, 42, 130, 132
Bungarus multicinctus, 115
Buspirone, 225, 246
Butaprost, 45
Butoxamine, 20
BW A868C, 45
BW 245C, 45

C-type natriuretic peptide, 326
Ca^{2+}-activated nonspecific cation channel, 54
Ca^{2+} channels, 18–20, 27, 29, 30, 43, 47, 51, 56, 253, 259, 299
Ca^{2+} permeability
  AMPA receptor subtypes, 286
  NMDA receptor channel, 270
Calciseptine, 51
Calcitonin gene-related peptide (CGRP), 107
Calcitonin gene-related peptide receptor, 25
Calmidazolium, 53
Calmodulin, 53
Cannabinoid, 57
Cannabinoid receptor, 25
Carbocyanine dyes, 53
5-Carboxamidotryptamine (5-CT), 225
Cardiomyopathy, 162
Catecholamines, 138, 167
Cation permeability, 285
Caudate, 235
CCK receptors, 26
CCK/gastrin, 57
CEC, 18
β-Cells, 53
Cerebral ischaemia
  AMPA receptor antagonists, 281
  NMDA receptor antagonists, 281
Cetiedil, 53
CGP20712A, 20
CGP35348, 29
CGP36742, 29
CGP37849, 31, 275
CGP42112A, 21
CGRP, 25
CGS19755, 31, 275
CGS21680, 17
Charybdotoxin, 53

Chemoreceptor theory, 3
Chimeric receptors, 13
Chirality of ligands, 204
CHO cells, 20
Cholecystokinin receptor, 26, 342
Choroid plexus
  5-HT_{1c} binding sites, 234
Chromaffin cells
  K^+ channels, 53
Chronic neurodegenerative conditions, 282
CI977, 43, 301
CI988, 26
Cicaprost, 45
Cingulate cortex
  5-$HT_{1B/1D}$ binding sites, 231
Cirazoline, 18
CL-218872, 185
Cl^- channel, 29, 33
Clonidine, 19
Clozapine, 27
CNP, 22
Codeine, 297
α-Conotoxin, 132
μ-Conotoxin, 56
ω-Conotoxin, 51
Cooperativity, 7, 62, 79, 83–85
  negative, 79, 84
  positive, 79, 83
Corynanthine, 18
CP55940, 25
CP66713, 17
CP93129, 35
CP99994, 48
CPPene, 275
Cs^+, 52, 53
CTAP, 43
CTOP, 301
Curare, 3, 113
CV6209, 44
Cyanopindolol, 230
2-Cl-N^6-Cyclopentyladenosine, 17
N^6-Cyclopentyladenosine, 17
8-Cyclopentyltheophylline, 17
D-Cycloserine, 282

DALDA, 301
DAMGO, 43, 301
Decamethonium, 53
Delayed rectifier
  scorpion venom, phalloidin, 52
Deltorphin, 43

Dementia
   NMDA receptors, 282
Dendrotoxin, 52
Desensitization, 46, 99, 100, 152, 236, 242, 285
Devazepide, 26
Dexetimide, 201
Diabetes insipidus
   vasopressin receptors, 49
Diacylglycerol, 236
5,7-Dichlorokynurenate, 31
Dihydropyridine, 51
5,7-Dihydroxytryptamine, 231
Dimaprit, 34
Direct binding plot, 75
Dissociation rate constant, 63, 64, 74, 87, 88
Dizocilpine, 31
DMCM, 29, 185
DOB, 235
DOI, 235
Domoate, 31
Domperidone, 27
Dopamine, 57
Dopamine receptors, 8, 27, 251
   biochemical characterisation, 257
   cellular responses, 259
   characterisation, 257
   D_1, 252, 254, 256
   D_2, 252
   D_3, 254, 256
   D_4, 254, 256
   D_5, 254, 256
   D_1-like, 254, 256
   D_2-like, 254, 256
   dopamine receptor activation, 259
   dopamine receptors to G-proteins, 258
   inhibition, 259
   mechanism of ligand binding, 260
   regulation, 262
   subtypes defined from molecular biological studies, 252, 254
   subtypes defined from physiological, pharmacological and biochemical studies, 252, 253
Dorsal horn interneurons
   glutamate receptors, 268
Dorsal raphe (DR)
   5-HT synthesis, 225
Down-regulation, 102, 103, 105, 106, 158
DPCPX, 17
DPDPE, 43, 301

Drosophila, 132
DSBULET, 43
Dynorphin, 43, 300

Eadie–Hofstee plot, 91
Ectonucleotidases, 46
EGF, 9
Eicosanoids, 37
Electric ray, 42, 115, 118
Electrophorus electricus, 115, 118
Endogenous opiates, 5
Endorphin, 43, 300
Endothelin, 28, 58
Endothelin receptors, 28
Endothelium-derived relaxing factor (EDRF), 288
Enkephalin, 5, 299, 300
[Leu]Enkephalin, 43
[Met]Enkephalin, 43
Enprostil, 45
Entorhinal cortex
   5-HT_3 binding site, 240
Epidermal growth factor, 105
Epilepsy
   NMDA antagonists, 280
   therapeutic index, 280
Epinephrine, 138
Equilibration, 67, 86, 96
   incomplete, 81
Equilibrium, 73
   binding, 73
   constants, 63
   dialysis, 94, 95
Ethanol, 193
Ethyl β-carboline, 185
Ethylketocyclazocine, 43, 298
Etorphine, 298
EXP985, 21
EXP31274, 21

Fenoldopam, 27
First messenger
   triune receptor model, 6
Flumazenil, 29
Flunarizine, 51
Flunitrazepam, 29, 186
2-(*m*-Fluorophenyl)histamine, 34
5-Fluorowillardine, 31
Fluoxetine, 228, 246
Fluprostenol, 45
FMRF, 55

Focal brain ischaemia
  NMDA receptor antagonists, 281
Forskolin, 52, 229
Free ligand
  binding studies, 72
Funnel web spider toxin (FTX)
  Ca^{2+} channels, 51

G-proteins, 9–13, 25, 100, 105, 137, 140, 163, 164, 223, 252, 258, 283, 299
  G_i, 8, 141, 165
  G_o, 141
  G_s, 8, 104, 140, 158, 165, 168
  $\beta\gamma$-subunits, 141
GABA, 6, 9, 29, 58, 107, 123, 183
$GABA_A$ receptors, 7, 9, 29, 123, 183
$GABA_B$ receptors, 29, 183
GABA/benzodiazepine receptors, 5
Galanin, 30
Galanin receptor, 30
Galantide, 30
Gallamine, 39, 42
Gardos effect
  Ca^{2+}-activated K^+ channel, 53
Gastrin receptor, 26
Gastrin releasing peptide (GRP), 23
Gene regulation, 286
Gephyrin, 188
Ginkgolide B, 44
Glia
  Na^+ channels, 56
Global brain ischaemia
  NMDA receptor antagonists, 281
Glu_2 receptor
  expression cloning, 287
  G-protein-linked receptors, 287
Glucocorticoids, 160
Glu_G receptor, 268, 269, 273, 279, 287
  adenylate cyclase, 288
  heterogeneity, 273
  intracellular second-messenger systems, 273
  role in LTP, 274
  pharmacology, 280
  phosphatidyl inositol (PI) turnover, 273, 288
GluR1–4 proteins
  alternately spliced variant, 286, 287
Glutamate, 58, 107
Glutamate receptors, 7, 9, 11, 31, 268, 269, 288
  heterogeneity, 288
  ionotropic, 31

  metabotropic, 32
  subtypes, 268
L-Glutamic acid diethylester, 268
γ-Glutamyl transpeptidase, 37
Glutathione-S-transferase, 37
Glycine, 6, 33, 107, 123, 183
  binding, 33
  co-agonist site, 274, 277
Glycine antagonist
  L-365 260, 26
  L-659 877, 48
  L-659 989, 44
  L-687 414, 277
  L-689 560, 31, 277
Glycine receptors, 12, 29, 33, 121, 183
  mutant mouse spastic, 195
GMP, 6, 325
GR32191, 45
GR64349, 48
GR82334, 48
GR94800, 48
GR113808, 36
Granisetron, 36
Growth hormone releasing factor (GRF), 50
GRP, 23
GTI, 35
Guanylate cyclase, 234
Guanylin, 326, 329
Guanylyl cyclase, 9, 10, 22, 325
Guanylyl cyclase receptors
  membrane-bound guanylyl cyclases, 325
  peptide hormones, 326
  receptor-linked enzymes, 325
  soluble guanylyl cyclase, 325
GYKI52466, 31, 279

H-current, 243
HA966, 31
Haloperidol, 53
Head trauma
  NMDA receptor antagonists, 281
Heart failure
  β-adrenoceptors, 162
HEAT, 18
Heat-shock protein, 17
Heat-stable enterotoxins of *Escherichia coli*, 329
Helix-M2 model, 127
Heme, 331
12R-HETE, 37
Hexahydrosiladifenidol, 39

Hexamethonium, 42, 53
Hill equation, 84
Hill plot, 86
Himbacine, 39
Hippocampus
  galanin receptors, 30
  5-HT receptors, 226
Histamine receptors, 34
HOE-140, 24
5-HT, 58
5-HT receptors, 35, 36, 221
  D-receptor, 222
  desensitization, 236
  EPSP, 238
    fast, 242
    slow, 239
  heteroreceptor, 232
  5-HT$_2$, 222
  5-HT$_3$, 239
  5-HT$_4$, 243
  5-HT autoreceptors, 226
  5-HT$_{1A}$, 222
  5-HT$_{1B}$, 222
  5-HT$_{1C}$, 222
  5-HT$_{1D}$, 223
  M-receptor, 222
Hydropathy plots, 11
8-Hydroxy-2-(di-n-propylamino)tetralin
  (8-OH-DPAT), 225
5-Hydroxyindoleacetic acid, 223
5-Hydroxytryptamine (5-HT, serotonin), 221
5-Hydroxytryptophan, 223
Hyperthyroidism
  β-adrenoceptors, 166
Hypothalamus
  galanin receptors, 30
Hypothyroidism
  β-adrenoceptors, 166

I-BOP, 45
Iberatoxin, 53
IC$_{50}$, 65
ICI118551, 20
ICI174864, 43, 301
ICI197067, 43
ICI198615, 37
ICS 205-930, 239
Iloprost, 45
Imetit, 34
Impromidine, 34
Inhibition constant, 65, 88, 89

Inhibition experiments, 64, 72, 88–91
  Cheng–Prusoff equation, 88
  logit–log plot, 89
Inositol-1,4,5-triphosphate (IP3), 236
Insulin, 9, 105
Insulin-like growth factor, 105
Insulin receptor, 105
  endocytosis, down-regulation, 106
Insulin receptor serine kinase (IRSK), 106
Intracellular level of cAMP, 216
Inward rectifier (I_{IR})
  Gaboon viper venom, 52
Iodocyanopindolol, 20
Iodophenpropit, 34
Ion channel, 7, 9, 114
Ion channel family, 283
Ionotropic receptors, 9
Ipsapirone, 225
IPSP, 222
Isoguvacine, 29
Isomerization, 214
Isoproterenol, 138

K$^+$ channels, 19, 25, 27, 29, 30, 35, 39, 43,
  47, 50, 52–55, 229, 253, 259, 299
43k protein, 121
Kainate, 31, 269
Kainate-activated channels, 272
  Na$^+$/K$^+$ permeability, 272
Kainate binding protein, 31
Kainate receptor, 272, 287
  regional localization, 272
Kallidin, 24
Ketanserin, 235, 238
KFI7837, 17
K_i, see Inhibition constant
Kinetic methods
  experiments, 64, 85–88
  muscarinic agonists, 215
  nonradioactive ligands, 219
  radioactive antagonists, 213
  radioligand binding, 212

L calcium channel, 230
Leiurotoxin, 53
Leukotriene, 58
Leukotriene receptors, 37
Levocabastine, 41
Levonantradol, 25
Ligand, 62, 67–72
  classification, 67

handling, 69–71
heterogeneity of, 80
homogeneity, 70
mode of action, 68
pharmacological model, 71
purity, 70
Ligand affinity, 63, 68
Ligand binding
  association, 85
  association rate constant, 63, 64, 86, 87
  dissociation, 85
  dissociation rate constant, 63, 64, 87, 88
  thermodynamics, 64
Ligand-gated ion-channel receptors, 107
Ligand-gated ion channels, 9–12, 42, 114, 124, 132, 183
Ligand–ligand interactions, 80
Limbic system
  opioid receptors, 299
Lipid- and calcium-dependent protein kinase (PKC), 103
Lipoxygenase, 55
Lophotoxin, 132
Lorglumide, 26
Losartan, 21
LTB_4, 37
LTC_4, 37
LTD_4, 37
LTP, 271, 282, 286, 289
Luzindole, 38
LY53857, 36
LY97241, 52
LY170680, 37
LY255283, 37
LY262691, 26
Lysergic acid diethylamide, 222

M2, 11, 123
M current, 54
Mast cell degranulating peptide, 52
Maxi-K channel, 53
McCune–Albright syndrome
  β-adrenergic system, 168
Mecamylamine, 42
Median raphe
  5-HT synthesis, 225
Melatonin receptor, 38
Mepyramine, 34
Mesulergine, 36
Methoctramine, 39
Methoxamine, 18

5-Methoxytryptamine, 36
N-Methyl-4-piperidinyl benzilate, 201
Methylcaconitine, 42
Methylcarbamylcholine, 42
cis-[^{3}H]Methyldioxolane, 201
α,β-Methylene ATP, 46
Methylenedioxymethamphetamine (MDMA), 225
α-Methylpropranolol, 20
(−)-N-Methylscopolamine, 39
5-Methylurapadil, 18
MK571, 37
MK678, 47
MK801, 31, 278
MNQX, 31
Monoamine oxidase (MAO), 223
Monosynaptic transmission, 271
Morphine, 43, 297, 298
MPTP, 282
Muscarinic acetylcholine receptors, 11, 39, 58, 199, 205, 208, 217
  agonists, 203
  antagonists, 203
  electrophysiological responses, 200
  lectin chromatography, 208
  organ-bath assay procedures, 200
  subtypes, 205
  three-dimensional structure, 208
Muscimol, 186
Muscle relaxant, 6
Myasthenia gravis
  AChR, 126
Myocardial infarction
  β-adrenergic system, 164
Myriceron caffeoyl ester, 28

N-0437, 27
N calcium channel, 230
Na^+ channels, 51, 56
Nabilone, 25
Naloxone, 297, 298
Naltrexone, 43
Naltrindole, 43, 298, 301
NAN-190, 225
Narcotic drugs, 297
NBQX, 279
NECA, 15
α-Neo-dynorphin, 43
Neosurugatoxin, 132
Neurokinin A (NKA), 48
Neurokinin B (NKB), 48

Neuromedin B (NMB), 23
Neuromedin L, 48
Neuromedin N, 41
Neuromodulators, 4
Neuromuscular endplate
  AChR, 113, 129
Neuromuscular junction
  AChR, 113, 129
Neuronal bungarotoxin
  AChR, 40, 130, 132
Neuropeptide K, 48
Neuropeptide receptor
  oncogene, 350
  tumorigenesis, 350
Neuropeptide receptors, 339
  adenylate cyclase, 348
  binding sites, 345
  extracellular domain, 345
  G-proteins, 348
  inositol phospholipid turnover, 348
  intracellular loop, 347
  neurotensin, 41, 58, 306, 342
  opioids, 300
  palmitoylation, 349
  phosphorylation, 347
  potassium and calcium channels, 348
  sequence identity, 345
  topologies, 345
  transmembrane domains, 345
Neuropeptide Y (NPY), 40
Neurosteroids, 184
α-Neurotoxin-binding site, 125
α-Neurotoxins, 115
Nicotine, 42, 113, 130
Nicotinic acetylcholine receptors (NAChR), 4, 7, 10, 12, 42, 107
Niguldipine, 18
(+)-Niguldipine, 18
Nitrendipine, 53
Nitric oxide, 325
NKA, 48
NMB, 23
NMDA, 7, 31, 269
NMDA receptors, 33, 271, 274, 275, 277, 283
  antagonists, 281, 282
  Ca^{2+} imaging studies, 270
  cognitive function, 282
  dissociative anaesthetics, 277
  distribution, 271
  expression cloning, *Xenopus* oocytes, 284
  function, 270
  heterogeneity, 271
  in-situ hybridization, 284
  localization, 284
  negative slope conductance, 270
  nitric oxide, 288
  open channel blockers, 277
  pharmacology, 274
  postsynaptic density, 271
  stimulation, biochemical sequelae, 271
  structure–activity relationships, 275
  subunits, 284
  voltage-dependent channel block by Mg^{2+}, 270
  voltage-dependent conductance, 270
NMDAR1
  alternately spliced forms, NMDAR2, 284, 285
NO, 9, 288, 289
  astrocytes, 289
  guanylate cyclase, 289
  NADPH diaphorase, 289
  neurons, 289
  NMDA-induced neurotoxicity, 289
  synthetase, 289, 331
  transcellular messenger, 289
Nonspecific binding, 65, 79, 81–83, 92
  evaluation, 82
  minimization, 81
Nootropics, 6
Nor-binaltorphimine, 43, 298, 301
Noradrenaline, 20
Norepinephrine, 138
Noxiustoxin, 52, 53
Nucleus accumbens
  $5-HT_2$ receptors, 235

Octanol, 51
Odorant receptor, 8
Odorants, 13
Off-rate, 85, 88, 92
  evaluation, 88
8-OH-DPAT, 35
Olfactory bulb
  $5-HT_2$ receptors, 235
On-rate, 85–87
  evaluation, 87
Ondansetron, 36, 239, 246
ONO-LB457, 37
OPC21268, 49
Opioid receptors, 43
  affinity labelling, 314

classification, 302
cloning, 316
coupling to effector systems, 307
endogenous ligands, 302
interactions with G-proteins, 306
molecular biology, 316
NG108-15, 306
purification, 309
solubilisation, 309
states, 308
Opioids, 58
  addiction, 299
  agonists, 298
  antagonists, 298
  ligands, 300
  tolerance, 299
Opium, 297
Organ-bath assay procedures, 200
Orphan receptors, 4
Oxoquazepam, 185
Oxotremorine, 39, 201
Oxymetazoline, 19
Oxytocin (OT), 49, 59, 335

PACAP, 50
PAF, 58
PAF receptor, 44
Pain
  opioids, 299
Palmitate, 104
Palmitoylation, 104
Pandinus imperator toxin, 52
PAPA-APEC, 17
Parkinson's disease, 251, 281
  AMPA receptor antagonists, 282
  NMDA receptor antagonists, 282
Partial agonists, 69, 203
Patch-clamp, 113, 130
PC12 cells, 130
PCP, see Phencyclidine
PD123177, 21
PD140548, 26
PDGF, 9
Peptide histidine isoleucinamide (PHI), 50
Peptide histidine methionineamide (PHM), 50
Peptide hormones, 13
Peptide YY (PYY), 40
Pertussis toxin
  Ca^{2+} conductance, 5-HT, 230
  inhibitory post-synaptic potential, 5-HT, 228

Phencyclidine (PCP), 31, 52, 278, 300
  psychosis, 278
Phenylalkylamines, 51
Phenylephrine, 18
Phosphatidylinositide (PI), 223, 234
Phospholipase, 216
Phospholipase C (PLC), 236, 253
Phosphorylation, 12, 13, 101–107, 208
Photoaffinity labeling, 13, 128
Photoreceptor disk, 6
PI-response, 8
Picrotoxin, 183
Pilocarpine, 3, 201
Pimozide, 53
Pirenzepine, 39
Pituitary adenylyl cyclase activating polypeptide (PACAP), 50
PKA, 103, 104, 106, 107, 128, 158
PKC, 103, 105–107, 128, 155, 193, 236
PL017, 43
Platelet-derived growth factor, 105
Platelets, 45, 46
Polyamines, 278
  NMDA receptor, 278
Prazosin, 18, 19
Procaterol, 20
Propylbenzilylcholine mustard, 210
N-Propylbenzilylcholine mustard, 201
Prostacyclin, 45
Prostaglandin, 45
Prostanoid, 59
Prostanoid receptors, 45
Protein kinases, 153–155, 193, 236
Purinoceptors, 17
Purkinje cells
  Ca^{2+} channels, 51
Putamen
  5-HT$_2$ receptors, 235

Quinacrine, 53
Quinidine, 52
Quinine, 53, 54
Quinoxalinediones, 31
Quinuclidinyl benzilate, 39, 201
Quisqualate receptor, 268

Radioligand, 67, 69–72, 79, 201
  iodinated compounds, 69
  technical requirements, 69, 71, 72
  tritiated compounds, 69
Ranitidine, 34

Rauwolscine, 19
Receptive substance, 3, 113
Receptor occupancy, 6
Receptor–ligand complex, 73
 isomerization, 81
Receptors, 65–67, 72, 113, 123
 acetylcholine, 113
 binding site, 62
 classification, 9, 268
 concept, 8
 criteria, 5
 crosstalk, 7
 definition, 4
 diversity, 8
 effector (E), 6
 expression, 157
 expression cloning, 342
 families, 8
 general principles, 3
 history, 3
 IUPHAR rules, 9
 neurotensin, 342
 nicotinic acetylcholine, 113
 nomenclature, 3, 9, 15
 preformed states, 8
 regulation, 7, 13, 99
 solubilized, 93, 95
 structural features, 11
 subtypes, 8
 transducer (T), 6
 triune model, 6, 8
Red cell
 Ca^{2+}-activated K^+ channels, 53
Reserpine, 222
Reverse genetics, 4
Rhodopsin, 6, 146, 149
Ritanserin, 36
mRNA, 102
mRNA editing, 9
Ro194603, 29
RP67580, 48
RS2359190, 36
RTE model, 125
RU-5135, 185
RU-24969, 230
RX821002, 19

Saclofen, 29
Salivary gland
 Ca^{2+}-activated K^+ channels, 53
Sarafotoxin, 28

Sarcoplasmic reticulum channel, 53
Saturation curve, 75
Saturation experiments, 63, 73–85
 graphical analysis, 91, 92
 limits due to the ligand, 79
 numerical analysis, 91, 92
Saxitoxin, 56
SC19220, 45
SC41930, 37
Scatchard plot, 74–81
 nonlinear, 79–81
SCH23390, 27
SCH39166, 27
Schizophrenia
 dopamine receptors, 251
Scopolamine, 201
Scrapie prion protein, 130
Scyllatoxin, 53
Second messenger, 6
Secretin, 50
Senktide, 48
Separation of bound and free ligands, 72, 77, 92–95
 centrifugation, 94
 charcoal adsorption, 95
 comparison of techniques, 93
 equilibrium dialysis, 94, 95
 filtration, 92
 gel filtration, 94
 precipitation by polyethyleneglycol, 95
Sequestration, 101, 103
Serine–borate complex, 37
Serotonin, 35
Signal converters, 6
Signal sequence, 13
Signal transduction, 114
Site-directed mutagenesis, 13, 210, 261, 285
SKF38393, 27
SKF83566, 27
SKF104353, 37
SKF108566, 21
Slow delayed rectifier, 52
Somatostatin, 47, 59
 cerebellum, 352
 external granule cell layer, 352
 granule cell layers, 353
 Purkinje cell layers, 353
 synaptogenesis, 352
Somatostatin receptors, 47, 343
 astrocytes, 346
 somatostatin-14, 343

somatostatin-28, 343
Specific binding, 65, 82
  criteria, 66
Spiroperidol, 222
SQ29548, 45
Sr^{2+}, 52
SR48692, 41
SR48968, 48
SR95531, 29
SRI63072, 44
SS-bridges, 210
STA, 45
Stroke
  NMDA receptor antagonists, 281
Strychnine, 29, 33, 185
Substance K, 48
Substance K receptor, 340
Substance P, 48
Substantia nigra
  5-$HT_{1B/1D}$ receptors, 231
Sufentanyl, 43
(−)Sulpiride, 27
Sumatriptan, 35, 230, 246
Superfamilies, 10, 11, 114, 124, 130
Supersensitivity, 128

T calcium channel, 230
Tachykinin receptors, 48, 59
Talcum, 5
TBPS, 185
TEA, 52–54
Telenzepine, 39
Terminal heteroreceptor
  5-HT receptors, 232
Ternary agonist–antagonist–receptor complex, 215
Tetrahydroaminoacridine, 52
(−)-Δ^9 Tetrahydrocannabinol, 25
Tetrodotoxin, 56, 130
TFMPP, 230
Thioperamide, 34
Thromboxane, 45
Thyroid hormones, 160, 166
Tiotidine, 34
1TM, 10
4TM, 10
Torpedo, 42, 115, 118
Toxin I, 52
Transmembrane segment, 285
Transmitter
  classification, 6
  inhibition, 6
Trifluoperazine, 53

Triprolidine, 34
Trophic factor, 128
Tropicamide, 39
Tropisetron, 36
Tryptophan hydroxylase, 223
Tubocurarine, 53
D-Tubocurarine, 115, 132
Two-site receptor model, 216, 217
Two-step consecutive binding mechanism, 213
Type C natriuretic peptide (CNP), 22
Type I benzodiazepine recognition site, 185
Type I receptors, 9, 11
Type II benzodiazepine recognition site, 185
Type II receptors, 9, 12
Type III receptors, 9, 10
Tyrosine kinase receptors
  lymphocytes, 106
Tyrosine kinase receptors (TKR), 10, 105, 128

U46619, 45
U69593, 43, 298, 301
U75302, 37
UK14304, 19
Up-regulation, 160
UTP, 46

V_1 protein, 118
Vasoactive intestinal peptide (VIP), 50
Vasopressin (AVP), 49, 59
Vasopressin and oxytocin receptors, 49, 343
VIP receptors, 50

WAY100135, 35
WB4101, 18
WEB2086, 44
WIN55212-2, 25

Xamoterol, 20
Xenopus, 28
Xenopus oocytes, 4, 9, 56, 192, 235, 343
  B_2 bradykinin, 344
  electrophysiological testing, 344
  neurotensin, 344
  oxytocin, 344
  substance P, 344
  receptors, 344
  thyrotropin-releasing hormone, 344

YM091512, 27
Yohimbine, 19

Zacopride, 36
ZK110841, 45
ZK93426, 29
Zn^{2+}, 52
Zolpidem, 29